Methods in Molecular Biology

Series Editor
John M. Walker
School of Life and Medical Sciences
University of Hertfordshire
Hatfield, Hertfordshire, AL10 9AB, UK

For further volumes:
http://www.springer.com/series/7651

Zygotic Genome Activation

Methods and Protocols

Edited by

Kiho Lee

Department of Animal and Poultry Sciences,
Virginia Tech, Blacksburg, VA, USA

Humana Press

Editor
Kiho Lee
Department of Animal and Poultry Sciences
Virginia Tech
Blacksburg, VA, USA

ISSN 1064-3745 ISSN 1940-6029 (electronic)
Methods in Molecular Biology
ISBN 978-1-4939-6986-9 ISBN 978-1-4939-6988-3 (eBook)
DOI 10.1007/978-1-4939-6988-3

Library of Congress Control Number: 2017937534

Printed on acid-free paper

This Humana Press imprint is published by Springer Nature
The registered company is Springer Science+Business Media LLC
The registered company address is: 233 Spring Street, New York, NY 10013, U.S.A.

Preface

Proper embryogenesis requires well-orchestrated events. After fertilization, initially maternal factors stored in the egg lead the development and the zygotic genome is dormant. Then, zygotic genome controls the development by initiating its own transcription. Successful transition into this event, zygotic genome activation (ZGA), is critical for embryo survival. Previous studies have demonstrated that dramatic degradation of maternal mRNA occurs and activation of specific zygotic genes is involved during ZGA. However, specific pathways and factors involved in the process have not been fully elucidated. One of the main obstacles to investigating the process is limited tools available for molecular analyses of the event. Specifically, due to the limited amount of samples (DNA, RNA, and protein) available from early stage embryos, assessing the global profile of gene expression at the RNA and protein level has been a challenge. Similarly, following specific changes in epigenetic marks such as DNA methylation and histone codes during ZGA has been difficult. Recent technological advancements in molecular analyses now allow us to follow these changes at higher accuracy. Advanced next-generation sequencing technology allows the expression profile of transcripts during ZGA to be detected and analyzed. In addition, advancement in data processing allows us to effectively utilize mass data analysis approaches to investigate gene expression patterns during ZGA. Sensitivity of quantitative PCR is sufficient to assess the level of mRNA, small RNA, and long noncoding RNA. Immunocytochemistry, based on either antibody or fluorescence in situ hybridization (FISH), can now visualize the presence of specific epigenetic marks or RNA. The ability to alter genes during embryogenesis has not been widely available to study ZGA, at least in mammals. This is due to difficulty in generating and maintaining genetically modified animals for embryo collection. The application of siRNA technology now allows us to alter the level of transcripts during embryogenesis and the use of gene editing technology such as CRISPR/Cas9 system allows us to completely remove the function of target genes during embryogenesis. These technological advancements can overcome traditional barriers we have had that discourage us from investigating events of ZGA. This volume of the Methods in Molecular Biology series provides an overview of ZGA and use of the recent tools that can be used to elucidate the events during ZGA. We expect that new findings will emerge as now more practical approaches are available to monitor the changes we see during ZGA.

Blacksburg, VA, USA *Kiho Lee*

Contents

Contributors

MALAVIKA K. ADUR • *Department of Animal Science, Iowa State University, Ames, IA, USA*
KATIA ANCELIN • *Unité de Génétique et Biologie du Développement, Institut Curie, PSL Research University, CNRS UMR 3215, INSERM U934, Paris, France*
KEISUKE AOSHIMA • *Laboratory of Comparative Pathology, Graduate School of Veterinary Medicine, Hokkaido University, Sapporo, Japan*
VERONIKA BENESOVA • *Laboratory of Developmental Biology, Institute of Animal Physiology and Genetics, Academy of Science of Czech Republic, v.v.i., Libechov, Czech Republic; Faculty of Science, Charles University in Prague, Prague, Czech Republic*
DMITRY S. BOGOLYUBOV • *Institute of Cytology RAS, St. Petersburg, Russia*
IRINA O. BOGOLYUBOVA • *Institute of Cytology RAS, St. Petersburg, Russia*
ALI H. BRIVANLOU • *Laboratory of Molecular Vertebrate Embryology, The Rockefeller University, New York, NY, USA*
WILLIAM F. BURKHOLDER • *Microfluidics Systems Biology Laboratory, Institute of Molecular and Cell Biology, Agency for Science, Technology and Research (A*STAR), Singapore, Singapore*
LIH FENG CHEOW • *Microfluidics Systems Biology Laboratory, Institute of Molecular and Cell Biology, Agency for Science, Technology and Research (A*STAR), Singapore, Singapore*
JESÚS GARCÍA-LÓPEZ • *Department of Cellular and Molecular Biology, Centro de Investigaciones Biológicas (CSIC), Madrid, Spain; Oncology Department, St. Jude Children's Research Hospital, Memphis, TN, USA*
JÖRG GROSSHANS • *Institute for Developmental Biochemistry, Medical School, University of Göttingen, Göttingen, Germany*
BENJAMIN J. HALE • *Department of Animal Science, Iowa State University, Ames, IA, USA*
NOBUHIKO HAMAZAKI • *Department of Stem Cell Biology and Medicine, Graduate School of Medical Sciences, Kyushu University, Fukuoka, Japan*
KATSUHIKO HAYASHI • *Department of Stem Cell Biology and Medicine, Graduate School of Medical Sciences, Kyushu University, Fukuoka, Japan*
EDITH HEARD • *Unité de Génétique et Biologie du Développement, Institut Curie, PSL Research University, CNRS UMR 3215, INSERM U934, Paris, France*
PATRICIA HEYN • *Max Plank Institute of Molecular Cell Biology and Genetics, Dresden, Germany; MRC Human Genetics Unit, IGMM, University of Edinburgh, Edinburgh, UK*
TAKUYA IMAMURA • *Department of Stem Cell Biology and Medicine, Graduate School of Medical Sciences, Kyushu University, Fukuoka, Japan*
JIRI KANKA • *Laboratory of Developmental Biology, Institute of Animal Physiology and Genetics, Academy of Science of Czech Republic, v.v.i., Libechov, Czech Republic*
KYEOUNG-HWA KIM • *Department of Biomedical Sciences, Institute of Reproductive Medicine, College of Life Science, CHA University, Pan-Gyo, South Korea*
TAKASHI KIMURA • *Laboratory of Comparative Pathology, Graduate School of Veterinary medicine, Hokkaido University, Sapporo, Japan*

VERONIKA KINTEROVA • *Laboratory of Developmental Biology, Institute of Animal Physiology and Genetics, Academy of Science of Czech Republic, v.v.i., Libechov, Czech Republic; Department of Veterinary Sciences, Czech University of Life Sciences in Prague, Prague, Czech Republic*

EDUARDO LARRIBA • *Department of Cellular and Molecular Biology, Centro de Investigaciones Biológicas (CSIC), Madrid, Spain*

KIHO LEE • *Department of Animal and Poultry Sciences, Virginia Tech, Blacksburg, VA, USA*

KYUNG-AH LEE • *Department of Biomedical Science, Institute of Reproductive Medicine, College of Life Science, CHA University, Pan-Gyo, South Korea*

SU-YEON LEE • *Department of Biomedical Science, Institute of Reproductive Medicine, College of Life Science, CHA University, Pan-Gyo, South Korea*

YAN LI • *Human Reproduction Unit, Northern Clinical School, Sydney Medical School, University of Sydney, Sydney, NSW, Australia*

KA YI LING • *Developmental Epigenetics and Disease Laboratory, Institute of Molecular and Cell Biology, Agency for Sciences, Technology and Research (A*STAR), Singapore, Singapore*

BOYANG LIU • *Institute for Developmental Biochemistry, Medical School, University of Göttingen, Göttingen, Germany*

ZOLTAN MACHATY • *Department of Animal Sciences, Purdue University, West Lafayette, IN, USA*

ANTONIO MARCO • *School of Biological Sciences, University of Essex, Colchester, UK*

JESÚS DEL MAZO • *Department of Cellular and Molecular Biology, Centro de Investigaciones Biológicas (CSIC), Madrid, Spain*

DANIEL M. MESSERSCHMIDT • *Developmental Epigenetics and Disease Laboratory, Institute of Molecular and Cell Biology, Agency for Science, Technology and Research (A*STAR), Singapore, Singapore*

KINICHI NAKASHIMA • *Department of Stem Cell Biology and Medicine, Graduate School of Medical Sciences, Kyushu University, Fukuoka, Japan*

KARLA M. NEUGEBAUR • *Molecular Biophysics and Biochemistry, Yale University, New Haven, CT, USA*

CHRISTOPHER O'NEILL • *Human Reproduction Unit, Northern Clinical School, Sydney Medical School, University of Sydney, Sydney, NSW, Australia*

YUKI OKADA • *Laboratory of Pathology and Development, Institute of Molecular and Cellular Biosciences, University of Tokyo, Tokyo, Japan*

IKUHIRO OKAMOTO • *Department of Anatomy and Cell Biology, Graduate School of Medicine, Kyoto University, Kyoto, Japan*

STEPHEN R. QUAKE • *Department of Bioengineering and Applied Physics, Stanford University, Stanford, CA, USA; Howard Hughes Medical Institute, Stanford, CA, USA*

NOÉMIE RANISAVLJEVIC • *Unité de Génétique et Biologie du Développement, Institut Curie, PSL Research University, CNRS UMR 3215, INSERM U934, Paris, France*

ALESSANDRO ROSA • *Department of Biology and Biotechnology 'Charles Darwin', Sapienza University of Rome, Rome, Italy; Laboratory of Molecular Vertebrate Embryology, The Rockefeller University, New York, NY, USA*

JASON W. ROSS • *Department of Animal Sciences, Iowa State University, Ames, IA, USA*

JUNGHYUN RYU • *Department of Animal and Poultry Sciences, Virginia Tech, Blacksburg, VA, USA*

TEREZA TORALOVA • *Laboratory of Developmental Biology, Institute of Animal Physiology and Genetics, Academy of Science of Czech Republic, v.v.i., Libechov, Czech Republic*
KYUNGJUN UH • *Department of Animal and Poultry Science, Virginia Tech, Blacksburg, VA, USA*
HAOYI WANG • *The Jackson Laboratory, Bar Harbor, MA, USA; State Key Laboratory of Stem Cell and Reproductive Biology, Institute of Zoology, Chinese Academy of Sciences, Beijing, China*
WENBO WANG • *The Jackson Laboratory, Bar Harbor, MA, USA; The University of Pennsylvania Perelman School of Medicine, Philadelphia, PA, USA*
LU ZHANG • *Department of Animal Sciences, Purdue University, West Lafayette, IN, USA*
YINGFAN ZHANG • *The Jackson Laboratory, Bar Harbor, MA, USA*

Chapter 1

Clearance of Maternal RNAs: Not a Mummy's Embryo Anymore

Antonio Marco

Abstract

Until the zygotic genome is activated, early development relies on the products deposited by the mother. Once the zygotic genome starts to be transcribed, most maternal products are not needed anymore by the developing embryo. This emancipation from the maternal genome occurs during the Zygotic Genome Activation (ZGA). Although the process by which the maternal content is replaced with zygotic products differs from species to species, there is a common theme to all of them: maternal transcripts are actively degraded. Here, a review of how the degradation of maternal RNAs is regulated during early development and discussions on some computational tools that may be of use in this research area are outlined.

Key words RNA degradation, Deadenylation, RNA-binding proteins, microRNAs, Maternal-to-zygotic transition, Zygotic genome activation

1 The Discovery of Maternal RNA Degradation

Generous mothers provide invaluable gene products to the unfertilized egg. These products will be crucial for the formation of the embryo. Indeed, early embryologists already noticed the importance of maternal products in the first stages of development. The first case of an enucleated sea urchin embryo undergoing cleavage was reported as early as in 1898 [1] (see discussion in [2]). In a classic experiment, Briggs and collaborators activated frog (*Rana pipiens*) eggs with X-ray-treated sperm [3]. These chromosome-free embryos underwent segmentation (although slower than nucleated embryos) and even initiated gastrulation [3]. These experiments indicated that the genetic information provided by the mother was enough to start the developmental programme. Parallel to the developments in embryology, geneticists also found early in the twentieth Century the so-called maternal-effect genes (*see* [4] and reference within), providing further evidence of a maternal contribution independent of the zygotic genome. In the fruit fly (*Drosophila melanogaster*) maternally deposited products

Kiho Lee (ed.), *Zygotic Genome Activation: Methods and Protocols*, Methods in Molecular Biology, vol. 1605,
DOI 10.1007/978-1-4939-6988-3_1, © Springer Science+Business Media LLC 2017

were necessary to establish the polarity of the embryo during early development [5]. When the anterior of the egg was irradiated with UV light, presumably destroying maternally deposited RNAs, the embryo did not develop a head [5, 6]. By the beginning of the molecular era, it was well established that important maternal products were loaded into the developing egg, and had a function during early development. Multiple experiments demonstrated that not only messenger RNAs, but also other gene products such as ribosomes or tRNAs, are maternally deposited (reviewed in [7]).

In the early 1970s, it was found, in sea urchins, that maternal RNAs were poly(A)-rich [8, 9]. It was also proposed that poly(A) tails may have a function other than transcript transportation [8]. At that time, poly(A) tails were believed to participate in nucleous-to-cytoplasm transport [10], probably because these tails had not been detected in histones. Further experiments confirmed that poly(A) tails were a common characteristic of maternal RNAs in sea urchin [11, 12] and starfish [13]. Also, it was observed that maternal RNAs tend to disappear from the polysomes as development progresses [11], and some authors suggested that that may be due to stochastic decay due to replacement of maternal RNA by zygotic transcripts [11]. But a few years later, a study in *Xenopus* suggested that maternal RNAs stability depended on the presence of poly(A) tails, and that maternal RNAs may be selectively destroyed during gastrulation [14]. This hypothesis was confirmed thanks to the development of new RNA labeling techniques, showing that there is specific (active) degradation of maternal RNAs in mouse embryos [15]. These results were later on corroborated (e.g., [16, 17]): there was indeed a maternal RNA degradation machinery.

How maternal RNAs were selectively degraded was not known, since gene regulation at the post-transcriptional level was not well understood. A major breakthrough in molecular biology was the discovery of AU-Rich Elements (ARE), short motifs in the RNA that control the stability of messenger RNAs [18]. Equipped with this conceptual toolkit, Duval et al. compared the 3′ UTRs of maternal RNAs deadenylated during *Xenopus* development, and detected motifs that may serve as signals for deadenylation/degradation [19]. By the early 1990s, all indicated that maternal RNA degradation was a regulated process, involving the action of RNA Binding Proteins (RBP). In the next section, I review the various molecular mechanisms behind maternal transcript degradation.

2 The Zygotic and Maternal Pathways of Maternal Transcript Degradation

The first insights on the molecular mechanisms behind maternal RNA degradation came from Howard Lipshitz's lab, when they studied the degradation of maternal transcripts in *Drosophila* [20].

In this species, eggs are mechanically activated during deposition, independently of fertilization. They found that the levels of specific maternal transcripts decreased with time in unfertilized eggs, and that this degradation did not occur if specific fragments were removed from the 3′ UTR [20]. Strikingly, by injecting *Drosophila* constructs into *Xenupos* oocytes, they showed that the specific regulatory sequences were also recognized by the *Xenopus* clearance machinery. This suggests that there is a conserved maternal pathway of RNA degradation. On the other hand, the degradation of some maternal RNAs was faster if there was fertilization, suggesting a second pathway encoded in the zygotic genome. These two pathways, the maternal and the zygotic, were supported by microarray experiments in other organisms such as mouse, zerafish, *Caenorhabditis elegans*, and humans (reviewed in [21, 22]).

The RNA-binding protein Smaug (SMG) was first identified in *Drosophila*, where it regulates the translation of transcripts during early development [23–25]. SMG binds to transcripts via specific RNA motifs, the Smaug Recognition Elements (SRE). In *Drosophila*, the maternal transcript from *Hsp83* is recognized by SMG, which subsequently recruits the CCR4/POP2/NOT deadenylation complex and triggering transcript degradation [26]. SMG is translated from maternal transcripts; thus, SMG-dependent transcript clearance seemed to be the maternal pathway proposed a few years before [20]. The translation of SMG transcript is a tightly regulated mechanism itself, which requires activation by the Pan GU (PGU) kinase [27]. SMG can also block translation by recruiting the Cup-eIF4E complex [28], or by interacting with AGO1 [29]. Other studies using microarrays showed that SMG triggers the degradation of two thirds of the unstable maternal RNAs [27]. Co-immunoprecipitation assays revealed that over 300 transcripts are the direct target of SMG, and also that SMG represses the translation of about 3000 genes [30]. These findings provided a mechanistic explanation for the maternal pathway of transcript degradation.

Parallel to these developments in *Drosophila*, the analysis of Dicer mutants in zebrafish revealed that microRNAs may be involved in the zygotic pathway of RNA transcript degradation [31]. MicroRNAs are short regulatory RNAs that bind to gene transcripts by pairwise complementarity, inducing translational repression or degradation [32]. MicroRNAs are involved in virtually any biological process, are crucial during development [33] and are very often clustered in the genome and transcribed as polycitronic molecules [34]. The biogenesis of microRNAs requires the action of a RNase called Dicer (reviewed in [35]). However, in zebrafish, Dicer mutants develop well into day 10 [36]. Giraldez and collaborators suggested that maternal Dicer action may be compensating the lack of zygotic Dicer, as this is crucial during early development. Therefore, they generated zebrafish with neither maternal nor zygotic Dicer [31]. These mutants had an almost normal early development but showed severe errors during organogenesis. The authors even suggested a role of a cluster composed by microRNAs of the mir-430 family, highly expressed during early zygotic development, in maternal transcript degradation. In a follow-up paper they showed that indeed mir-430 accelerated transcript decay by inducing transcript deadenylation, suggesting that microRNAs may act in the zygotic pathway of maternal clearance [37]. These mechanisms are conserved in *Xenopus*, where the family mir-427 (presumably an ortholog of mir-430) is also involved in maternal transcript degradation [38].

In *Drosophila*, where the maternal pathway seemed to be controlled by SMG, it was suggested that microRNAs, like in zebrafish, may also be involved in the zygotic degradation pathway [27]. This was based on the fact that degraded transcripts were enriched for target sites for several microRNA families [27]. Confirmation of a role of microRNAs in the zygotic degradation pathway was found soon after in *Drosophila* [39]. The microRNA cluster *mir-309* (formed by eight precursor microRNAs) encode mature microRNAs that, when zygotically expressed, target maternal transcripts that will undergo degradation [39]. However, this scenario was a bit more complex. First, there is a significant overlap between SMG targets and mir-309 targets [39]. Second, microRNA-mediated transcript degradation often depends on SMG activity [40]. Thus, the microRNA and non-microRNA pathways seemed to be related.

Further experiments showed that the zygotic pathway was more complex. The expression profiling of multiple chromosomal deletions in *Drosophila* showed that this pathway had multiple players, some of which were probably RNA-binding proteins other than SMAUG [41]. This work showed evidence that AU-rich elements (ARE, see above) as well as a new motif that they called Bicoid Stabilizing Factor (BSF) may be involved in the selection of transcripts for degradation [41]. ARE-binding proteins are probably involved in maternal transcript degradation in *Xenopus*, *C. elegans*, and mouse (reviewed in [22, 42]). Other computational analyses of sequence motifs detected ARE and SRE motifs in both the zygotic and the maternal degradation pathways, and another type of element, the Pumilio-like Binding Site (PBS), mostly present in transcripts that undergo degradation by the zygotic pathway [43].

A role of microRNAs in the maternal pathway has not been demonstrated. However, a study found that, in *Drosophila*, destabilized transcripts were enriched in target sites for maternally deposited microRNAs [44]. In particular, this paper suggests that the microRNA mir-9c may be involved in maternal transcript degradation. Indeed, the maternal loss of mir-9c affects the number of germ cells [45]. This kind of maternal product degrading other maternal products seems intuitively nonsense. However, in *Caenorhabditis elegans*, maternal microRNAs trigger the deadenylation of maternal transcripts [46], although degradation has

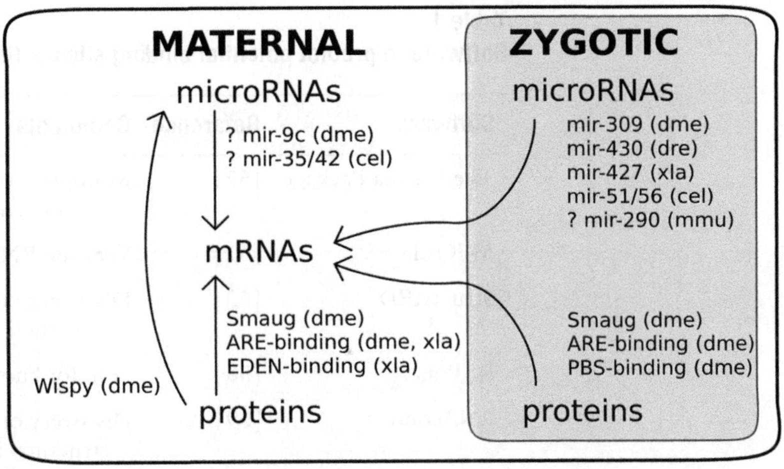

Fig. 1 Mechanism of maternal transcript clearance. The cartoon shows the two pathways described in the main text, the maternal and zygotic pathway. Species names are in brackets: Drosophila melanogaster (dme); Danio rerio (dre); Xenopus laevis (xla); Caenorhabditis elegans (cel); Mus musculus (mmu)

not been observed. Also, maternal microRNAs themselves are cleared from the egg by a maternal protein, Wispy [47]. In the light of these observations, we cannot discard a role of microRNAs in the maternal pathway of transcript clearance. Figure 1 summarizes the difference mechanisms by which maternal transcripts are cleared from the embryo.

3 Finding and Predicting Targets for Degradation

The perception that transcripts are long nucleotide strings freely floating in the cytoplasm is misleading. RNA molecules form complex structures [48]. Typically, RNA-binding proteins bind to double-stranded chains. Therefore, binding sites at the single-stranded transcripts require that these molecules fold into hairpin-like structures. A well-studied model is that of yeast Vts1p, which bind to specific hairpin motifs to regulate translation [49]. The folding properties of RNA molecules have been studied in great detail, giving rise to multiple computational tools that predict local structures from primary sequences. Among the most popular are the Vienna Package [50] and MFOLD [51], both having standalone and web server versions (Table 1).

To search for SMG recognition elements (SRE), for instance, transcripts are scanned for the motif CNGG, and then a RNA-folding prediction program is run to detect those motifs in the loop of a hairpin. The prediction of hairpin structures around SRE motifs has been done with the Vienna Package [30] and with MFOLD [27]. The chromosomal ablation experiments discussed

Table 1
Software to predict potential binding sites in RNA sequences

Software	Reference	Comments
The Vienna Package	[57]	Multiple tools for RNA folding and thermodynamics
MFOLD	[51]	Versatile RNA-folding prediction
StructRED	[52]	Discovery of novel binding sites using structural information
RBPmap	[58]	Scan for known RNA-binding motifs
RNAcontext	[59]	Discovery of novel binding sites using structural information
MEMERIS	[60]	Incorporates structural information to the popular MEME

above, which identified several RNA motifs, did not use any folding predictions and their results were based only on statistical overrepresentation of sequence motifs [41]. A similar approach was used by Thomsen et al. [43]. Although this strategy has been proved successful, the power to detect bona fide RNA-binding motifs is lower. Using a more sophisticated approach to discover structured regulatory elements, Foat and Stormo found SRE to be associated with maternal transcript degradation [52], in agreement with previous studies [27]. These SRE were remarkably similar to the yeast Vts1p-binding sites (Vst1p is a homolog of Smaug in yeast). This indicates that the mechanism of action of Vts1p/Smaug is highly conserved, and predates its role in maternal RNA clearance. This algorithm is implemented in the software StructRED (Table 1). Using StructRED, other novel RNA-binding motifs have been discovered [52], stressing the potential of RNA structure-aware software to study maternal clearance. Table 1 lists additional software to predict RNA-binding bites that may be of use in future research endeavors.

The other big players in maternal transcript degradation are the microRNAs. MicroRNA target sites are very short (often between 6 and 8 nucleotides) [32], which creates and obvious technical limitation as multiple false positives are expected. For that reason, different programs use different strategies, among them, evolutionary conservation is a common approach to filter out false positives. However, as reported by Giraldez et al. [37], target sites for mir-430 (see above) are not preferentially conserved. As a matter of fact, if we expect maternal transcript degradation to be an evolutionarily dynamic mechanism [44], we will expect that target site conservation has a minor importance. Thus, it is recommended that evolutionary conservation is not used to study maternal RNA clearance. One strategy consists in scanning transcripts for

Table 2
Software to predict microRNA target sites

Software	Reference	Comments
seedVicious	[53]	Canonical seeds and other features. Custom data analysis via web interface
TargetScan	[61]	Canonical seeds plus evolutionary conservation
miRanda	[62]	Combines hybridization energy with other features
RNAhybrid	[63]	Prioritize folding/hybridization energy
Sylammer	[54]	MicroRNA-unaware detection of enriched motifs

canonical seed target sites [32], and filters out target sites with a high binding energy, and/or considers only transcripts with multiple sites. This strategy is implemented in the program (also available as a web-server) seedVicious [53]. An alternative approach is that implemented in Sylammer [54], which compares the word distribution of RNA sequences from different experiments. Other microRNA target prediction algorithms have been reviewed elsewhere [32, 55]. Table 2 lists some useful microRNA target predictions tools.

4 Why Degrading Maternal Products?

Detailed discussion on the possible roles of maternal clearance has been published elsewhere [21, 22]. In summary, they describe a permissive function, in which the elimination of a broadly expressed maternal transcript allows its zygotic counterpart to have a more restricted (spatially) expression profile. They also describe an instructive function, in which maternal transcripts are removed to restrict their function. For instance, in *Drosophila* development maternal transcripts encode cell cycle regulators that, upon degradation, the cell cycle slows down, which is essential during the last syncytial nuclear divisions [27]. Other roles include removing transcripts that are no longer needed [22] prevent abnormal mRNA dosages in the embryo [21], or the spatial elimination of maternal transcripts that are otherwise stable in specific organs [21]. According to these authors, maternal clearance may have multiple functions. An alternative idea has been suggested by Giraldez and collaborators [42, 56]. In their view, maternal clearance is required to delete the old, highly differentiated, program that will be replaced by the pluripotent zygotic program. This is proposed in a context of cellular reprogramming. The idea is original and certainly attractive. Interestingly, maternal clearance shows parallelisms with the artificial reprogramming of somatic cells (reviewed in [42]).

On the other hand, an alternative possibility exists: maternal clearance is a by-product of other maternal and zygotic activities.

It is evident the potential of maternal clearance as a regulatory mechanism, and the fact that it is evolutionarily conserved may indicate a function. On the other hand, the Dicer mutants described in zebrafish progress until organogenesis with no major issues, and mir-309 mutants in *Drosophila* do not show any defect in patterning. SMG mutants in *Drosophila* are indeed lethal [25], but SMG is required for several different functions (including protein folding and degradation and basic metabolism [30]), as it is more likely that these mutants suffer from massive pleiotropic effects. In summary, despite the existing evidence and the different regulatory roles proposed, it has not been proved yet whether maternal clearance has a well-defined function.

5 Conclusion

The clearance of maternal RNAs is a mechanism that operates in early development. Whether maternal clearance has a well-defined function or not, can only be found by a fine dissection of the molecular details of this process. Thanks to the advances in high-throughput expression analysis and computational biology, there has been significant progress during the last decade. Current developments in Next-Generation Sequencing, as well as the emergence of novel gene-editing techniques such as CRISPR/Cas9, indicate that we are now equipped to study maternal clearance at an unprecedented level of accuracy. After four decades of research in maternal clearance, there are still important open questions, and the coming developments in this field promise to be very exciting.

References

1. Ziegler HE (1898) Experimentelle Studien über die Zelltheilung. Arch Für Entwicklungsmechanik Org 6:249–293. doi:10.1007/BF02152958
2. Chambers R (1924) The physical structure of protoplasm as determined by microdissection and injection. In: Cowdry EV (ed) General cytology. The University of Chicago Press, Chicago, IL, pp 237–309
3. Briggs R, Green EU, King TJ (1951) An investigation of the capacity for cleavage and differentiation in Rana pipiens eggs lacking "functional" chromosomes. J Exp Zool 116:455–499. doi:10.1002/jez.1401160307
4. Redfield H (1926) The maternal inheritance of a sex-limited lethal effect in DROSOPHILA MELANOGASTER. Genetics 11:482–502
5. Kalthoff K, Sander K (1968) Der Entwicklungsgang der Mißbildung "Doppelabdomen" im partiell UV-bestrahlten Ei von Smittia parthenogenetica (Dipt., Chironomidae). Wilhelm Roux Arch Für Entwicklungsmechanik Org 161:129–146. doi:10.1007/BF00585968
6. Gilbert SF, Singer SR, Tyler MS, Kozlowski RN (2006) Developmental biology. Sinauer Associates, Sunderland, MA
7. Davidson EH (1986) Gene activity in early development, 3rd revised edn. Academic Press Inc, Orlando, FL
8. Slater I, Gillespie D, Slater DW (1973) Cytoplasmic adenylylation and processing of maternal RNA. Proc Natl Acad Sci U S A 70:406–411
9. Wilt FH (1973) Polyadenylation of maternal RNA of sea urchin eggs after fertilization. Proc Natl Acad Sci 70:2345–2349
10. Watson JD (1976) Molecular biology of the gene, 3rd edn. Benjamin-Cummings Publishing Co, Menlo Park, CA
11. Hough-Evans BR, Wold BJ, Ernst SG et al (1977) Appearance and persistence of maternal RNA sequences in sea urchin development.

Dev Biol 60:258–277. doi:10.1016/0012-1606(77)90123-3
12. Wilt FH (1977) The dynamics of maternal poly(A)-containing mRNA in fertilized sea urchin eggs. Cell 11:673–681
13. Jeffery WR (1977) Polyadenylation of maternal and newly-synthesized RNA during starfish oocyte maturation. Dev Biol 57:98–108
14. Sagata N, Shiokawa K, Yamana K (1980) A study on the steady-state population of poly(A)+RNA during early development of Xenopus laevis. Dev Biol 77:431–448
15. Bachvarova R, De Leon V (1980) Polyadenylated RNA of mouse ova and loss of maternal RNA in early development. Dev Biol 74:1–8
16. Pikó L, Clegg KB (1982) Quantitative changes in total RNA, total poly(A), and ribosomes in early mouse embryos. Dev Biol 89:362–378
17. De Leon V, Johnson A, Bachvarova R (1983) Half-lives and relative amounts of stored and polysomal ribosomes and poly(A) + RNA in mouse oocytes. Dev Biol 98:400–408
18. Shaw G, Kamen R (1986) A conserved AU sequence from the 3' untranslated region of GM-CSF mRNA mediates selective mRNA degradation. Cell 46:659–667
19. Duval C, Bouvet P, Omilli F et al (1990) Stability of maternal mRNA in Xenopus embryos: role of transcription and translation. Mol Cell Biol 10:4123–4129. doi:10.1128/MCB.10.8.4123
20. Bashirullah A, Halsell SR, Cooperstock RL et al (1999) Joint action of two RNA degradation pathways controls the timing of maternal transcript elimination at the midblastula transition in Drosophila melanogaster. EMBO J 18:2610–2620. doi:10.1093/emboj/18.9.2610
21. Tadros W, Lipshitz HD (2009) The maternal-to-zygotic transition: a play in two acts. Development 136:3033–3042. doi:10.1242/dev.033183
22. Walser CB, Lipshitz HD (2011) Transcript clearance during the maternal-to-zygotic transition. Curr Opin Genet Dev 21:431–443. doi:10.1016/j.gde.2011.03.003
23. Smibert CA, Wilson JE, Kerr K, Macdonald PM (1996) smaug protein represses translation of unlocalized nanos mRNA in the Drosophila embryo. Genes Dev 10:2600–2609. doi:10.1101/gad.10.20.2600
24. Smibert CA, Lie YS, Shillinglaw W et al (1999) Smaug, a novel and conserved protein, contributes to repression of nanos mRNA translation in vitro. RNA 5:1535–1547
25. Dahanukar A, Walker JA, Wharton RP (1999) Smaug, a novel RNA-binding protein that operates a translational switch in Drosophila. Mol Cell 4:209–218. doi:10.1016/S1097-2765(00)80368-8
26. Semotok JL, Cooperstock RL, Pinder BD et al (2005) Smaug recruits the CCR4/POP2/NOT deadenylase complex to trigger maternal transcript localization in the early Drosophila Embryo. Curr Biol 15:284–294. doi:10.1016/j.cub.2005.01.048
27. Tadros W, Goldman AL, Babak T et al (2007) SMAUG is a major regulator of maternal mRNA destabilization in Drosophila and its translation is activated by the PAN GU kinase. Dev Cell 12:143–155. doi:10.1016/j.devcel.2006.10.005
28. Nelson MR, Leidal AM, Smibert CA (2004) Drosophila Cup is an eIF4E-binding protein that functions in Smaug-mediated translational repression. EMBO J 23:150–159. doi:10.1038/sj.emboj.7600026
29. Pinder BD, Smibert CA (2013) microRNA-independent recruitment of Argonaute 1 to nanos mRNA through the Smaug RNA-binding protein. EMBO Rep 14:80–86. doi:10.1038/embor.2012.192
30. Chen L, Dumelie JG, Li X et al (2014) Global regulation of mRNA translation and stability in the early Drosophila embryo by the Smaug RNA-binding protein. Genome Biol 15:R4. doi:10.1186/gb-2014-15-1-r4
31. Giraldez AJ, Cinalli RM, Glasner ME et al (2005) MicroRNAs regulate brain morphogenesis in zebrafish. Science 308:833–838. doi:10.1126/science.1109020
32. Bartel DP (2009) MicroRNAs: target recognition and regulatory functions. Cell 136:215–233. doi:10.1016/j.cell.2009.01.002
33. Sayed D, Abdellatif M (2011) MicroRNAs in development and disease. Physiol Rev 91:827–887. doi:10.1152/physrev.00006.2010
34. Marco A, Ninova M, Griffiths-Jones S (2013) Multiple products from microRNA transcripts. Biochem Soc Trans 41:850–854. doi:10.1042/BST20130035
35. Axtell MJ, Westholm JO, Lai EC (2011) Vive la différence: biogenesis and evolution of microRNAs in plants and animals. Genome Biol 12:221. doi:10.1186/gb-2011-12-4-221
36. Wienholds E, Koudijs MJ, van Eeden FJM et al (2003) The microRNA-producing enzyme Dicer1 is essential for zebrafish development. Nat Genet 35:217–218. doi:10.1038/ng1251
37. Giraldez AJ, Mishima Y, Rihel J et al (2006) Zebrafish MiR-430 promotes deadenylation

and clearance of maternal mRNAs. Science 312:75–79. doi:10.1126/science.1122689

38. Lund E, Liu M, Hartley RS et al (2009) Deadenylation of maternal mRNAs mediated by miR-427 in Xenopus laevis embryos. RNA 15:2351–2363. doi:10.1261/rna.1882009
39. Bushati N, Stark A, Brennecke J, Cohen SM (2008) Temporal reciprocity of miRNAs and their targets during the maternal-to-zygotic transition in Drosophila. Curr Biol 18:501–506. doi:10.1016/j.cub.2008.02.081
40. Benoit B, He CH, Zhang F et al (2009) An essential role for the RNA-binding protein Smaug during the Drosophila maternal-to-zygotic transition. Development 136:923–932. doi:10.1242/dev.031815
41. Renzis SD, Elemento O, Tavazoie S, Wieschaus EF (2007) Unmasking activation of the zygotic genome using chromosomal deletions in the Drosophila Embryo. PLoS Biol 5:e117. doi:10.1371/journal.pbio.0050117
42. Lee MT, Bonneau AR, Giraldez AJ (2014) Zygotic genome activation during the maternal-to-zygotic transition. Annu Rev Cell Dev Biol 30:581–613. doi:10.1146/annurev-cellbio-100913-013027
43. Thomsen S, Anders S, Janga SC et al (2010) Genome-wide analysis of mRNA decay patterns during early Drosophila development. Genome Biol 11:R93. doi:10.1186/gb-2010-11-9-r93
44. Marco A (2015) Selection against maternal microRNA target sites in maternal transcripts. G3 5:2199–2207. doi:10.1534/g3.115.019497
45. Kugler J-M, Chen Y-W, Weng R, Cohen SM (2013) Maternal loss of miRNAs leads to increased variance in primordial germ cell numbers in Drosophila melanogaster. G3 3:1573–1576. doi: 10.1534/g3.113.007591
46. Wu E, Thivierge C, Flamand M et al (2010) Pervasive and cooperative deadenylation of 3′UTRs by embryonic MicroRNA families. Mol Cell 40:558–570. doi:10.1016/j.molcel.2010.11.003
47. Lee M, Choi Y, Kim K et al (2014) Adenylation of maternally inherited microRNAs by wispy. Mol Cell 56:696–707. doi:10.1016/j.molcel.2014.10.011
48. Elliott D (2011) Molecular biology of RNA. Oxford University Press, Oxford
49. Oberstrass FC, Lee A, Stefl R et al (2006) Shape-specific recognition in the structure of the Vts1p SAM domain with RNA. Nat Struct Mol Biol 13:160–167. doi:10.1038/nsmb1038
50. Hofacker IL (2003) Vienna RNA secondary structure server. Nucleic Acids Res 31:3429–3431
51. Zuker M (2003) Mfold web server for nucleic acid folding and hybridization prediction. Nucleic Acids Res 31:3406–3415
52. Foat BC, Stormo GD (2009) Discovering structural cis-regulatory elements by modeling the behaviors of mRNAs. Mol Syst Biol 5:268. doi:10.1038/msb.2009.24
53. Marco A (2017) seedVicious: a versatile microRNA target site prediction tool with evolutionary applications. http://seedvicious.essex.ac.uk/
54. van Dongen S, Abreu-Goodger C, Enright A (2008) Detecting microRNA binding and siRNA off-target effects from expression data. Nat Methods 5(1025):1023
55. Alexiou P, Maragkakis M, Papadopoulos GL et al (2009) Lost in translation: an assessment and perspective for computational microRNA target identification. Bioinformatics 25(23):3049–3055. doi:10.1093/bioinformatics/btp565
56. Giraldez AJ (2010) microRNAs, the cell's Nepenthe: clearing the past during the maternal-to-zygotic transition and cellular reprogramming. Curr Opin Genet Dev 20:369–375. doi:10.1016/j.gde.2010.04.003
57. Lorenz R, Bernhart SH, Höner zu Siederdissen C et al (2011) ViennaRNA package 2.0. Algorithms. Mol Biol 6:26. doi:10.1186/1748-7188-6-26
58. Paz I, Kosti I, Ares M et al (2014) RBPmap: a web server for mapping binding sites of RNA-binding proteins. Nucleic Acids Res 42:W361–W367. doi:10.1093/nar/gku406
59. Kazan H, Ray D, Chan ET et al (2010) RNAcontext: a new method for learning the sequence and structure binding preferences of RNA-binding proteins. PLoS Comput Biol 6:e1000832. doi:10.1371/journal.pcbi.1000832
60. Hiller M, Pudimat R, Busch A, Backofen R (2006) Using RNA secondary structures to guide sequence motif finding towards single-stranded regions. Nucleic Acids Res 34:e117. doi:10.1093/nar/gkl544
61. Agarwal V, Bell GW, Nam J-W, Bartel DP (2015) Predicting effective microRNA target sites in mammalian mRNAs. Elife. doi:10.7554/eLife.05005
62. Enright A, John B, Gaul U et al (2003) MicroRNA targets in Drosophila. Genome Biol 5:R1. doi:10.1186/gb-2003-5-1-r1
63. Kruger J, Rehmsmeier M (2006) RNAhybrid: microRNA target prediction easy, fast and flexible. Nucleic Acids Res 34:W451–W454. doi:10.1093/nar/gkl243

Chapter 2

Link of Zygotic Genome Activation and Cell Cycle Control

Boyang Liu and Jörg Grosshans

Abstract

The activation of the zygotic genome and onset of transcription in blastula embryos is linked to changes in cell behavior and remodeling of the cell cycle and constitutes a transition from exclusive maternal to zygotic control of development. This step in development is referred to as mid-blastula transition and has served as a paradigm for the link between developmental program and cell behavior and morphology. Here, we discuss the mechanism and functional relationships between the zygotic genome activation and cell cycle control during mid-blastula transition with a focus on *Drosophila* embryos.

Key words Cell cycle, Mid-blastula transition, Zygotic genome activation

1 Introduction

In most animals, from nematodes to chordates, embryogenesis starts with a series of rapid cleavage cell cycles after fertilization. These fast divisions lead to an exponentially increasing number of cells without an accompanied growth of the embryo. After a species-specific number of divisions, the cell cycle slows down and finally enters a pause. Subsequently, the embryo enters gastrulation with its characteristic morphogenetic movements, loss of symmetry, and cell type-specific differentiation. Mammalian embryogenesis is special in that it begins with differentiation of inner cell mass (ICM) and trophoblast, and the fast embryonic cleavage cycles eventually arise at late blastocyst stage [1–3]. Maternally supplied materials, including proteins, RNAs, and conceivably also metabolites contribute to the initial developmental processes. Maternal products exclusively control development during this first period, as the zygotic genome starts expression only with a delay after fertilization. Following zygotic genome activation (ZGA), both maternal and zygotic factors contribute to developmental control. The switch from maternal to zygotic control is especially prominent in species with large, externally deposited eggs. ZGA coincides with striking changes in cell behavior and molecular processes, including

Kiho Lee (ed.), *Zygotic Genome Activation: Methods and Protocols*, Methods in Molecular Biology, vol. 1605,
DOI 10.1007/978-1-4939-6988-3_2, © Springer Science+Business Media LLC 2017

cell cycle, DNA replication, maternal RNAs degradation, chromatin structure, metabolite composition, and status of DNA checkpoint. This morphologically visible switch in early development during the blastula stage was first described 120 years ago in sea urchin *Echinus microtuberculat* and *Sphaerechinus granularis*, and later has been referred to as mid-blastula transition (MBT) [4, 5].

1.1 MBT in Model Organisms

Many model organisms are well studied in terms of MBT. Amphibian *Xenopus laevis*, for instance, undergoes 12 short and synchronized cleavage cycles with a lack of gap phases, 35 min each and proceeds with a series of progressively longer and less synchronized divisions from cycles 13 to 15. The transition period is defined as the MBT [5–8]. S phase progressively lengthens, and the cell cycle pauses in G1 or G2 phases during the MBT [9]. Concomitantly, maternal transcripts are deadenylated and degraded. The first zygotic transcripts are detected at cycle 7 and transcription rate increases up to and beyond MBT [10]. During the MBT, developmental control is handed over from maternal to zygotic factors (maternal-zygotic transition, MZT).

In zebrafish *Danio rerio* embryo, 9 rapid cycles with approximately 15 min each are followed by gradually longer cell cycles [11]. MBT begins at cycle 10, and the cell cycle loses synchrony with acquisition of a G1 phase in cycle 11 [12]. Similar to *Xenopus*, ZGA is regulated by the nuclear-cytoplasmic ratio, but DNA damage checkpoint acquisition is independent of zygotic transcription [13]. Maternal factors Nanog, Pou5f1, and SoxB1 are required for de novo zygotic transcription as well as inducing maternal clearance by activating the microRNA *miR-430* expression [14].

In the nematode *Caenorhabditis elegans* (*C. elegans*), zygotic transcription is already activated in the 4-cell stage. Multiple mechanisms and maternal factors, including OMA-1 and OMA-2, are involved and regulated by phosphorylation, nuclear shuttling, and protein destabilization [15, 16]. In contrast to the other species discussed above, cells divide asynchronously and asymmetrically following fertilization in *C. elegans* embryos [17, 18].

1.2 MBT in Drosophila

MBT is observed in embryos of *Drosophila melanogaster* at about 2 h post fertilization. Embryonic development starts with 13 rapid and meta-synchronized nuclear divisions, with extraordinary short S phases and no gap phases [19]. The extraordinary speed of about 10 min per pre-blastoderm cell cycle is achieved by fast replication of DNA and the absence of cytokinesis [20–22]. The syncytial mode of early development is a special feature of insect embryogenesis [23]. Due to the absence of cytokinesis, the early cell cycles are often referred to as nuclear cycles (NC). The onset of the embryonic cell cycle is regulated by *pan gu*, *plutonium*, and *giant nuclei* [24–27]. From NC8 to 9, the nuclei move from the interior of the egg toward the periphery, forming the syncytial blastoderm. From NC10 to 13, nuclei undergo four more divisions at the egg cell cortex, until the nuclei number reaches approximately 6000. Some nuclei remain in the interior egg to differentiate into polyploid yolk nuclei. After mitosis 13, the cell cycle mode changes with the introduction of a long G2 phase, and the embryo enters into cellularization stage [19]. Following NC11, the cell cycle gradually slows down from 10 min in NC11 to 21 min in NC13 and an hour-long G2 pause in interphase 14 (25 °C) [19]. The S phase lengthens and by cycle 14 a difference between early and late replicating euchromatin and the satellite DNA becomes obvious. In addition, the usage of replication origins changes [28].

Interphase 14 corresponds to the MBT in *Drosophila*. Interphase 14 is the stage when the cell cycle pauses in a G2 phase, zygotic transcription strongly increases, and DNA replication switches to a slow replication mode. During interphase 14, visible morphology changes from the syncytial to cellular blastoderm, in a process called cellularization. Cellularization is the first morphological process that depends on zygotic gene products [29, 30].

However, the first signs of MBT are already visible earlier. As mentioned above, the extending interphases in NC11–14 depend on zygotic transcription. The first transcripts and activated RNA polymerase II (Pol II) can be already detected in pre-blastoderm stages. Transcription slowly increases until cycle 12. In cycle 13 many zygotic genes are clearly expressed [31]. Genome-wide analysis showed that gene expression is initiated at different time points throughout early development [32, 33], suggesting that rather than a sharp switch, MZT is likely regulated by multiple and diverse mechanisms [9, 34, 35]. The timing of these multiple and diverse mechanisms depends, to a certain degree, on the ratio of nuclear and cytoplasmic content (N:C ratio). This is further discussed in Subheading 5.

Approximately, two-thirds of all genes are contained in *Drosophila* eggs as maternal mRNAs [34, 36]. A third of all maternal transcripts are eliminated in stages leading to MBT in three ways [36]: First, maternally encoded factors activate mRNA degradation of over 20% of maternal transcripts after egg activation in a ZGA-independent manner [34, 37–39]. The RNA-binding protein Smaug is such a factor, acting together with the CCR4/POP2/NOT deadenylase complex [38, 40, 41]. Another RNA-binding protein, Brain Tumor, functions in a similar way [42]. Second, 15% of maternal mRNAs are eliminated depending on zygotic transcription during MBT [43, 44]. Third, microRNAs induce maternal RNA degradation. More than 100 maternal transcripts are degraded depending on zygotically expressed microRNAs from the *miR-309* cluster, which is activated by the early zygotic transcription factor Vielfältig/Zelda [45–47].

2 Mechanism of Zygotic Genome Activation

Transcription of the zygotic genome only begins shortly after fertilization [48]. The highly dynamic transcription profile was characterized by number of methods, including high-throughput strategies, global run-on sequencing (GRO-seq), and fluorescent labeling of nascent RNA [14, 49–52]. In general, the initiation of low-level zygotic transcription, mostly of signaling and patterning genes, already appears before NC10 ahead of large-scale ZGA [31, 53]. These include small and intron-less genes, as well as genes with TAGteam DNA motif in the control region [36]. A comparable profile is also observed in that of the zebrafish [54]. Full activation of zygotic transcription is observed during MBT, when thousands of genes are transcriptionally activated and transcribed in high levels. Taken together, the activation of the zygotic genome is a gradual process rather than a single sharp switch. This suggests that ZGA is triggered by multiple and diverse events [9, 34, 35].

A contribution to ZGA is intrinsically provided by the division of nuclei and doubling of DNA with every nuclear cycle. Even with a constant activity of the individual zygotic transcription units, the total number of transcripts would exponentially increase. In general, zygotic transcription is quantified in relation to the number of embryos, total mass of embryos (protein or total RNA content), or in comparison to an abundant maternal RNA, such as ribosomal RNA. Most of the older data are based on samples prepared from mixed stages comprising several nuclear division cycles. Alternatively, zygotic transcription may be normalized to the number of nuclei in an embryo. Given recent technological advances, transcription profiling can be conducted with few or even single *Drosophila* embryos, allowing highly accurate staging according to the nuclear division cycle [33, 55]. Such normalization is important to reveal the actual transcriptional activity of a locus.

This hypothesis was tested with normalized transcriptional profiles of selected early zygotic genes (Fig. 1) based on a data set from manually staged embryos [56]. Normalization to the number of nuclei was performed with the assumption of a doubling with every cell cycle. In case of a doubling transcript number from one cycle to the next, this results in a zero value. An increase in transcript number higher than a factor two results in a positive number, whereas an increase less than a factor two, in a negative number (Fig. 1). This simple and exemplary calculation indicates that both the increasing number of nuclei and an increased activity of the transcription units contribute to the overall increase in zygotic transcripts per embryo. There is, however, also transcript-dependent variation. A similar finding was reported recently for dorsoventrally patterning genes [57]. This indicates that depending on the zygotic gene, both an increased activity of individual transcription units and an increased number of transcription units/nuclei contribute to ZGA.

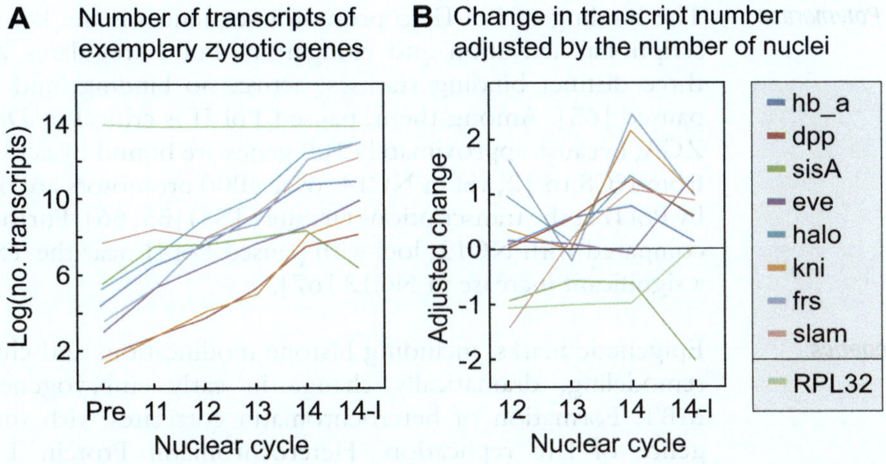

Fig. 1 Zygotic transcription and number of nuclei. (**a**) Number of selected zygotic transcripts based on NanoString analysis with extracts from manually staged embryos plotted on a logarithmic scale. (**b**) The number of transcripts was normalized to the number of nuclei that double with every cycle. Plotted is the difference of log2 of the number of transcripts from one cycle to the previous cycle minus 1. The number of transcripts in pre-blastoderm stages is not included. Transcripts for the ribosomal protein L32 serve as a reference. Staging by the nuclear cycle, pre-blastoderm stage (Pre) and late cellularization (14-l). Data are from Sung et al. [56]

2.1 Vielfältig/Zelda Functions in ZGA Regulation

The zinc-finger protein Vielfältig/Zelda (Vfl/Zld) plays a major role in ZGA. Vfl/Zld specifically binds to TAGteam elements in the early *Drosophila* embryo. The TAGteam CAGGTAG sequence was identified by genome-wide studies as a general *cis*-regulatory element and as the most highly enriched regulatory motif in genes involved in anterior-posterior patterning [36, 58, 59]. Vfl/Zld is an essential transcriptional activator during early zygotic gene expression, as demonstrated by the strongly reduced (but not absent) expression of many early zygotic genes in embryos from females with *Vfl/Zld* mutant germline [60]. *Vfl/Zld* is maternally deposited and uniformly distributed throughout the egg and early embryo. The Vfl/Zld protein levels increase coincidently with the activation of zygotic genome during pre-blastoderm stage, prior to large-scale transcription [49, 61].

Vfl/Zld consists of a cluster of four zinc fingers and a low-complexity activation domain, both of which are required for promoting DNA binding and mediating transcriptional activation [62]. Vfl/Zld binding to promoters is detected already in NC8 for particular genes and roughly a thousand genes during NC10 [63, 64]. The DNA binding is maintained at least until NC14 [49]. During ZGA, Vfl/Zld-binding sites are highly enriched specifically in regions of accessible chromatin, allowing transcription factors to subsequently bind and drive zygotic transcription [63, 64]. Thus, Vfl/Zld acts as a co-activator during MZT. Vfl/Zld also controls the accurate temporal and spatial expression of microRNAs [46].

2.2 RNA Polymerase II Pausing

The binding of Pol II to promotor sequences is the key to transcriptional activation and elongation. Pol II regulates ZGA by three distinct binding statuses: active, no binding, and stalled/paused [65]. Among them, paused Pol II is critical in *Drosophila* ZGA, because approximately 100 genes are bound by active Pol II from NC8 to 12, yet in NC14, over 4000 promotors are occupied by Pol II at the transcription start site (TSS) [55, 66]. Furthermore, compared with NC12, loci with paused Pol II near the TSS show a significant increase in NC13 [67].

2.3 Epigenetics and ZGA

Epigenetic marks, including histone modifications and chromatin remodeling, dramatically change in early embryogenesis and MBT. Formation of heterochromatin correlates with the emergence of late replication. Heterochromatin Protein 1 (HP1) together with histone modifications on H3K9 and H3K4 is involved in establishing of tightly packed chromatin structure [68, 69]. Modifications of lysine acetylation and methylation in histones H3 and H4 appear during MZT. In zebrafish, a striking change in histone modification correlates with ZGA [70]. An increase in histone methylation during MZT matches high level of zygotic transcription [70, 71]. In *Xenopus* embryo, maternally provided histones H3/H4 and their modification states control the regulation of transcriptional activation and cell cycle lengthening [72, 73]. Similarly, during *Drosophila* early development, genome-wide studies showed that domains of histone methylation H3K4me1, H3K4me3, H3K27me3, and H3K36me3 increased from undetectable to widespread level at NC14 [48, 55, 74]. Levels of acetylation on H3K9 appear correspondingly to methylation marks, whereas H3K18ac, H3K27ac, and H4K8ac levels are evidently precocious at NC12 [48]. These early appearing acetylation marks are strongly correlated with maternal DNA-binding protein Vfl/Zld, demonstrating that Vfl/Zld may regulate transcriptional activation by recruiting histone acetylation, thus allowing opening of genome state [34, 48]. In contrast, the mark H4K5ac, whose level was previously shown to bookmark active transcription in mammalian cells, decreases from NC8 with the slowdown of the cleavage cycles [48, 75]. In addition to histone modifications, remodeling of nucleosomes and linker histones with histone variants may contribute to ZGA. *Drosophila* maternal-specific linker histone H1 dBigH1 is replaced by somatic H1 in early development [76]. dBigH1 seems to suppress ZGA, since increased levels of activated Pol II and expression of zygotic genes are observed in embryos with reduced dBigH1 levels [76].

Both histone modification and Vfl/Zld DNA binding ultimately affect transcriptional activation by altering chromatin accessibility. Highly accessible chromatin regions are locally and globally marked by H3/H4 acetylation and Vfl/Zld enrichment from NC8 to 12 in *Drosophila* [77]. In NC13, however, thousands of enhancers and promotors with nucleosome-free regions accumulate additional transcription factors in a cascade way [48, 78]. This phenomenon has also been observed in zebrafish [79].

2.4 Other Regulators

Drosophila zygotic transcription is modulated by multiple factors including *cis*-regulatory elements. For instance, TATA-dependent promoters, as well as enhancers, are central in transcriptional regulation [80, 81]. Distinct enhancer-core-promoter specificities ensure that developmental and housekeeping genes are activated precisely across the entire genome [81]. Likewise, the post-transcriptional regulation of TATA-binding protein (TBP) affects transcription pattern together with the earliest transcribed genes during the MZT [55]. Smaug may involve ZGA regulation through maternal clearance of transcription factor *tramtrack* mRNA, which is involved in triggering transcription of transcripts depending on the N:C ratio [38, 53].

3 Switch in Cell Cycle Mode During the MBT

The cell cycle switch from a fast syncytial mode to a mode with slow replication and extended G2 phase is the most obvious aspect of MBT in morphological terms. A long-standing question is the functional relationship of the cell cycle switch with ZGA. According to one model, the cell cycle switch allows for the strong increase in zygotic transcription (Fig. 2) [82]. In the opposing model, zygotic

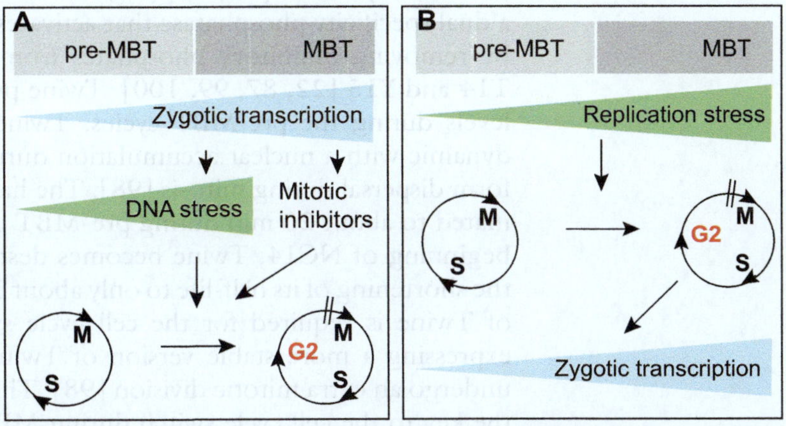

Fig. 2 Models for the control of cell cycle remodeling during MBT. (**a**) The onset of zygotic transcription leads to the activation of the DNA checkpoint due to interference of transcription and replication as well as expression of mitotic inhibitors. These two processes lead to the cell cycle remodeling. (**b**) Activation of the DNA checkpoint, caused by limiting amounts of replication factors, for example, triggers a slowdown and subsequent pause of the cell cycle. The longer interphase promotes zygotic transcription

transcription triggers the remodeling of the cell cycle [56, 67]. Depending on the experimental system, strong experimental evidence speaks in favor of the first or the second model. A synthesis has not been achieved, yet.

3.1 Cell Cycle Regulation in Drosophila Early Embryogenesis

Cyclin and its partner cyclin-dependent kinase (Cdk) are essential for cell cycle control. In *Drosophila*, cyclin A/B/B3:Cdk1 complexes regulate entry into M phase [20, 83]. The rapid S phases in pre-MBT cycles are maternally controlled, and the catalytic activity level of cyclin:Cdk1 complexes determines the timing for mitotic entry [21, 84]. Distinct mechanisms regulate cyclin:Cdk1 complexes in pre-MBT: First, during each nuclear division, Cyclin A, B and B3 proteins are synthesized in S phase by maternally supplied mRNA [85, 86], and degraded in mitosis by the ubiquitin pathway [87, 88]. Cyclin A, B, and B3 fulfill a redundant but essential function, as RNAi-mediated depletion stops the syncytial cycles [20, 89]. Cyclin B levels also contribute to the cell cycle switch as changes in *cyclin B* gene dose affect the number of nuclear divisions [90]. Second, the inhibitory phosphorylation of T14Y15 sites of Cdk1 are pairwise regulated by maternally supplied kinases Wee1/Myt1 and phosphatase Cdc25/Twine [85, 91–95]. Therefore, Cdk1 is timely activated and inactivated by controlling T14Y15 inhibitory phosphorylation sites [96].

3.2 Cdc25/Twine Degradation at the MBT

In NC14 and to a certain degree already in NC12 and 13, S phase lengthens and a G2 phase is introduced. Central to these changes is the induced inactivation and final degradation of the phosphatase Cdc25/Twine [97, 98] (Fig. 3). *Drosophila* Cdc25/Twine is a dual specificity phosphatase that activates cyclin:Cdk1 complexes by removing inhibitory phosphates from the ATP-binding sites T14 and Y15 [22, 87, 99, 100]. Twine protein is present in high levels during the pre-MBT cycles. Twine protein localization is dynamic with a nuclear accumulation during interphases and uniform dispersal during mitosis [98]. The half-life of Twine was estimated to about 20 min during pre-MBT cycles [98]. Yet with the beginning of NC14, Twine becomes destabilized as indicated by the shortening of its half-life to only about 5 min [98]. Degradation of Twine is required for the cell cycle switch because embryos expressing a more stable version of Twine protein (Twine$^{106\text{-}180}$) undergo an extra mitotic division [98]. The rapid destabilization is the key to the cell cycle switch during MBT, as it depends on the N:C ratio and on zygotic transcription [98].

Prior to MBT, the steady-state level of Twine is relatively stable due to balanced synthesis and degradation. The link of zygotic transcription and the switch-like decrease in the half-life of Twine suggests that zygotic factors may be involved. One of these factors is the pseudokinase Tribbles [101–103], as RNAi-mediated depletion of *tribbles* accelerates Twine degradation [97].

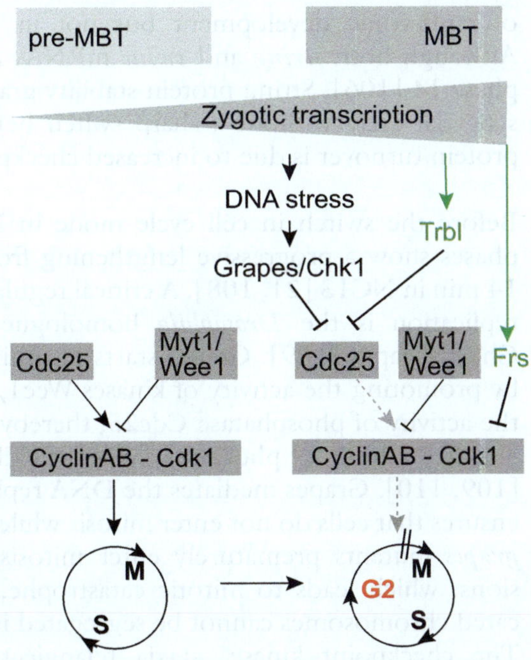

Fig. 3 Model of cell cycle remodeling in Drosophila. Cyclin:Cdk1 is activated by the phosphatase Cdc25 and inactivated by the kinases Myt1/Wee1. In pre-MBT Cyclin:Cdk1 activity is high and promotes fast cell cycles. During MBT the balance of Cyclin:Cdk1 control is shifted toward low activity. Cdc25 is inhibited by the DNA checkpoint, which is activated by DNA stress caused by interference of DNA replication and zygotic transcription. In addition, the zygotic mitotic inhibitors, Tribbles and Frühstart, promote Cdc25 degradation and inhibition of the Cyclin:Cdk1 complexes, respectively

However, *tribbles* is not essential for the cell cycle switch, since embryos deficient for maternal and zygotic *tribbles* do not undergo an extra nuclear cycle [101, 102]. The mechanism for how *tribbles* induces Twine degradation remains unknown, but in other organisms such as yeast, *Xenopus*, and human cells, Cdc25 (or Cdc25C) degradation is induced by phosphorylation due to multiple pathways [56, 104, 105]. In addition to induced destabilization of Cdc25/Twine at NC14, additional mechanisms control pre-MBT levels and activity of Twine. The number of pre-MBT cell cycles is rather insensitive to changes in *twine* gene dose. A tripling of *twine* gene dose to 6×*twine*[+] induces an extra nuclear division in only a few embryos [106], suggesting that mechanisms exist that make Twine protein levels independent of gene dose.

The second *Drosophila* homologue of Cdc25, String, has distinct developmental functions in cell cycle control [84, 107]. String but not Twine is required for mitotic entry in zygotically controlled cycles 14–16. In contrast to these later stages, *string* is not required for progression of the syncytial cell cycles [84]. Premature expression of *string* is sufficient to trigger mitotic entry during later stages of embryonic development but not in pre-MBT stages [84]. Although both *string* and *twine* mRNAs are destructed in interphase 14 [106], String protein stability gradually decreases during syncytial cycle without a sharp switch before MBT [97]. String protein turnover is due to increased checkpoint activity [98].

3.3 DNA Replication Checkpoint at NC13

Before the switch in cell cycle mode in NC14 in *Drosophila*, S phases show a progressive lengthening from 3.4 min in NC8 to 14 min in NC13 [21, 108]. A critical regulator of the slowdown of replication is the *Drosophila* homologue of checkpoint kinase Chk1, Grapes [109]. Grapes starts to inhibit cyclin:Cdk1 activity by promoting the activity of kinases Wee1/Myt1 and suppressing the activity of phosphatase Cdc25, thereby shifting the balance to T14Y15 inhibitory phosphorylation of Cdk1 from NC11 onward [109, 110]. Grapes mediates the DNA replication checkpoint and ensures that cells do not enter mitosis while replication is ongoing. *grapes* mutants prematurely enter mitosis during syncytial divisions, which leads to mitotic catastrophe, as incompletely replicated chromosomes cannot be segregated in anaphase [109, 110]. The checkpoint kinase, ataxia telangiectasia and Rad3-related (ATR, Mei-41 in *Drosophila*), acts upstream and activates Chk1/Grapes similar as in *Xenopus* [111, 112]. *mei-41* mutants show a similar phenotype during syncytial divisions as *grapes*, indicating a functional replication checkpoint is required at the MBT [67].

In Drosophila the DNA checkpoint is triggered by ZGA. Blocking transcription by α-amanitin in *Drosophila* pre-MBT embryos does not suppress lethality of *mei-41* mutant [67]. Nonetheless, embryos from *mei-41 Vfl/Zld* double mutant mothers could partially suppress the mitotic catastrophe, indicating that replication has been finished in time [67]. These observations are consistent with the model that zygotic transcription reduces replication speed and induces DNA stress, leading to DNA checkpoint activation at ZGA [56, 67].

3.4 Other Regulators

In *Drosophila*, cyclin-dependent kinase inhibitor (CKI) Frühstart is another zygotic regulator, which functions to inhibit cyclin:Cdk1 activity by binding the hydrophobic patch of cyclins, thereby interfering with Cdk1 substrate recognition [101, 113, 114]. Together with large-scale ZGA, *frühstart* starts transcription immediately after mitosis 13, and generates a uniform cell cycle pause in cycle 14 [114]. In the absence of Frühstart, embryos enter an extra round of nuclear division especially in embryos with extra copies of *twine*[+] [114]. The expression of Frühstart depends on the N:C ratio, suggesting that Frühstart is involved in the link of N:C with cell cycle regulation [115]. Wee1 and Myt1 kinases are Cdk1 inhibitors that oppose functions to Cdc25 phosphatases [91–93, 116, 117] (Fig. 3). Wee1 can be activated by Grapes, and inhibits Cdk1 activity by adding inhibitory phosphorylation at T14 and Y15 sites [9, 118, 119]. Cyclin:Cdk1 activity is also influenced by some other factors such as mitotic kinase Aurora-A and acquisition of late-replicating heterochromatin domains [95, 120].

In summary, the switch of the cell cycle from a fast syncytial mode to a slow embryonic mode is controlled on two levels of inhibition: (1) indirectly by interference of zygotic transcription with DNA replication and subsequent activation of the DNA checkpoint, (2) directly by expression of zygotic genes encoding mitosis inhibitors.

4 What Is the Trigger for MBT?

The MBT cell cycle switch depends on ZGA (Fig. 2). First, injection of α-amanitin, a Pol II inhibitor, before MBT induces an extra synchronized mitotic division, indicating that widespread zygotic transcription is required for the cell cycle switch in *Drosophila* [106]. Second, ZGA correlates with DNA stress. About 80% of the RpA-70-GFP-binding sites in early MBT cycles also have RNA Pol II bound [67]. RpA70-GFP marks sites of DNA stress [121]. This indicates that ZGA causes DNA stress and activates the DNA checkpoint [67]. Third, a precocious onset of zygotic transcription is sufficient for an earlier MBT [56]. Fourth, dependent on ZGA, Tribbles and other factors trigger Twine destruction in NC14, resulting in inhibition of Cdk1 activation, thereby pausing the cell cycle [101, 102].

The essential role of the DNA checkpoint for triggering MBT was initially shown by the analysis of the checkpoint mutants, *grapes*/Chk1 and *mei-41*/ATR, in *Drosophila* [109, 111]. Embryos from *grapes* females do not switch the cell cycle mode and do not enter MBT, indicating that the DNA checkpoint is required for MBT in *Drosophila* [67, 109]. Based on the observation that *grapes* embryos would not express zygotic genes, the authors concluded that the checkpoint would be upstream of ZGA [109]. Recent data clearly show, however, that ZGA is normal in checkpoint-deficient embryos and that the initial observation was probably due to technical difficulties in detecting expression of early zygotic genes [67].

An alternative source for checkpoint activation beside interference of replication and transcription are limiting amounts of replications factors. Experiments from mostly *Xenopus* support this model (Fig. 2). In *Xenopus* embryos slowdown of DNA replication has been proposed to be upstream of ZGA [82]. The replications factors Cut5, RecQ4, Treslin, and Drf1 become limiting in MBT, which leads to an activation of the DNA checkpoint, slowdown of the cell cycle, and ZGA [82].

In summary, in vivo and genetic experiments provide strong evidence for the model that ZGA is the trigger for MBT in *Drosophila*. ZGA acts upstream of cell cycle control, including the DNA checkpoint and degradation of Cdc25/Twine. First, ZGA is required for MBT and timely cell cycle pause; second, ZGA is associated with induction of replication stress in time and space (on the chromosome); third, precocious ZGA leads to precocious MBT. In other organisms experimental evidence mainly in *Xenopus* speaks in favor of the alternative model, i.e., that cell cycle control acts upstream ZGA. However not all three criteria are fulfilled in vivo: the mechanism should be necessary, sufficient, and temporally and spatially associated with MBT.

5 What Is the Timer for MBT?

A central unresolved question concerning MBT is the timing mechanism for the associated processes including ZGA and number of pre-MBT cell cycles. Tight control of the cell cycle is important for further embryonic development, since the number of divisions determines the cell number and size. Too few cells may be incompatible with the formation of stripes of pair-rule gene expression, for example, as stripes should be at least one cell wide.

5.1 Molecular Clocks

With the onset of embryonic development, fertilization may trigger a molecular clock, on which MBT and its associated processes may depend. A conceivable mechanism is translation of certain maternal mRNAs, which would lead to a time-dependent accumulation of the product following onset after fertilization. Translational regulators such as FMRP are required for MBT regulation in *Drosophila*, through dynamically regulating RNA metabolism and controlling the availability of specific transcripts, as well as mediating the *frühstart* mRNA activation level [122, 123]. A target for translational regulation may be Vfl/Zld, whose protein level increases during blastoderm concomitantly with activation of zygotic transcription [34, 124].

Maternal RNA degradation may represent a second such a mechanism constituting a molecular clock. A large fraction of these maternal RNAs is degraded following egg activation and independent of zygotic transcription. For some RNAs at least, the degradation proceeds with a constant speed [38, 56], and may in this manner constitute a molecular clock. It has been proposed that the speed of RNA degradation affects the number of nuclear divisions, as expression levels of *smaug* affect the timing of MBT [40, 125]. Distinct from Vfl/Zld, Smaug reaches its peak expression level at NC10, and performs downregulation at the MBT [38, 125]. Smaug is functional to mRNA clearance, and times the ZGA through inducing the destruction of maternal transcriptional inhibitor [27].

5.2 N:C Ratio as a Clock

In contrast to a molecular clock as an absolute timer, more evidence speaks in favor of a regulatory process. The morphologically visible MBT depends on genome ploidy, because haploid embryos undergo one more division and tetraploid embryos, one less division [11]. It has been proposed that the N:C ratio represents the timer for MBT. Nuclear content is determined by the amount of DNA or chromatin, which doubles with every cell cycle, whereas cytoplasmic content remains constant during cleavage divisions. The embryo may measure the N:C in that the increasing amount of chromatin titrates a constant cytoplasmic factor until this becomes rate-limiting [6, 53]. Potential cytoplasmic factors are repressors of transcription, replication, or the cell cycle, for example. In *Xenopus* embryos, DNA content is important for MBT [5, 7]. Injection of purified DNA leads to precocious onset of zygotic transcription, as measured by total transcription rate [7]. However, the amount of DNA seems not to be the only determinant, since an increased or decreased nuclear volume, while keeping the DNA content unchanged, leads to a precocious or delayed MBT including zygotic activation and corresponding cell cycle remodeling [126]. Similar findings come from zebrafish that the timing of ZGA is governed by the N:C ratio [13].

It is unclear what is titrated by the exponentially increasing amount of DNA and chromatin, but maternal histones proteins H3/H4 may be a central factor [72]. Depletion and overexpression of H3/H4 delay the cell cycle switch, and also induce premature transcriptional activation [72]. In *Drosophila* embryos, the maternal form of the linker histone H1 dBigH1 has been implicated in the timing of MBT [76]. Maternal dBigH1 is replaced by the somatic form in early embryogenesis. Embryos with half of the maternal contribution and lacking zygotic expression show increased levels of activated Pol II and zygotic gene expression. However, the link of dBigH1 to MBT remains unclear as mutant defects and embryonic genotypes were not analyzed with sufficiently high temporal resolution and with respect to MBT and ZGA.

The replication factors Cut5, RecQ4, Treslin, and Drf1 have been found to be limiting for replication initiation during MBT in *Xenopus* embryos [82]. Titration of the maternal pool of these replication factors by the exponentially increasing chromatin leads to slower replication initiation, ZGA, longer interphases, and DNA checkpoint activation.

Other cytoplasmic factors may also be titrated, such as metabolites. It has been proposed that deoxynucleotides may serve as a marker for the cytoplasm [127]. The maternal pool may be incorporated in the exponentially increasing amounts of DNA. The existence of such a maternal pool is well known, as inhibition of zygotic synthesis by hydroxyurea (HU), which inhibits the NDP reductase, causes a cell cycle arrest only briefly before MBT [127].

Although it is clear that ploidy determines the number of pre-MBT cell cycles in model organisms, it is much less clear whether all of the MBT-associated processes, including ZGA, cell cycle, RNA degradation, are controlled by the N:C ratio. Haploid *Drosophila* embryos switch the cell cycle mode only after an extra division 14 in NC15 [115, 128]. In contrast, ZGA does not depend on the N:C ratio in *Drosophila*. Although older data indicated a link of ploidy and ZGA in *Drosophila* [53], genome-wide analysis of embryonic transcripts with carefully staged *Drosophila* embryos revealed that the majority of zygotic transcripts (127 out of 215 genes) show an expression profile comparable between haploid and diploid embryos [115]. These data suggest that ZGA timing is controlled by a molecular clock in *Drosophila*. However, a small set of zygotic transcripts (88 out of 215 genes) shows clearly delayed expression in haploid embryos [115]. This small gene set includes genes encoding mitotic inhibitors such as Frühstart [114], which are involved in the MBT-associated remodeling of the cell cycle.

6 Conclusions

Recent years brought striking advances in our understanding of zygotic genome activation and its relation to MBT. This is mainly due to improved technology now allowing to analyze transcriptional activity and chromosome status with high resolution and importantly with very little material, down to single embryos. In this way, the variation and limited temporal resolution of mixtures of many embryos can be overcome. Despite this progress, there is no unifying model for zygotic genome activation, MBT, and cell cycle control. Conclusion on central questions and favored models depend on the experimental system. Strong evidence supports the model that DNA replication onset triggers MBT and ZGA in *Xenopus*. However, the alternative model is supported by convincing experiments from *Drosophila*, where ZGA triggers MBT and cell cycle remodeling. It will be the task for future work to reconcile these opposing views. Having the new technologies available and standardized, we can expect new and surprising findings to come.

Acknowledgment

BL was supported by China Scholarship Council. The work in JG's laboratory was in part supported by the German Research Council (Deutsche Forschungsgemeinschaft (DFG) GR1945/3-1, SFB937/TP10).

References

1. Hiiragi T, Solter D (2004) First cleavage plane of the mouse egg is not predetermined but defined by the topology of the two apposing pronuclei. Nature 430(6997):360–364. doi:10.1038/nature02595
2. O'Farrell PH, Stumpff J, Su TT (2004) Embryonic cleavage cycles: how is a mouse like a fly? Curr Biol 14(1):R35–R45
3. O'Farrell PH (2015) Growing an embryo from a single cell: a hurdle in animal life. Cold

Spring Harb Perspect Biol 7(11):a019042. doi:10.1101/cshperspect.a019042

4. Boveri T (1893) An organism produced sexually without characteristics of the mother. Am Soc Nat 27(315):222–232
5. Gerhart JC (1980) Mechanisms regulating pattern formation in the amphibian egg and early embryo. In: Goldberger R (ed) Biological regulation and development, vol 2. Springer, Boston, MA, pp 133–316
6. Newport J, Kirschner M (1982) A major developmental transition in early Xenopus embryos: I. Characterization and timing of cellular changes at the midblastula stage. Cell 30(3):675–686
7. Newport J, Kirschner M (1982) A major developmental transition in early Xenopus embryos: II. Control of the onset of transcription. Cell 30(3):687–696
8. Newport JW, Kirschner MW (1984) Regulation of the cell cycle during early Xenopus development. Cell 37(3):731–742
9. Farrell JA, O'Farrell PH (2014) From egg to gastrula: how the cell cycle is remodeled during the Drosophila mid-blastula transition. Annu Rev Genet 48:269–294. doi:10.1146/annurev-genet-111212-133531
10. Collart C, Owens ND, Bhaw-Rosun L, Cooper B, De Domenico E, Patrushev I, Sesay AK, Smith JN, Smith JC, Gilchrist MJ (2014) High-resolution analysis of gene activity during the Xenopus mid-blastula transition. Development 141(9):1927–1939. doi:10.1242/dev.102012
11. Kane DA, Kimmel CB (1993) The zebrafish midblastula transition. Development 119(2):447–456
12. Zamir E, Kam Z, Yarden A (1997) Transcription-dependent induction of G1 phase during the zebra fish midblastula transition. Mol Cell Biol 17(2):529–536
13. Zhang M, Kothari P, Mullins M, Lampson MA (2014) Regulation of zygotic genome activation and DNA damage checkpoint acquisition at the mid-blastula transition. Cell Cycle 13(24):3828–3838. doi:10.4161/15384101.2014.967066
14. Lee MT, Bonneau AR, Takacs CM, Bazzini AA, DiVito KR, Fleming ES, Giraldez AJ (2013) Nanog, Pou5f1 and SoxB1 activate zygotic gene expression during the maternal-to-zygotic transition. Nature 503(7476):360–364. doi:10.1038/nature12632
15. Robertson S, Lin R (2015) The maternal-to-zygotic transition in C. elegans. Curr Top Dev Biol 113:1–42. doi:10.1016/bs.ctdb.2015.06.001
16. Guven-Ozkan T, Nishi Y, Robertson SM, Lin R (2008) Global transcriptional repression in C. elegans germline precursors by regulated sequestration of TAF-4. Cell 135(1):149–160. doi:10.1016/j.cell.2008.07.040
17. Sulston JE, Schierenberg E, White JG, Thomson JN (1983) The embryonic cell lineage of the nematode Caenorhabditis elegans. Dev Biol 100(1):64–119
18. Rose L, Gonczy P (2014) Polarity establishment, asymmetric division and segregation of fate determinants in early C. elegans embryos. WormBook:1–43. doi:10.1895/wormbook.1.30.2
19. Foe VE, Alberts BM (1983) Studies of nuclear and cytoplasmic behaviour during the five mitotic cycles that precede gastrulation in Drosophila embryogenesis. J Cell Sci 61:31–70
20. McCleland ML, O'Farrell PH (2008) RNAi of mitotic cyclins in Drosophila uncouples the nuclear and centrosome cycle. Curr Biol 18(4):245–254. doi:10.1016/j.cub.2008.01.041
21. Shermoen AW, McCleland ML, O'Farrell PH (2010) Developmental control of late replication and S phase length. Curr Biol 20(23):2067–2077. doi:10.1016/j.cub.2010.10.021
22. Farrell JA, Shermoen AW, Yuan K, O'Farrell PH (2012) Embryonic onset of late replication requires Cdc25 down-regulation. Genes Dev 26(7):714–725. doi:10.1101/gad.186429.111
23. Rabinowitz M (1941) Studies on the cytology and early embryology of the egg of Drosophila melanogaster. J Morphol 69(1):1–49
24. Axton JM, Shamanski FL, Young LM, Henderson DS, Boyd JB, Orr-Weaver TL (1994) The inhibitor of DNA replication encoded by the Drosophila gene plutonium is a small, ankyrin repeat protein. EMBO J 13(2):462–470
25. Fenger DD, Carminati JL, Burney-Sigman DL, Kashevsky H, Dines JL, Elfring LK, Orr-Weaver TL (2000) PAN GU: a protein kinase that inhibits S phase and promotes mitosis in early Drosophila development. Development 127(22):4763–4774
26. Lee LA, Van Hoewyk D, Orr-Weaver TL (2003) The Drosophila cell cycle kinase PAN GU forms an active complex with PLUTONIUM and GNU to regulate embryonic divisions. Genes Dev 17(23):2979–2991. doi:10.1101/gad.1132603
27. Laver JD, Marsolais AJ, Smibert CA, Lipshitz HD (2015) Regulation and function of maternal gene products during the maternal-to-zygotic transition in Drosophila. Curr Top Dev Biol 113:43–84. doi:10.1016/bs.ctdb.2015.06.007

28. Schubeler D, Scalzo D, Kooperberg C, van Steensel B, Delrow J, Groudine M (2002) Genome-wide DNA replication profile for Drosophila melanogaster: a link between transcription and replication timing. Nat Genet 32(3):438–442. doi:10.1038/ng1005
29. Edgar BA, Schubiger G (1986) Parameters controlling transcriptional activation during early Drosophila development. Cell 44(6): 871–877
30. Merrill PT, Sweeton D, Wieschaus E (1988) Requirements for autosomal gene activity during precellular stages of Drosophila melanogaster. Development 104(3):495–509
31. Ali-Murthy Z, Lott SE, Eisen MB, Kornberg TB (2013) An essential role for zygotic expression in the pre-cellular Drosophila embryo. PLoS Genet 9(4):e1003428. doi:10.1371/journal.pgen.1003428
32. Graveley BR, Brooks AN, Carlson JW, Duff MO, Landolin JM, Yang L, Artieri CG, van Baren MJ, Boley N, Booth BW, Brown JB, Cherbas L, Davis CA, Dobin A, Li R, Lin W, Malone JH, Mattiuzzo NR, Miller D, Sturgill D, Tuch BB, Zaleski C, Zhang D, Blanchette M, Dudoit S, Eads B, Green RE, Hammonds A, Jiang L, Kapranov P, Langton L, Perrimon N, Sandler JE, Wan KH, Willingham A, Zhang Y, Zou Y, Andrews J, Bickel PJ, Brenner SE, Brent MR, Cherbas P, Gingeras TR, Hoskins RA, Kaufman TC, Oliver B, Celniker SE (2011) The developmental transcriptome of Drosophila melanogaster. Nature 471(7339):473–479. doi:10.1038/nature09715
33. Lott SE, Villalta JE, Schroth GP, Luo S, Tonkin LA, Eisen MB (2011) Noncanonical compensation of zygotic X transcription in early Drosophila melanogaster development revealed through single-embryo RNA-seq. PLoS Biol 9(2):e1000590. doi:10.1371/journal.pbio.1000590
34. Harrison MM, Eisen MB (2015) Transcriptional activation of the zygotic genome in Drosophila. Curr Top Dev Biol 113:85–112. doi:10.1016/bs.ctdb.2015.07.028
35. Lee MT, Bonneau AR, Giraldez AJ (2014) Zygotic genome activation during the maternal-to-zygotic transition. Annu Rev Cell Dev Biol 30:581–613. doi:10.1146/annurev-cellbio-100913-013027
36. De Renzis S, Elemento O, Tavazoie S, Wieschaus EF (2007) Unmasking activation of the zygotic genome using chromosomal deletions in the Drosophila embryo. PLoS Biol 5(5):e117. doi:10.1371/journal.pbio.0050117
37. Tadros W, Houston SA, Bashirullah A, Cooperstock RL, Semotok JL, Reed BH, Lipshitz HD (2003) Regulation of maternal transcript destabilization during egg activation in Drosophila. Genetics 164(3):989–1001
38. Tadros W, Goldman AL, Babak T, Menzies F, Vardy L, Orr-Weaver T, Hughes TR, Westwood JT, Smibert CA, Lipshitz HD (2007) SMAUG is a major regulator of maternal mRNA destabilization in Drosophila and its translation is activated by the PAN GU kinase. Dev Cell 12(1):143–155. doi:10.1016/j.devcel.2006.10.005
39. Tadros W, Lipshitz HD (2009) The maternal-to-zygotic transition: a play in two acts. Development 136(18):3033–3042. doi:10.1242/dev.033183
40. Semotok JL, Cooperstock RL, Pinder BD, Vari HK, Lipshitz HD, Smibert CA (2005) Smaug recruits the CCR4/POP2/NOT deadenylase complex to trigger maternal transcript localization in the early Drosophila embryo. Curr Biol 15(4):284–294. doi:10.1016/j.cub.2005.01.048
41. Chen L, Dumelie JG, Li X, Cheng MH, Yang Z, Laver JD, Siddiqui NU, Westwood JT, Morris Q, Lipshitz HD, Smibert CA (2014) Global regulation of mRNA translation and stability in the early Drosophila embryo by the Smaug RNA-binding protein. Genome Biol 15(1):R4. doi:10.1186/gb-2014-15-1-r4
42. Laver JD, Li X, Ray D, Cook KB, Hahn NA, Nabeel-Shah S, Kekis M, Luo H, Marsolais AJ, Fung KY, Hughes TR, Westwood JT, Sidhu SS, Morris Q, Lipshitz HD, Smibert CA (2015) Brain tumor is a sequence-specific RNA-binding protein that directs maternal mRNA clearance during the Drosophila maternal-to-zygotic transition. Genome Biol 16:94. doi:10.1186/s13059-015-0659-4
43. Walser CB, Lipshitz HD (2011) Transcript clearance during the maternal-to-zygotic transition. Curr Opin Genet Dev 21(4):431–443. doi:10.1016/j.gde.2011.03.003
44. Bashirullah A, Halsell SR, Cooperstock RL, Kloc M, Karaiskakis A, Fisher WW, Fu W, Hamilton JK, Etkin LD, Lipshitz HD (1999) Joint action of two RNA degradation pathways controls the timing of maternal transcript elimination at the midblastula transition in Drosophila melanogaster. EMBO J 18(9): 2610–2620. doi:10.1093/emboj/18.9.2610
45. Bushati N, Stark A, Brennecke J, Cohen SM (2008) Temporal reciprocity of miRNAs and their targets during the maternal-to-zygotic transition in Drosophila. Curr Biol 18(7): 501–506. doi:10.1016/j.cub.2008.02.081
46. Fu S, Nien CY, Liang HL, Rushlow C (2014) Co-activation of microRNAs by Zelda is essential for early Drosophila development. Development 141(10):2108–2118. doi:10.1242/dev.108118

47. Huntzinger E, Izaurralde E (2011) Gene silencing by microRNAs: contributions of translational repression and mRNA decay. Nat Rev Genet 12(2):99–110. doi:10.1038/nrg2936
48. Li XY, Harrison MM, Villalta JE, Kaplan T, Eisen MB (2014) Establishment of regions of genomic activity during the Drosophila maternal to zygotic transition. Elife 3. doi:10.7554/eLife.03737
49. Harrison MM, Li XY, Kaplan T, Botchan MR, Eisen MB (2011) Zelda binding in the early Drosophila melanogaster embryo marks regions subsequently activated at the maternal-to-zygotic transition. PLoS Genet 7(10):e1002266. doi:10.1371/journal.pgen.1002266
50. Lecuyer E, Yoshida H, Parthasarathy N, Alm C, Babak T, Cerovina T, Hughes TR, Tomancak P, Krause HM (2007) Global analysis of mRNA localization reveals a prominent role in organizing cellular architecture and function. Cell 131(1):174–187. doi:10.1016/j.cell.2007.08.003
51. Saunders A, Core LJ, Sutcliffe C, Lis JT, Ashe HL (2013) Extensive polymerase pausing during Drosophila axis patterning enables high-level and pliable transcription. Genes Dev 27(10):1146–1158. doi:10.1101/gad.215459.113
52. Ferraro T, Lucas T, Clemot M, De Las Heras Chanes J, Desponds J, Coppey M, Walczak AM, Dostatni N (2016) New methods to image transcription in living fly embryos: the insights so far, and the prospects. Wiley Interdiscip Rev Dev Biol 5(3):296–310. doi:10.1002/wdev.221
53. Pritchard DK, Schubiger G (1996) Activation of transcription in Drosophila embryos is a gradual process mediated by the nucleocytoplasmic ratio. Genes Dev 10(9):1131–1142
54. Heyn P, Kircher M, Dahl A, Kelso J, Tomancak P, Kalinka AT, Neugebauer KM (2014) The earliest transcribed zygotic genes are short, newly evolved, and different across species. Cell Rep 6(2):285–292. doi:10.1016/j.celrep.2013.12.030
55. Chen K, Johnston J, Shao W, Meier S, Staber C, Zeitlinger J (2013) A global change in RNA polymerase II pausing during the Drosophila midblastula transition. Elife 2:e00861. doi:10.7554/eLife.00861
56. Sung HW, Spangenberg S, Vogt N, Grosshans J (2013) Number of nuclear divisions in the Drosophila blastoderm controlled by onset of zygotic transcription. Curr Biol 23(2):133–138. doi:10.1016/j.cub.2012.12.013
57. Sandler JE, Stathopoulos A (2016) Quantitative single-embryo profile of Drosophila genome activation and the dorsal-ventral patterning network. Genetics 202(4):1575–1584. doi:10.1534/genetics.116.186783
58. ten Bosch JR, Benavides JA, Cline TW (2006) The TAGteam DNA motif controls the timing of Drosophila pre-blastoderm transcription. Development 133(10):1967–1977. doi:10.1242/dev.02373
59. Li XY, MacArthur S, Bourgon R, Nix D, Pollard DA, Iyer VN, Hechmer A, Simirenko L, Stapleton M, Luengo Hendriks CL, Chu HC, Ogawa N, Inwood W, Sementchenko V, Beaton A, Weiszmann R, Celniker SE, Knowles DW, Gingeras T, Speed TP, Eisen MB, Biggin MD (2008) Transcription factors bind thousands of active and inactive regions in the Drosophila blastoderm. PLoS Biol 6(2):e27. doi:10.1371/journal.pbio.0060027
60. Liang HL, Nien CY, Liu HY, Metzstein MM, Kirov N, Rushlow C (2008) The zinc-finger protein Zelda is a key activator of the early zygotic genome in Drosophila. Nature 456(7220):400–403. doi:10.1038/nature07388
61. Staudt N, Fellert S, Chung HR, Jackle H, Vorbruggen G (2006) Mutations of the Drosophila zinc finger-encoding gene vielfaltig impair mitotic cell divisions and cause improper chromosome segregation. Mol Biol Cell 17(5):2356–2365. doi:10.1091/mbc.E05-11-1056
62. Hamm DC, Bondra ER, Harrison MM (2015) Transcriptional activation is a conserved feature of the early embryonic factor Zelda that requires a cluster of four zinc fingers for DNA binding and a low-complexity activation domain. J Biol Chem 290(6):3508–3518. doi:10.1074/jbc.M114.602292
63. Foo SM, Sun Y, Lim B, Ziukaite R, O'Brien K, Nien CY, Kirov N, Shvartsman SY, Rushlow CA (2014) Zelda potentiates morphogen activity by increasing chromatin accessibility. Curr Biol 24(12):1341–1346. doi:10.1016/j.cub.2014.04.032
64. Schulz KN, Bondra ER, Moshe A, Villalta JE, Lieb JD, Kaplan T, McKay DJ, Harrison MM (2015) Zelda is differentially required for chromatin accessibility, transcription factor binding, and gene expression in the early Drosophila embryo. Genome Res 25(11):1715–1726. doi:10.1101/gr.192682.115
65. Zeitlinger J, Stark A, Kellis M, Hong JW, Nechaev S, Adelman K, Levine M, Young RA (2007) RNA polymerase stalling at developmental control genes in the Drosophila melanogaster embryo. Nat Genet 39(12):1512–1516. doi:10.1038/ng.2007.26

66. Boettiger AN, Levine M (2009) Synchronous and stochastic patterns of gene activation in the Drosophila embryo. Science 325(5939): 471–473. doi:10.1126/science.1173976
67. Blythe SA, Wieschaus EF (2015) Zygotic genome activation triggers the DNA replication checkpoint at the midblastula transition. Cell 160(6):1169–1181. doi:10.1016/j.cell.2015.01.050
68. Rudolph T, Yonezawa M, Lein S, Heidrich K, Kubicek S, Schafer C, Phalke S, Walther M, Schmidt A, Jenuwein T, Reuter G (2007) Heterochromatin formation in Drosophila is initiated through active removal of H3K4 methylation by the LSD1 homolog SU(VAR)3-3. Mol Cell 26(1):103–115. doi:10.1016/j.molcel.2007.02.025
69. Yuan K, O'Farrell PH (2016) TALE-light imaging reveals maternally guided, H3K9me2/3-independent emergence of functional heterochromatin in Drosophila embryos. Genes Dev. doi:10.1101/gad.272237.115
70. Lindeman LC, Andersen IS, Reiner AH, Li N, Aanes H, Ostrup O, Winata C, Mathavan S, Muller F, Alestrom P, Collas P (2011) Prepatterning of developmental gene expression by modified histones before zygotic genome activation. Dev Cell 21(6):993–1004. doi:10.1016/j.devcel.2011.10.008
71. Vastenhouw NL, Zhang Y, Woods IG, Imam F, Regev A, Liu XS, Rinn J, Schier AF (2010) Chromatin signature of embryonic pluripotency is established during genome activation. Nature 464(7290):922–926. doi:10.1038/nature08866
72. Amodeo AA, Jukam D, Straight AF, Skotheim JM (2015) Histone titration against the genome sets the DNA-to-cytoplasm threshold for the Xenopus midblastula transition. Proc Natl Acad Sci U S A 112(10):E1086–E1095. doi:10.1073/pnas.1413990112
73. Hontelez S, van Kruijsbergen I, Georgiou G, van Heeringen SJ, Bogdanovic O, Lister R, Veenstra GJ (2015) Embryonic transcription is controlled by maternally defined chromatin state. Nat Commun 6:10148. doi:10.1038/ncomms10148
74. Boettiger AN, Bintu B, Moffitt JR, Wang S, Beliveau BJ, Fudenberg G, Imakaev M, Mirny LA, Wu CT, Zhuang X (2016) Super-resolution imaging reveals distinct chromatin folding for different epigenetic states. Nature 529(7586):418–422. doi:10.1038/nature16496
75. Zhao R, Nakamura T, Fu Y, Lazar Z, Spector DL (2011) Gene bookmarking accelerates the kinetics of post-mitotic transcriptional re-activation. Nat Cell Biol 13(11):1295–1304. doi:10.1038/ncb2341
76. Perez-Montero S, Carbonell A, Moran T, Vaquero A, Azorin F (2013) The embryonic linker histone H1 variant of Drosophila, dBigH1, regulates zygotic genome activation. Dev Cell 26(6):578–590. doi:10.1016/j.devcel.2013.08.011
77. Li XY, Thomas S, Sabo PJ, Eisen MB, Stamatoyannopoulos JA, Biggin MD (2011) The role of chromatin accessibility in directing the widespread, overlapping patterns of Drosophila transcription factor binding. Genome Biol 12(4):R34. doi:10.1186/gb-2011-12-4-r34
78. Thomas S, Li XY, Sabo PJ, Sandstrom R, Thurman RE, Canfield TK, Giste E, Fisher W, Hammonds A, Celniker SE, Biggin MD, Stamatoyannopoulos JA (2011) Dynamic reprogramming of chromatin accessibility during Drosophila embryo development. Genome Biol 12(5):R43. doi:10.1186/gb-2011-12-5-r43
79. Zhang Y, Vastenhouw NL, Feng J, Fu K, Wang C, Ge Y, Pauli A, van Hummelen P, Schier AF, Liu XS (2014) Canonical nucleosome organization at promoters forms during genome activation. Genome Res 24(2):260–266. doi:10.1101/gr.157750.113
80. Juven-Gershon T, Kadonaga JT (2010) Regulation of gene expression via the core promoter and the basal transcriptional machinery. Dev Biol 339(2):225–229. doi:10.1016/j.ydbio.2009.08.009
81. Zabidi MA, Arnold CD, Schernhuber K, Pagani M, Rath M, Frank O, Stark A (2015) Enhancer-core-promoter specificity separates developmental and housekeeping gene regulation. Nature 518(7540):556–559. doi:10.1038/nature13994
82. Collart C, Allen GE, Bradshaw CR, Smith JC, Zegerman P (2013) Titration of four replication factors is essential for the Xenopus laevis midblastula transition. Science 341(6148): 893–896. doi:10.1126/science.1241530
83. Lehner CF, O'Farrell PH (1990) Drosophila cdc2 homologs: a functional homolog is coexpressed with a cognate variant. EMBO J 9(11):3573–3581
84. Edgar BA, O'Farrell PH (1990) The three postblastoderm cell cycles of Drosophila embryogenesis are regulated in G2 by string. Cell 62(3):469–480
85. Edgar BA, Sprenger F, Duronio RJ, Leopold P, O'Farrell PH (1994) Distinct molecular mechanism regulate cell cycle timing at successive stages of Drosophila embryogenesis. Genes Dev 8(4):440–452
86. Yuan K, O'Farrell PH (2015) Cyclin B3 is a mitotic cyclin that promotes the metaphase-anaphase transition. Curr Biol 25(6):811–816. doi:10.1016/j.cub.2015.01.053

87. Sigrist S, Ried G, Lehner CF (1995) Dmcdc2 kinase is required for both meiotic divisions during Drosophila spermatogenesis and is activated by the Twine/cdc25 phosphatase. Mech Dev 53(2):247–260
88. Glotzer M, Murray AW, Kirschner MW (1991) Cyclin is degraded by the ubiquitin pathway. Nature 349(6305):132–138. doi:10.1038/349132a0
89. Yuan K, Farrell JA, O'Farrell PH (2012) Different cyclin types collaborate to reverse the S-phase checkpoint and permit prompt mitosis. J Cell Biol 198(6):973–980. doi:10.1083/jcb.201205007
90. Ji JY, Squirrell JM, Schubiger G (2004) Both cyclin B levels and DNA-replication checkpoint control the early embryonic mitoses in Drosophila. Development 131(2):401–411. doi:10.1242/dev.00944
91. Jin Z, Homola EM, Goldbach P, Choi Y, Brill JA, Campbell SD (2005) Drosophila Myt1 is a Cdk1 inhibitory kinase that regulates multiple aspects of cell cycle behavior during gametogenesis. Development 132(18):4075–4085. doi:10.1242/dev.01965
92. Price D, Rabinovitch S, O'Farrell PH, Campbell SD (2000) Drosophila wee1 has an essential role in the nuclear divisions of early embryogenesis. Genetics 155(1):159–166
93. Stumpff J, Duncan T, Homola E, Campbell SD, Su TT (2004) Drosophila Wee1 kinase regulates Cdk1 and mitotic entry during embryogenesis. Curr Biol 14(23):2143–2148. doi:10.1016/j.cub.2004.11.050
94. Edgar BA, O'Farrell PH (1989) Genetic control of cell division patterns in the Drosophila embryo. Cell 57(1):177–187
95. Blythe SA, Wieschaus EF (2015) Coordinating cell cycle remodeling with transcriptional activation at the Drosophila MBT. Curr Top Dev Biol 113:113–148. doi:10.1016/bs.ctdb.2015.06.002
96. Ayeni JO, Varadarajan R, Mukherjee O, Stuart DT, Sprenger F, Srayko M, Campbell SD (2014) Dual phosphorylation of cdk1 coordinates cell proliferation with key developmental processes in Drosophila. Genetics 196(1):197–210. doi:10.1534/genetics.113.156281
97. Farrell JA, O'Farrell PH (2013) Mechanism and regulation of Cdc25/Twine protein destruction in embryonic cell-cycle remodeling. Curr Biol 23(2):118–126. doi:10.1016/j.cub.2012.11.036
98. Di Talia S, She R, Blythe SA, Lu X, Zhang QF, Wieschaus EF (2013) Posttranslational control of Cdc25 degradation terminates Drosophila's early cell-cycle program. Curr Biol 23(2):127–132. doi:10.1016/j.cub.2012.11.029
99. Edgar BA, Lehner CF (1996) Developmental control of cell cycle regulators: a fly's perspective. Science 274(5293):1646–1652
100. Alphey L, Jimenez J, White-Cooper H, Dawson I, Nurse P, Glover DM (1992) twine, a cdc25 homolog that functions in the male and female germline of Drosophila. Cell 69(6):977–988
101. Grosshans J, Wieschaus E (2000) A genetic link between morphogenesis and cell division during formation of the ventral furrow in Drosophila. Cell 101(5):523–531
102. Mata J, Curado S, Ephrussi A, Rorth P (2000) Tribbles coordinates mitosis and morphogenesis in Drosophila by regulating string/CDC25 proteolysis. Cell 101(5):511–522
103. Rorth P, Szabo K, Texido G (2000) The level of C/EBP protein is critical for cell migration during Drosophila oogenesis and is tightly controlled by regulated degradation. Mol Cell 6(1):23–30
104. Frazer C, Young PG (2012) Phosphorylation mediated regulation of Cdc25 activity, localization and stability. In: Huang C (ed) Protein phosphorylation in human health, Biochemistry, genetics and molecular biology. InTech, Rijeka, Croatia, pp 395–436. doi:10.5772/48315
105. Murphy JM, Nakatani Y, Jamieson SA, Dai W, Lucet IS, Mace PD (2015) Molecular mechanism of CCAAT-enhancer binding protein recruitment by the TRIB1 pseudokinase. Structure 23(11):2111–2121. doi:10.1016/j.str.2015.08.017
106. Edgar BA, Datar SA (1996) Zygotic degradation of two maternal Cdc25 mRNAs terminates Drosophila's early cell cycle program. Genes Dev 10(15):1966–1977
107. Chen F, Archambault V, Kar A, Lio P, D'Avino PP, Sinka R, Lilley K, Laue ED, Deak P, Capalbo L, Glover DM (2007) Multiple protein phosphatases are required for mitosis in Drosophila. Curr Biol 17(4):293–303. doi:10.1016/j.cub.2007.01.068
108. Blumenthal AB, Kriegstein HJ, Hogness DS (1974) The units of DNA replication in Drosophila melanogaster chromosomes. Cold Spring Harb Symp Quant Biol 38:205–223
109. Sibon OC, Stevenson VA, Theurkauf WE (1997) DNA-replication checkpoint control at the Drosophila midblastula transition. Nature 388(6637):93–97. doi:10.1038/40439
110. Fogarty P, Campbell SD, Abu-Shumays R, Phalle BS, Yu KR, Uy GL, Goldberg ML, Sullivan W (1997) The Drosophila grapes gene is related to checkpoint gene chk1/

rad27 and is required for late syncytial division fidelity. Curr Biol 7(6):418–426
111. Sibon OC, Laurencon A, Hawley R, Theurkauf WE (1999) The Drosophila ATM homologue Mei-41 has an essential checkpoint function at the midblastula transition. Curr Biol 9(6):302–312
112. Shimuta K, Nakajo N, Uto K, Hayano Y, Okazaki K, Sagata N (2002) Chk1 is activated transiently and targets Cdc25A for degradation at the Xenopus midblastula transition. EMBO J 21(14):3694–3703. doi:10.1093/emboj/cdf357
113. Gawlinski P, Nikolay R, Goursot C, Lawo S, Chaurasia B, Herz HM, Kussler-Schneider Y, Ruppert T, Mayer M, Grosshans J (2007) The Drosophila mitotic inhibitor Fruhstart specifically binds to the hydrophobic patch of cyclins. EMBO Rep 8(5):490–496. doi:10.1038/sj.embor.7400948
114. Grosshans J, Muller HA, Wieschaus E (2003) Control of cleavage cycles in Drosophila embryos by fruhstart. Dev Cell 5(2):285–294
115. Lu X, Li JM, Elemento O, Tavazoie S, Wieschaus EF (2009) Coupling of zygotic transcription to mitotic control at the Drosophila mid-blastula transition. Development 136(12): 2101–2110. doi:10.1242/dev.034421
116. Campbell SD, Sprenger F, Edgar BA, O'Farrell PH (1995) Drosophila Wee1 kinase rescues fission yeast from mitotic catastrophe and phosphorylates Drosophila Cdc2 in vitro. Mol Biol Cell 6(10):1333–1347
117. Bettencourt-Dias M, Giet R, Sinka R, Mazumdar A, Lock WG, Balloux F, Zafiropoulos PJ, Yamaguchi S, Winter S, Carthew RW, Cooper M, Jones D, Frenz L, Glover DM (2004) Genome-wide survey of protein kinases required for cell cycle progression. Nature 432(7020):980–987. doi:10.1038/nature03160
118. Fasulo B, Koyama C, Yu KR, Homola EM, Hsieh TS, Campbell SD, Sullivan W (2012) Chk1 and Wee1 kinases coordinate DNA replication, chromosome condensation, and anaphase entry. Mol Biol Cell 23(6):1047–1057. doi:10.1091/mbc.E11-10-0832
119. Royou A, McCusker D, Kellogg DR, Sullivan W (2008) Grapes(Chk1) prevents nuclear CDK1 activation by delaying cyclin B nuclear accumulation. J Cell Biol 183(1):63–75. doi:10.1083/jcb.200801153
120. Kang Q, Srividhya J, Ipe J, Pomerening JR (2014) Evidence toward a dual phosphatase mechanism that restricts Aurora A (Thr-295) phosphorylation during the early embryonic cell cycle. J Biol Chem 289(25):17480–17496. doi:10.1074/jbc.M113.527622
121. Zou L, Elledge SJ (2003) Sensing DNA damage through ATRIP recognition of RPA-ssDNA complexes. Science 300(5625): 1542–1548. doi:10.1126/science.1083430
122. Papoulas O, Monzo KF, Cantin GT, Ruse C, Yates JR 3rd, Ryu YH, Sisson JC (2010) dFMRP and Caprin, translational regulators of synaptic plasticity, control the cell cycle at the Drosophila mid-blastula transition. Development 137(24):4201–4209. doi:10.1242/dev.055046
123. Monzo K, Papoulas O, Cantin GT, Wang Y, Yates JR 3rd, Sisson JC (2006) Fragile X mental retardation protein controls trailer hitch expression and cleavage furrow formation in Drosophila embryos. Proc Natl Acad Sci U S A 103(48):18160–18165. doi:10.1073/pnas.0606508103
124. Nien CY, Liang HL, Butcher S, Sun Y, Fu S, Gocha T, Kirov N, Manak JR, Rushlow C (2011) Temporal coordination of gene networks by Zelda in the early Drosophila embryo. PLoS Genet 7(10):e1002339. doi:10.1371/journal.pgen.1002339
125. Benoit B, He CH, Zhang F, Votruba SM, Tadros W, Westwood JT, Smibert CA, Lipshitz HD, Theurkauf WE (2009) An essential role for the RNA-binding protein Smaug during the Drosophila maternal-to-zygotic transition. Development 136(6): 923–932. doi:10.1242/dev.031815
126. Jevtic P, Levy DL (2015) Nuclear size scaling during Xenopus early development contributes to midblastula transition timing. Curr Biol 25(1):45–52. doi:10.1016/j.cub.2014.10.051
127. Vastag L, Jorgensen P, Peshkin L, Wei R, Rabinowitz JD, Kirschner MW (2011) Remodeling of the metabolome during early frog development. PLoS One 6(2):e16881. doi:10.1371/journal.pone.0016881
128. Edgar BA, Kiehle CP, Schubiger G (1986) Cell cycle control by the nucleo-cytoplasmic ratio in early Drosophila development. Cell 44(2):365–372

Chapter 3

Role of MicroRNAs in Zygotic Genome Activation: Modulation of mRNA During Embryogenesis

Alessandro Rosa and Ali H. Brivanlou

Abstract

A fundamental process occurring during early development is the zygotic genome activation, i.e., the initiation of transcription from the embryonic genome. Before that step, cellular processes in the developing embryo are dictated by transcripts produced by the maternal genome and accumulated in the egg during oogenesis. The maternal-to-zygotic transition (MZT) involves both the clearance of maternal RNAs and the initiation of transcription of the embryonic genome and is a tightly regulated process. In some species, decay of maternal transcripts may be facilitated by the activity of microRNAs. These small RNAs can act pleiotropically, blocking translation and inducing destabilization of hundreds of different maternal targets. In this review, we will discuss the role of microRNAs during MZT, focusing on Drosophila melanogaster and vertebrate models, Xenopus laevis, Zebrafish and mouse, in which such a mechanism has been more extensively studied.

Key words microRNA, Zygotic genome activation, Maternal transcripts, miR-430, miR-427

1 Introduction: microRNAs Biogenesis and Mechanisms of Action

Several classes of noncoding RNAs play regulatory roles in pluricellular organisms. Among them, small RNAs such as microRNAs (miRNAs) and small interfering RNAs (siRNAs) are able to regulate gene expression at the post-transcriptional level. In metazoa, miRNAs play important roles during differentiation and development [1]. In most cases, they are generated by a canonical pathway that starts from the transcription of a long precursor by RNA Polymerase II [2] (Fig. 1). This transcript, named primary miRNA (pri-miRNA), is then recognized by a processing complex (Microprocessor) that contains the core components Drosha and DGCR8 (named Pasha in Drosophila) [3–6]. Other cofactors may promote the identification of the pri-miRNA [7]. Multiple miRNAs may be encoded in a cluster and released from a common precursor [8]. The pri-miRNA may be hosted in an intron (intronic miRNAs) or independently transcribed (intergenic miRNAs).

Kiho Lee (ed.), *Zygotic Genome Activation: Methods and Protocols*, Methods in Molecular Biology, vol. 1605,
DOI 10.1007/978-1-4939-6988-3_3, © Springer Science+Business Media LLC 2017

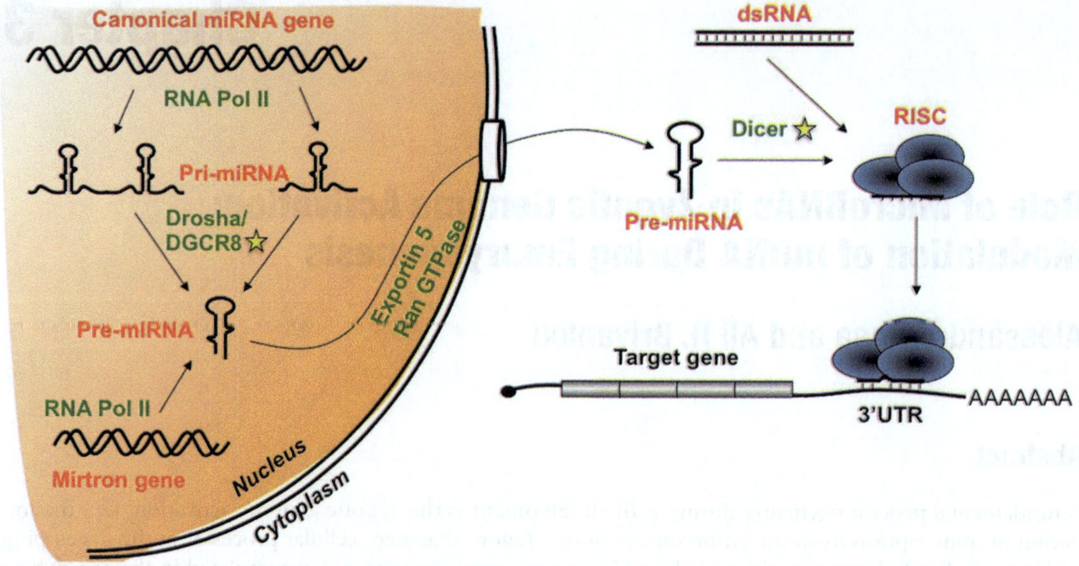

Fig. 1 The microRNA pathway. The figure depicts the canonical and noncanonical miRNA biogenesis pathways. Crucial factors for the production of mature miRNAs are indicated in green. Yellow stars mark factors that have been mutated to study the role of miRNAs in early vertebrate development by global loss of function analysis (see the text for details). Note that mutation in DGCR8 affects only the biogenesis of canonical miRNAs, while loss of Dicer function impairs also noncanonical miRNAs and double-stranded RNAs (dsRNAs, which generate siRNAs upon Dicer cleavage)

In some instances, the pri-miRNA is produced from the antisense strand of another gene. In any case, the biogenesis of miRNAs could be tightly regulated at the level of transcription by the same elements, in cis and trans, that control the expression of protein-coding genes. Several miRNAs are consequently highly enriched, or exclusively expressed, in specific developmental stages and/or cell types [9]. Drosha is an RNAse III able to cleave a double-stranded RNA. Upon Drosha cleavage of the pri-miRNA, a stem-loop shorter RNA (pre-miRNA) is released. In the case of clustered miRNAs, more than one pre-miRNA are produced by the Microprocessor. While Drosha is also involved in the processing of other transcripts, such as mRNAs and ribosomal RNA [10], DGCR8/Pasha is more specific for miRNAs. A relevant exception to this canonical pathway is represented by mirtrons, which are produced in a Microprocessor-independent way from debranched introns [11].

Upon export in the cytoplasm, the pre-miRNA is further processed by the RNAse III enzyme Dicer, releasing the mature miRNA as a single-stranded 20–25 nucleotide RNA from one of the arms of the stem-loop [12]. Dicer is also involved in the production of other small RNAs, such as siRNAs [10]. Products of DICER cleavage are then incorporated into the RNA-Induced

Silencing Complex (RISC), which contains Argonaute (Ago) family proteins and other cofactors [13]. Mature miRNAs serve as guides for RISC to bind by imperfect base-pairing target transcripts [14]. In metazoa, target sites (MRE, miRNA-responsive elements) are generally located in the 3′UTR and are not perfectly matched with cognate miRNAs, with perfect base-pairing limited to nucleotides 2–7 of the miRNA. This short sequence is referred to as the miRNA seed [15]. Binding of the miRNA on the MRE results in the acceleration of mRNA decay, mediated by induction of decapping, deadenylation and exonucleolytic trimming, and/or the inhibition of translation [16]. In rare cases, the outcome is an increase of RNA translation [17]. Notably, when the complementarity with the guide RNA (either a miRNA or a siRNA) is perfect, binding of the RISC results in the endonucleolytic cleavage of the target [13].

A miRNA family includes members with the same seed sequence, which usually share a large set of common targets. Multiple members of a miRNA family can be encoded in the same cluster. In this case, the coordinated production of several members of a family within a unique pri-miRNA could be a mechanism to increase the levels of miRNAs with the same seed, i.e., to amplify the repression on their targets. The mechanism of action of miRNAs is combinatorial, as single miRNAs are able to regulate simultaneously multiple targets. Moreover, a single mRNA can be targeted by multiple miRNAs [18–20].

Animal zygotes inherit a great number of maternal transcripts stored in the cytoplasm during oogenesis that are necessary to sustain the first cleavages upon fertilization. After the cleavage stage, one of the most striking examples of cellular "reprogramming" is that early embryos erase the whole maternal transcriptome and initiate transcription from the genome. Replacement of maternal transcripts with zygotic ones (maternal-to-zygotic transition, MZT) is necessary for normal development and must be accomplished in a tightly regulated fashion. On the other side, genes induced upon zygotic genome activation (ZGA), which is triggered by maternal factors, produce factors that inhibit the activity of maternally provided transcripts. Regulation of maternal gene expression can be only exerted at the level of mRNA stability and/or translation. In this view, two classes of factors may facilitate MZT: miRNAs and RNA-binding proteins. Both of them are able to recognize and bind specific transcripts, controlling their half-life.

Biological processes underlying MZT and ZGA have been recently extensively reviewed elsewhere [21–23]. Here we focus on the role played by miRNAs, with a particular emphasis on three classic model systems for molecular embryology: Drosophila melanogaster, Xenopus laevis, and Danio rerio.

2 Role of miRNAs in Developmental Transitions: Insights from Non-vertebrate Model Organisms

MiRNAs were first discovered in Ceanorabditis elegans from the analysis of mutants, named heterochronic, that fail to properly transit from a given developmental stage to the subsequent one [24, 25]. The founding members of this class of small RNAs, lin-4 and let-7 (initially named "small temporal RNAs," stRNAs), play a crucial role during nematode embryonic development, which is the silencing of transcripts that specify a previous phase when it is time to move to the next step. In this way, stRNAs define the boundaries between consecutive developmental stages. The discovery that let-7 was conserved in insects and vertebrates [26] opened the question of whether miRNAs might exert similar functions in other species. The shift between a maternal to a zygotic program represents one of the most studied examples of this kind of developmental transitions, characterized by a profound reshaping of the transcriptome.

In Drosophila melanogaster, the clearance of maternal mRNAs at MZT is ensured by two independent mechanisms. The first one is based on the activity of the RNA-binding protein (RBP) Smaug, which triggers transcript destruction by recruiting a deadenylase complex to target transcripts [21]. The second mechanism involves the activity of a group of miRNAs encoded by the miR-309 cluster [27]. This cluster contains eight miRNAs (miR-3, miR-4, miR-5, three copies of miR-6, miR-286, and miR-309) for a total of five different seeds (miR-286 and miR-309 have the same seed). These miRNAs were strongly induced at the onset of zygotic transcription. Seed-complementary sequences for miRNAs of the cluster were enriched in the 3′UTRs of maternal transcripts that showed a reciprocal profile of expression, i.e., were strongly decreased as miRNA levels rise. Mutant embryos, in which the genomic region comprising the miR-309 cluster was replaced with GFP, showed significant upregulation of maternal mRNAs. Experimental validation, by luciferase reporters fused to the 3′UTR of potential targets, confirmed that regulation of maternal mRNAs depended on the presence of miR-309 cluster MREs. Interestingly, some mRNAs targeted by the miR-309 cluster were also Smaug targets. Taken together, these observations point to an important role for miRNAs in the clearance of maternal transcript in drosophila embryos [27].

The maternally provided transcription factor Zelda induces several zygotically expressed miRNAs at ZGA, including the miR-309 cluster [28, 29]. Interestingly, expression of the miR-309 cluster is also dependent on Smaug, as levels of miRNAs encoded in this cluster were strongly reduced in Smaug mutants, in which a concomitant stabilization of their maternal targets was observed [30].

3 An Embryonic miRNA Family Promotes Clearance of Maternal mRNAs in Zebrafish

The teleost Danio rerio (zebrafish) and the amphibian Xenopus laevis represent valuable model systems for the study of molecular processes occurring during early vertebrate embryogenesis. Interestingly, the activity of a group of miRNAs playing crucial roles during embryonic development seems conserved in vertebrates. These miRNAs, generally undetectable in adult tissues, are particularly abundant in early embryos and belong to a greater family characterized by the AAGUGC seed sequence, hereafter referred to as the miR-430/427/302 family [1]. As discussed in this and the next session, the miR-430/427/302 family plays a central role in the clearance of maternal mRNAs in zebrafish and Xenopus and, more in general, is necessary for early development in vertebrates.

In zebrafish, most miRNAs are absent before segmentation, which occurs 12 hours post fertilization (hpf) [31]. The most relevant exception is represented by a group of more than 70 miRNAs, belonging to the miR-430/427/302 family and collectively referred to as miR-430 [32]. Interestingly, miR-430 levels rise at the onset of the ZGA (Fig. 2). Due to this peculiar timing of expression and their crucial functions during early development, miR-430 is considered the most important miRNA in the zebrafish embryo. This notion comes from the analysis of mutants in the miRNA processing machinery. About a decade ago, it was still unclear whether miRNAs played important roles during the earliest steps of vertebrate development. The first reported zebrafish

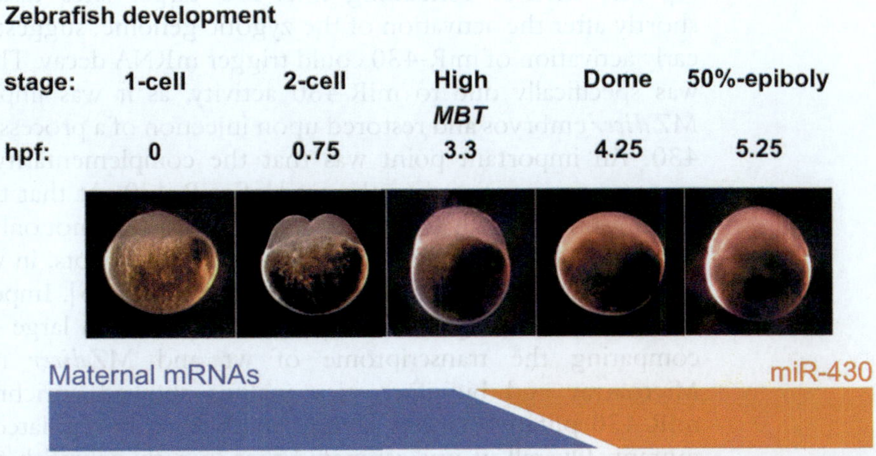

Fig. 2 miR-430 and zebrafish development. Schematic representation of the reciprocal expression of miR-430 and maternal mRNAs during early zebrafish development. Developmental stages and hours post fertilization (hpf) are indicated above each image of the zebrafish embryo

mutant for the miRNA processing enzyme Dicer (homozygous dicer1−/−) underwent growth arrest at 8 days post fertilization (dpf) and died at 14–15 dpf [33]. Strikingly, these embryos developed normally for the first week. However, the Dicer processing activity was still detectable up to 10 dpf in the mutants and inhibition of Dicer1 mRNA translation by antisense morpholino oligonucleotides in wild-type embryos produced a more severe phenotype than the null mutant [33]. These evidences suggested a contribution by maternal Dicer during the early phases of zebrafish development. In 2005, Giraldez et al. addressed this point by generating a maternal-zygotic Dicer mutant (MZdicer) [32], which was derived from zebrafish in which normal germ cells were replaced with dicer−/− germ cells. Depletion of both maternal and zygotic miRNA processing activity produced a more severe phenotype in these embryos, which displayed defective gastrulation, resulting in a reduced extension of the axis. Brain, somite, heart, and ear development were impaired. Strikingly, despite the fact that MZ dicer mutants lacked all mature miRNAs, most of these defects could be rescued by the injection of individual members of the miR-430 family. Mechanistically, as miRNAs act as pleiotropic factors, miR-430 could exert these pivotal functions by repressing a wide range of mRNAs. Relevant targets were found among components of the TGFbeta pathway. Target protectors, antisense modified oligonucleotides that prevent miRNA binding on specific mRNAs, were used to demonstrate that miR-430 target both agonists (Nodal-like ligands) and antagonists (Lefties) of the SMAD2/3 branch of the TGFbeta pathway, thus balancing this crucial signaling during early development [34].

Another crucial function of zebrafish miR-430 is the clearance of maternal mRNAs during MZT [35]. The authors showed that a reporter mRNA containing miR-430 target sites disappeared shortly after the activation of the zygotic genome, suggesting that early activation of miR-430 could trigger mRNA decay. This effect was specifically due to miR-430 activity, as it was impaired in MZ*dicer* embryos and restored upon injection of a processed miR-430. An important point was that the complementarity of the reporter was restricted to the seed of miR-430. At that time this result was not trivial, as the fact that miRNAs could not only inhibit translation but also destabilize their target transcripts, in vivo and by imperfect complementarity, was still debated [36]. Importantly, it allowed searching miR-430 target mRNAs in a large scale by comparing the transcriptome of wt and MZ*dicer* mutants. Microarray and bioinformatics analysis showed enrichment of miR-430 putative targets among transcripts up-regulated in the mutant. Overall, it was estimated that in early zebrafish embryos miR-430 targeted several hundreds of transcripts. Maternal mRNAs deposited into the egg before fertilization were significantly enriched in this target set. Levels of these maternal transcripts are high at 1.5 hpf and strongly decreased at 5 hpf, i.e., inversely correlated with miR-430 expression. Mechanistically, miR-430-induced destabilization of the transcripts by rapid shortening of the poly(A) tail and decay of mRNAs was preceded by translation repression at around 4 hpf [35, 37].

Individual members of the miR-430 family are produced by processing of a large pri-miR-430 polycistron, including 55 repeated pre-miR-430 hairpins. This gene is highly expressed in the first wave of zygotic transcription at sphere stage (4 hpf) [38]. Early zygotic genes, including the miR-430 cluster, are activated at ZGA by the combined action of three transcription factors, Nanog, Pou5fl, and SoxB1. The mammalian orthologs of these factors, respectively Nanog, Oct4, and Sox2, compose a core transcriptional regulatory circuitry in Embryonic Stem Cells (ESCs) [39]. A number of key pluripotency genes are positively regulated by these core factors in human and mouse ESCs, including miR-302 and miR-290-295 that belong to the miR-430/427/302 family [40, 41].

In summary, miR-430 induces deadenylation and decay of hundreds of target maternal mRNAs and thus promotes the ZGA transition in zebrafish [32, 35]. In turn, regulation of miR-430 activation at the onset of ZGA is under the control of key embryonic transcription factors and evolutionary conserved [38].

4 Evolutionary Conservation of miR-430/427/302 Activity in Xenopus laevis

Zygotic transcription starts in Xenopus laevis at the midblastula transition (MBT), stage 8.5, 7–8 hpf. Before MBT, embryos depend on translation of maternal mRNAs accumulated during oogenesis. MicroRNA profiling during Xenopus development revealed that some miRNAs were already present in oocytes and embryos before nuclear transcription and therefore could be considered maternal miRNAs [42]. A subset of miRNAs newly synthesized from the embryonic genome was transiently expressed in early stages and included the Xenopus miR-430/427/302 family member, miR-427 [42]. Similarly to its zebrafish ortholog miR-430, miR-427 is highly abundant in early embryos, absent in adult tissues, contains an AAGUGC seed, and regulates the Nodal pathway [42, 43]. Mature miR-427 can be first detected at 7–8 hpf [42, 43]. However, precursors of miR-427 accumulate as early as 6 hpf, before general activation of most zygotic transcripts [44]. These transcripts are produced by RNA polymerase II from about 1000 miR-427-coding regions, probably organized in tandem repeats arrays. Transcription of miR-427 is then rapidly turned off during gastrulation (9 hpf), even if elevated half-life allows persistence of mature miR-427 species until later stages [42–44]. MiRNAs expression might be regulated not only transcriptionally, but also at the maturation step. Some factors required for miRNA processing and nuclear export (Dicer, Exportin5) are already present at high levels in the oocyte cytoplasm and do not further

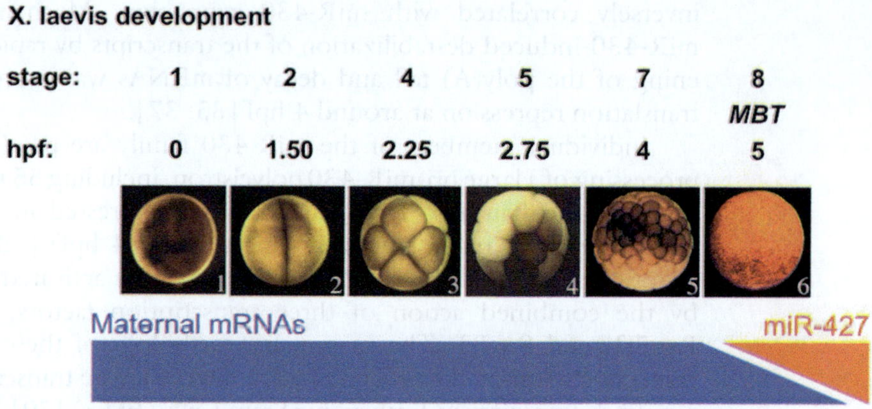

Fig. 3 miR-427 and X. laevis development. Schematic representation of the reciprocal expression of miR-427 and maternal mRNAs during early X. laevis development. Developmental stages and hours post fertilization (hpf) are indicated above each image of the X. laevis embryo

increase upon ZGA. Conversely, maternal levels of the RISC component Argonaute 2 (Ago2) are low in Xenopus embryos before MBT, while Ago2 zygotic mRNA and protein levels rise again after this stage [45]. However, reduced levels of Ago2 protein provided by maternal mRNA before ZGA are not limiting for the activity of miR-427, as a pre-miR-427 injected before transcription of zygotic Ago2 induced target decay [44].

As its zebrafish ortholog, miR-427 plays a role in the clearance of maternal mRNAs [44] (Fig. 3). In Xenopus embryos, maternal cyclin genes sustain cell division before ZGA. Cyclin A1 and B2 maternal transcripts are then rapidly deadenylated and destroyed after MBT [46, 47]. Cyclin A1 and B2 mRNA decay depends on a region on their 3′UTR, which is complementary to the miR-427 seed. Experimental validation demonstrated direct targeting by miR-427, which results in accelerated mRNA deadenylation and decay [44]. A single miR-427 MRE is necessary for this effect, while other elements on the cyclin B2 3′UTR, including a Pumilio protein-binding element, are not required for destabilization. Another maternal mRNA regulated by miR-427 is *DEADSouth*. This transcript, encoding for the RNA helicase DDX25, is present in the germ plasm, a portion of the cytoplasm localized in vegetal cortex of the fertilized egg and inherited by the germline. Clearance of maternal *DEADSouth* mRNA in somatic cells occurs via miR-427 targeting. In Primordial Germ Cells (PGCs), conversely, low levels of miR-427 expression and expression of zygotic *DEADSouth* after MBT result in high levels of DDX25 [48].

Despite not experimentally validated, other Xenopus mRNAs that are destabilized upon ZGA contain one or more potential MREs for miR-427 in the 3′UTR [44], suggesting that miR-427 might have a broader effect on maternal mRNAs clearance, as shown for its zebrafish ortholog [35].

5 The miR-430/427/302 Family Is Conserved in Mammals But Does Not Play a Role in Maternal-to-Zygotic Transition

Mammalian members of the miR-430/427/302 family are highly expressed in embryos and pluripotent stem cells [1]. Among them, the most relevant in mouse are the miRNAs encoded in the miR-290-295 cluster (hereafter, miR-290-295). In mouse embryos transcription of miR-290-295 is initiated at the onset of ZGA. Levels of these miRNAs are low in oocytes, zygotes, and two-cell embryos, while transition to the 4-cell stage led to a steep increase [49]. This early pattern of expression and the conserved AAGUGC seed sequence might suggest a role for the miR-290-295 during MZT, as previously described for the Xenopus and zebrafish orthologs. However, several lines of evidence suggest that members of the miR-430/427/302 family, and miRNAs in general, could not be involved in this process in mammals.

In vertebrates with external development, embryos undergo several cycles of cell division before ZGA at MBT. As already mentioned, this cleavage stage is entirely supported by maternal transcripts. The zygotic genome is activated very early in mammals. In the mouse, a first minor wave of activation is detectable in the zygote, while the major wave occurs at the 2-cell stage [22]. Maternal mRNA clearance begins during oocyte maturation, upon resumption of meiosis from the arrest in diplotene of the first meiotic prophase. This first wave of maternal mRNA degradation is followed by a second one, at fertilization, and a third one, triggered by ZGA. Levels of miR-290-295 are low until the embryo reaches the 4-cell stage [49]. At this stage most of the maternal mRNA load is already cleared. In general, among the classes of small RNAs (miRNAs, siRNAs, and piRNAs) detected in developing oocytes and pre-implantation embryos by next-generation sequencing, miRNAs represent a minor population [50]. Moreover, miRNA activity is suppressed during oocyte maturation [51]. Most remarkably, the activity of miRNAs is not required in mouse embryos until the stage of blastocyst. This notion comes from the analysis of mutants in which both maternal and zygotic Dgcr8 were deleted [52]. These mutants, generated by crossing conditional knockout mothers (Dgcr8$^{delta/flox}$; Zp3-Cre) with heterozygous knockout males, lacked both maternal and zygotic miRNAs generated by the canonical processing pathway. Surprisingly, maternal-zygotic Dgcr8$^{-/-}$ embryos developed normally to the blastocyst stage, suggesting that miRNAs generated by the Microprocessor are not required for pre-implantation development [52]. Notably, other small RNAs that do not require the Microprocessor for their biogenesis, such as miRNAs generated by mirtrons and endogenous siRNAs, are not affected by Dgcr8 loss. These small RNAs were conversely depleted in conditional Dicer null oocytes [49, 53] and their activity was impaired in early embryos in which Argonaute2 was knocked down [54]. Therefore, noncanonical miRNAs and/or endogenous siRNAs, but not miRNAs, could be responsible for the very severe developmental phenotypes generated by the depletion of either Dicer or Argonaute2, but not Dgcr8, in pre-implantation embryos [55].

For all these reasons, despite the miR-430/427/302 family being highly conserved in all vertebrates in terms of expression and seed sequence, it is highly unlikely that it has a role in maternal mRNAs clearance in mammals. However, other key functions during embryogenesis have been maintained for these miRNAs, such as the regulation of TGFbeta signaling by targeting Nodal antagonists [43].

6 Concluding Remarks

Maternal mRNAs abundance in embryos cannot be regulated at the level of transcription. Early in development, specific cell types need to build up their own specific transcriptome and get rid of the factors inherited from the egg cytoplasm. Post-transcriptional regulation of mRNA levels can be achieved by modulating their stability. RBPs and miRNAs represent two classes of factors able to exert this function. They act by recognizing and binding specific sequences on their targets, with the consequence of increasing or decreasing mRNA life. Both RBPs and miRNAs have been used by evolution to facilitate MZT.

An interesting feature of miRNAs is that their genes can be present in a huge number of clustered copies in the genome. This is very convenient during MZT, when a rapid and significant increase of "scavenger" factors is required. In vertebrates, miR-427 and miR-430 are among the earliest zygotic transcripts. The increase in their levels during early stages of embryonic development is remarkable. It has been estimated that 10^9 copies of miR-427 are present in a X. laevis embryo at MBT [44]. This is achieved by coordinated transcription of long precursors, which encode multiple miRNA copies. A similar polycistronic precursor produces multiple miR-430 copies in zebrafish [38].

Activation of miR-430/427 transcription at the onset of ZGA is sufficient to trigger the erasure of their maternal targets. The other components of the degradation machinery are already in place in the embryo before ZGA. Therefore, these miRNAs represent the only component of a very parsimonious switch from the maternal to the zygotic transcriptome. Regulation of the expression of miRNAs responsible for MZT occurs primarily at the level of transcription.

Both the pattern of expression and the seed conservation suggest an ancestral origin for this vertebrate miRNA family.

The strong conservation of the seed sequence in vertebrate miRNAs with a role in MZT might be suggestive of some sort of structural constraint. However, the evidence that other miRNAs worked as well as miR-427 in the clearance of a reporter bearing a "maternal" 3′UTR [44] points against this hypothesis. Interestingly, miRNAs deputed to the same function in Drosophila have a different seed sequence and therefore are not related to miR-430/427. This represents a remarkable example of convergent evolution between vertebrates and insects [56].

The conservation between miRNAs belonging to the miR-430/427/302 family is even broader. Evidences from zebrafish and ESCs suggest that these related miRNAs are under the control of embryonic core transcription factors belonging to the Oct4, Nanog, and Sox families. In zebrafish, miR-430 is activated by Nanog/Pou5f1/SoxB1 [38]. These factors have also a key role in the induction of the first wave of zygotic transcription. This represents a dual regulatory mechanism for these key TFs: they are able to activate transcription of zygotic genes and at the same time, by inducing the expression of scavenger miRNAs, elicit maternal mRNAs clearance. In Drosophila, a similar role is played by the transcription activator Zelda, which induces expression of many essential zygotic genes and, indirectly, promotes degradation of maternal transcripts by activating the miR-309 cluster [28].

In the case of miR-427 and cyclins A1 and B2, a single MRE is sufficient to induce consistent degradation of the target [44]. This is in contrast to the common situation in which greater mRNA destabilization is associated with multiple MREs [57]. The ability of a single MRE to direct rapid decay might be a consequence of the high concentration of RISCs loaded with miR-427 in Xenopus embryos [44], or it might be due to other still unidentified factors that increase the effect of this miRNA during MZT.

In conclusion, miRNAs hold important functions during early embryonic development in metazoa. In Drosophila, zebrafish, and Xenopus they facilitate the transition from an oocyte-inherited to an embryonic transcriptome by eliminating maternal mRNAs. Probably, this function has been established independently in insects and vertebrates during evolution and has been lost in mammals with the acquisition of internal development.

Acknowledgments

We thank Tomomi Haremaki for providing images of X. laevis and zebrafish development. This work has been supported by funding from Sapienza University to A.R. and Rockefeller University to A.H.B.

References

1. Rosa A, Brivanlou AH (2009) microRNAs in early vertebrate development. Cell Cycle 8:3513–3520
2. Lee Y, Kim M, Han J et al (2004) MicroRNA genes are transcribed by RNA polymerase II. EMBO J 23(20):4051–4060
3. Lee Y, Ahn C, Han J et al (2003) The nuclear RNase III Drosha initiates microRNA processing. Nature 425(6956):415–419
4. Denli AM, Tops BB, Plasterk RH et al (2004) Processing of primary microRNAs by the Microprocessor complex. Nature 432(7014): 231–235
5. Gregory RI, Yan KP, Amuthan G et al (2004) The Microprocessor complex mediates the genesis of microRNAs. Nature 432(7014): 235–240
6. Han J, Lee Y, Yeom KH et al (2004) The Drosha-DGCR8 complex in primary microRNA processing. Genes Dev 18(24): 3016–3027
7. Morlando M, Dini Modigliani S, Torrelli G et al (2012) FUS stimulates microRNA biogenesis by facilitating co-transcriptional Drosha recruitment. EMBO J 31:4502–4510
8. Lee Y, Jeon K, Lee JT et al (2002) MicroRNA maturation: stepwise processing and subcellular localization. EMBO J 21(17): 4663–4670
9. Landgraf P, Rusu M, Sheridan R et al (2007) A mammalian microRNA expression atlas based on small RNA library sequencing. Cell 129:1401–1414
10. Johanson TM, Lew AM, Chong MM (2013) MicroRNA-independent roles of the RNase III enzymes Drosha and Dicer. Open Biol 3(10):130144. doi:10.1098/rsob.130144
11. Ruby JG, Jan CH, Bartel DP (2007) Intronic microRNA precursors that bypass Drosha processing. Nature 448(7149):83–86
12. Grishok A, Pasquinelli AE, Conte D et al (2001) Genes and mechanisms related to RNA interference regulate expression of the small temporal RNAs that control C. elegans developmental timing. Cell 106(1):23–34
13. Hutvágner G, Zamore PD (2002) A microRNA in a multiple-turnover RNAi enzyme complex. Science 297(5589):2056–2060
14. Meister G (2013) Argonaute proteins: functional insights and emerging roles. Nat Rev Genet 14(7):447–459. doi:10.1038/nrg3462
15. Bartel DP (2009) MicroRNAs: target recognition and regulatory functions. Cell 136(2): 215–233. doi:10.1016/j.cell.2009.01.002
16. Jonas S, Izaurralde E (2015) Towards a molecular understanding of microRNA-mediated gene silencing. Nat Rev Genet 16(7):421–433. doi:10.1038/nrg3965
17. Vasudevan S, Tong Y, Steitz JA (2007) Switching from repression to activation: microRNAs can up-regulate translation. Science 318(5858):1931–1934
18. Brennecke J, Stark A, Russell RB et al (2005) Principles of microRNA-target recognition. PLoS Biol 3:e85
19. Lewis BP, Burge CB, Bartel DP (2005) Conserved seed pairing, often flanked by adenosines, indicates that thousands of human genes are microRNA targets. Cell 120:15–20
20. Rajewsky N (2006) microRNA target predictions in animals. Nat Genet 38:S8–13
21. Laver JD, Marsolais AJ, Smibert CA et al (2015) Regulation and function of maternal gene products during the maternal-to-zygotic transition in Drosophila. Curr Top Dev Biol 113:43–84. doi:10.1016/bs.ctdb.2015.06.007
22. Svoboda P, Franke V, Schultz RM (2015) Sculpting the transcriptome during the oocyte-to-embryo transition in mouse. Curr Top Dev Biol 113:305–349
23. Yartseva V, Giraldez AJ (2015) The maternal-to-zygotic transition during vertebrate development: a model for reprogramming. Curr Top Dev Biol 113:191–232
24. Lee RC, Feinbaum RL, Ambros V (1993) The C. elegans heterochronic gene lin-4 encodes small RNAs with antisense complementarity to lin-14. Cell 75:843–854
25. Reinhart BJ, Slack FJ, Basson M et al (2000) The 21-nucleotide let-7 RNA regulates developmental timing in Caenorhabditis elegans. Nature 403:901–906
26. Pasquinelli AE, Reinhart BJ, Slack F et al (2000) Conservation of the sequence and temporal expression of let-7 heterochronic regulatory RNA. Nature 408:86–89
27. Bushati N, Stark A, Brennecke J et al (2008) Temporal reciprocity of miRNAs and their targets during the maternal-to-zygotic transition in Drosophila. Curr Biol 18:501–506
28. Liang HL, Nien CY, Liu HY et al (2008) The zinc-finger protein Zelda is a key activator of the early zygotic genome in Drosophila. Nature 456(7220):400–403. doi:10.1038/nature07388
29. Fu S, Nien CY, Liang HL et al (2014) Co-activation of microRNAs by Zelda is essential for early Drosophila development. Development 141(10):2108–2118. doi:10.1242/dev.108118

30. Benoit B, He CH, Zhang F et al (2009) An essential role for the RNA-binding protein Smaug during the Drosophila maternal-to-zygotic transition. Development 136(6): 923–932
31. Wienholds E, Kloosterman WP, Miska E et al (2005) MicroRNA expression in zebrafish embryonic development. Science 309:310–311
32. Giraldez AJ, Cinalli RM, Glasner ME et al (2005) MicroRNAs regulate brain morphogenesis in zebrafish. Science 308:833–838
33. Wienholds E, Koudijs MJ, van Eeden FJM et al (2003) The microRNA-producing enzyme Dicer1 is essential for zebrafish development. Nat Genet 35:217–218
34. Choi W-Y, Giraldez AJ, Schier AF (2007) Target protectors reveal dampening and balancing of Nodal agonist and antagonist by miR-430. Science 318:271–274
35. Giraldez AJ (2006) Zebrafish MiR-430 promotes deadenylation and clearance of maternal mRNAs. Science 312:75–79
36. Pillai RS, Bhattacharyya SN, Artus CG et al (2005) Inhibition of translational initiation by Let-7 MicroRNA in human cells. Science 309:1573–1576
37. Bazzini AA, Lee MT, Giraldez AJ (2012) Ribosome profiling shows that miR-430 reduces translation before causing mRNA decay in zebrafish. Science 336:233–237. doi:10.1126/science.1215704
38. Lee MT, Bonneau AR, Takacs CM et al (2013) Nanog, Pou5f1 and SoxB1 activate zygotic gene expression during the maternal-to-zygotic transition. Nature 503:360–364. doi:10.1038/nature12632
39. Boyer LA, Lee TI, Cole MF et al (2005) Core transcriptional regulatory circuitry in human embryonic stem cells. Cell 122:947–956
40. Marson A, Levine SS, Cole MF et al (2008) Connecting microRNA genes to the core transcriptional regulatory circuitry of embryonic stem cells. Cell 134:521–533
41. Rosa A, Brivanlou AH (2011) A regulatory circuitry comprised of miR-302 and the transcription factors OCT4 and NR2F2 regulates human embryonic stem cell differentiation. EMBO J 30:237–248
42. Watanabe T, Takeda A, Mise K et al (2005) Stage-specific expression of microRNAs during Xenopus development. FEBS Lett 579:318–324
43. Rosa A, Spagnoli FM, Brivanlou AH (2009) The miR-430/427/302 family controls mesendodermal fate specification via species-specific target selection. Dev Cell 16:517–527
44. Lund E, Liu M, Hartley RS et al (2009) Deadenylation of maternal mRNAs mediated by miR-427 in Xenopus laevis embryos. RNA 15:2351–2363. doi:10.1261/rna.1882009
45. Lund E, Sheets MD, Imboden SB et al (2011) Limiting Ago protein restricts RNAi and microRNA biogenesis during early development in Xenopus laevis. Genes Dev 25:1121–1131. doi:10.1101/gad.2038811
46. Audic Y, Anderson C, Bhatty R et al (2001) Zygotic regulation of maternal cyclin A1 and B2 mRNAs. Mol Cell Biol 21:1662–1671
47. Audic Y, Garbrecht M, Fritz B et al (2002) Zygotic control of maternal cyclin A1 translation and mRNA stability. Dev Dyn 225:511–521
48. Yamaguchi T, Kataoka K, Watanabe K et al (2014) Restriction of the Xenopus DEADSouth mRNA to the primordial germ cells is ensured by multiple mechanisms. Mech Dev 131:15–23. doi:10.1016/j.mod.2013.11.002
49. Tang F, Kaneda M, O'Carroll D et al (2007) Maternal microRNAs are essential for mouse zygotic development. Genes Dev 21:644–648
50. Ohnishi Y, Totoki Y, Toyoda A et al (2010) Small RNA class transition from siRNA/piRNA to miRNA during pre-implantation mouse development. Nucleic Acids Res 38:5141–5151
51. Ma J, Flemr M, Stein P et al (2010) MicroRNA activity is suppressed in mouse oocytes. Curr Biol 20:265–270
52. Suh N, Baehner L, Moltzahn F et al (2010) MicroRNA function is globally suppressed in mouse oocytes and early embryos. Curr Biol 20:271–277
53. Murchison EP, Stein P, Xuan Z et al (2007) Critical roles for Dicer in the female germline. Genes Dev 21:682–693
54. Lykke-Andersen K, Gilchrist MJ, Grabarek JB et al (2008) Maternal Argonaute 2 is essential for early mouse development at the maternal-zygotic transition. Mol Biol Cell 19:4383–4392. doi:10.1091/mbc.E08-02-0219
55. Svoboda P, Flemr M (2010) The role of miRNAs and endogenous siRNAs in maternal-to-zygotic reprogramming and the establishment of pluripotency. EMBO Rep 11:590–597
56. Giraldez AJ (2010) microRNAs, the cell's Nepenthe: clearing the past during the maternal-to-zygotic transition and cellular reprogramming. Curr Opin Genet Dev 20: 369–375
57. Grimson A, Farh KK-H, Johnston WK et al (2007) MicroRNA targeting specificity in mammals: determinants beyond seed pairing. Mol Cell 27:91–105

Chapter 4

Gene Expression Analysis in Mammalian Oocytes and Embryos by Quantitative Real-Time RT-PCR

Kyeoung-Hwa Kim, Su-Yeon Lee, and Kyung-Ah Lee

Abstract

This chapter describes methods for preparing oocytes and embryos to analyze their gene expression at specific developmental stages. We illustrate how to collect germinal vesicles (GVs) and mature metaphase II (MII) stage oocytes, as well as how to collect embryos at specific developmental stages from the pronucleus (PN) to the blastocyst stage from female mice. We also describe how to prepare mRNAs from these precious cells to analyze the expression of the target genes. The materials and methods in this chapter are used mainly for mouse oocytes and embryos, but with subtle modifications, they may be applicable for most mammalian species.

Key words Oocytes, Pronuclear stage embryos, Zygotes, Gene expression, qRT-PCR

1 Introduction

Mature mammalian oocytes are imperative for fertilization and further embryo development because the initial phases of embryogenesis are controlled by maternal factors that accumulate in the oocytes during the oocyte growth period without de novo transcription. Various mRNAs and proteins are produced and stored in oocytes during oocyte growth, which is characterized by folliculogenesis and the subsequent maturation and acquisition of competence for fertilization. We studied the changes in gene expression during oocyte maturation using an innovative technology called annealing control primer-polymerase chain reaction (ACP-PCR) or Gene Fishing [1]. The accumulation of abundant mRNAs and proteins in oocytes is very important to drive normal fertilization and the early stages of embryonic development [2]; therefore, an understanding of the molecular profile of the oocytes is crucial for an understanding of the molecular signature of the quality of oocytes.

The dynamics of the degradation of maternal factors has been known to occur during oocyte maturation, and the selective disappearance of factors is associated with meiotic arrest at the GV stage and the progression of oocyte maturation, such as oxidative

Kiho Lee (ed.), *Zygotic Genome Activation: Methods and Protocols*, Methods in Molecular Biology, vol. 1605,
DOI 10.1007/978-1-4939-6988-3_4, © Springer Science+Business Media LLC 2017

phosphorylation, energy production, and protein synthesis and metabolism. Meanwhile, transcripts encoding factors essential for maintaining MII arrest, such as those involved in signaling pathways, are stable [3]. These maternal factors that accumulate in oocytes are in charge of regulating the completion of meiosis, fertilization, embryonic cell division, zygotic gene activation (ZGA), and early embryogenesis until implantation occurs [4]. When fertilization succeeds, embryonic development begins, zygotic transcription commences, and the degradation of maternal mRNAs and proteins begins to decline concomitantly. This stage of embryonic development, in which development is controlled by the zygotic genome, is also called the maternal to zygotic transition (MZT) and is defined as a period of development that begins just after fertilization when maternal transcripts begin to be eliminated [5].

Studying the molecular signature of the oocytes or embryos after targeting specific gene expression at specific stages has been the major purpose of our projects, and we revealed the functions of several maternal factors for oocyte maturation as well as for embryo development. We used RNA interference (RNAi) as a main technique for studying the loss of function of a specific gene for the maternal factors and quantitative real-time RT-PCR or microarray analysis to profile the changes in gene expression after RNAi of the target gene in the oocytes or embryos [6, 7]. In this chapter, we describe the details of quantitative real-time RT-PCR using a small quantity of cells, such as 10 or 20 oocytes or embryos, and illustrate how to prepare oocytes and embryos for molecular gene expression studies.

2 Materials

2.1 Collection of Oocytes and Embryos

2.1.1 Superovulation and Mating

1. Four-week-old female and 8-week-old male ICR mice: Maintain and breed all mice on a 16:8 h light:dark cycle.
2. Phosphate-buffered saline (PBS): Store in a sterile container at 4 °C.
3. 5 IU/0.1 ml pregnant mare serum gonadotrophin (PMSG; Sigma-Aldrich, St. Louis, MO): Use 50 IU/ml in ice-cold PBS or sterile saline and aliquot into microcentrifuge tubes placed on ice. Store at −20 °C.
4. 0.2 mM 3-Isobutyl-1-methyl-xanthine (IMBX; Sigma-Aldrich): Use 200× IBMX (40 mM) in DMSO and aliquot into microcentrifuge tubes placed on ice. Store at −20 °C.
5. Dimethyl sulfoxide (DMSO; Sigma-Aldrich).
6. 5 IU/0.1 ml human chorionic gonadotrophin (hCG; Sigma-Aldrich): Use 50 IU/ml in ice-cold PBS or sterile saline and aliquot into microcentrifuge tubes placed on ice. Store at −20 °C.
7. 27-gauge needle: Attach to a 1 ml syringe.

2.1.2 Isolation of Mouse Oocytes/Embryos

1. M2 medium (Sigma-Aldrich): Add 100× antibiotics to 100 ml M2 medium and mix well. Sterilize by passing through a 0.22 μm filter and store in a sterile container at 4 °C (*see* **Note 1**).
2. 100× antibiotics (Invitrogen): Store at −20 °C.
3. Millex-GV Syringe Filter Unit, 0.22 μm filter (Millipore, Bedford, MA).
4. Hyaluronidase (Sigma-Aldrich): Use 300 U/ml in M2 medium and aliquot into microcentrifuge tubes. Store at −20 °C.
5. Micro-dissecting forceps (curved and straight; Sigma-Aldrich) and scissors (Sigma-Aldrich): Clean with alcohol and sterilize by autoclaving.
6. Two-well dish (SPL, Korea).
7. 35 mm petri dish (SPL).
8. Dissecting microscope (Olympus, Japan).
9. Mouth pipette (Sigma-Aldrich).
10. Alcohol burner and 100% ethyl alcohol (DUKSAN, Korea).
11. Capillary tube (Thermo Fisher Scientific, San Jose, CA): To pull the capillary to the required diameter, rotate the center of the tube over a flame of the alcohol burner until it begins to melt. Promptly withdraw the tube from the flame while pulling both ends sharply in opposite directions, stretching the center of the melted tube into a fine narrow diameter. Prepare the capillary tube with a diameter that is slightly larger than that of the oocytes/embryos.

2.1.3 Washing and Storage

1. 35 mm petri dish (SPL).
2. PBS-PVA: Dissolve 0.5 mg of Polyvinyl alcohol (PVA; Sigma-Aldrich) in 500 ml of PBS and sterilize by passing the solution through a 0.22 μm filter. Store in a sterile container at 4 °C.
3. Millex-GV Syringe Filter Unit, 0.22 μm filter (Millipore).
4. Liquid nitrogen.
5. 1.7 ml ultra-clear tube (Axygen, Union City, CA).

2.2 Isolation of mRNA from the Mouse Oocytes/Embryos

1. Mouse oocytes/embryos: Store 10 or 20 oocytes/embryos at −70 °C.
2. Dynal MPC-S Magnetic Particle Concentrator (Invitrogen).
3. RNase-free 1.5 ml tube (Ambion, Austin, TX).
4. Dynabeads® mRNA DIRECT™ Kit (Invitrogen).
   (a) Dynabeads® Oligo $(dT)_{25}$.
   (b) Lysis/Binding Buffer: 100 mM Tris-HCl [pH 7.5], 500 mM LiCl, 10 mM EDTA [pH 8], 1% LiDS, 5 mM DTT.
   (c) Washing Buffer A: 10 mM Tris-HCl [pH 7.5], 0.15 M LiCl, 1 mM EDTA, 0.1% SDS.

(d) Washing Buffer B: 10 mM Tris-HCl [pH 7.5], 0.15 M LiCl, 1 mM EDTA.

(e) Elution Buffer: 10 mM Tris-HCl [pH 7.5].

2.3 Reverse Transcription

1. C1000 PCR system (Bio-Rad, Hercules, CA).
2. Prepare mRNAs from the oocytes or embryos.
3. Oligo dT_{25}: Purchase a synthesized 25 bp deoxythymidine nucleotide.
4. M-MLV Reverse transcriptase (200 U/μl) (Promega, Madison, WI).
5. M-MLV Reverse Transcriptase 5× Reaction Buffer: 50 mM Tris-HCl [pH 8.3], 75 mM KCl, 3 mM $MgCl_2$, 10 mM DTT.
6. RNasin® Ribonuclease Inhibitors (20 U/μl) (Promega).
7. dNTP stocks: 100 mM each of dATP (Takara, Japan), dGTP (Takara), dCTP (Takara), and dTTP (Takara). Store at −20 °C.
8. 10 mM dNTP mix: Mix 100 μl each of the 100 mM stocks of dATP, dGTP, dCTP, and dTTP and deionized H_2O to a total volume of 600 μl. Store in 100 μl aliquots at −20 °C.

2.4 Quantitative Real-Time RT-PCR

1. iCycler iQ thermal cycler (Bio-Rad, Model: MyiQ-iCycler).
2. 96-Well iQ Optical PCR plates (Bio-Rad).
3. Microseal 'B' Seal (Bio-Rad).
4. iQ SYBR Green Supermix (Bio-Rad): 100 mM KCl, 40 mM Tris-HCl [pH 8.4], 0.4 mM each dNTP, 50 U/ml iTaq DNA polymerase, 6 mM $MgCl_2$, SYBR Green I, 20 nM fluorescein, and stabilizers.
5. Oligonucleotides: 10 pmol/μl.
6. Deionized H_2O.

3 Methods

3.1 Collection of Oocytes and Embryos (See Note 2)

3.1.1 Superovulation and Mating

1. Inject 4-week-old female ICR mice with 5 IU/0.1 ml PMSG (*see* **Note 3**).
2. To obtain germinal vesicle (GV) oocytes, sacrifice female mice at 46 h after PMSG.
3. To obtain metaphase II (MII) oocytes, inject PMSG primed female mice with 5 IU/0.1 ml hCG 46 h after PMSG. Obtain superovulated MII oocytes from the oviducts 16 h after hCG injection.
4. To obtain early developmental stage of embryos, cage female mice overnight with 8-week-old males from the same strain at a 1:1 ratio of males to females.

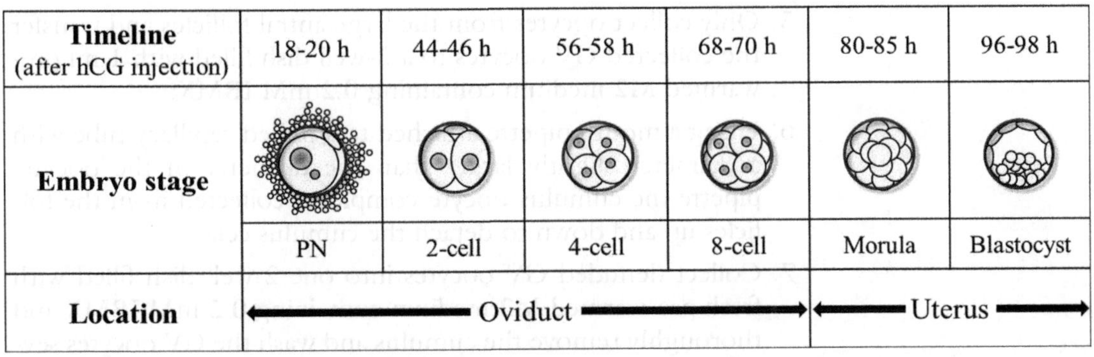

Fig. 1 Timeline of mouse embryo development after hCG injection. Embryos at the various developmental stages were obtained at the specific time points after hCG injection, which are listed in the column above the picture of each stage of embryo. After fertilization, PN embryos undergo a series of cleavage divisions and concurrently move through the oviducts into the lumen of the uteri. Therefore, embryos from the PN to the 8-cell stage are collected from the oviducts; meanwhile, the morula and blastocysts are collected from the uteri

5. Confirm successful mating by the presence of vaginal plugs in the female mice. The next morning, remove the males after vaginal plug confirmation [8].
6. Obtain mouse embryos at specific time points after hCG injection as follows: pronuclear (PN) zygote stage embryos at 16–18 h, 2-cell stage embryos at 40–42 h, 4-cell stage embryos at 52–54 h, 8-cell stage embryos at 64–66 h, morula stage embryos at 76–78 h, and blastocyst stage embryos at 88–90 h by flushing the oviducts or uterus (Fig. 1) [8, 9].

3.1.2 Isolation of Oocytes (See Note 4)

Isolation of GV Oocytes

1. Prepare the 35 mm petri dish and 2-well dish filled with 2.5 ml or 1 ml of M2 medium containing 0.2 mM IBMX, respectively, and warm it at 37 °C. We used IBMX (*see* **Note 5**), an inhibitor of phosphodiesterase, to maintain a high concentration of cyclic AMP (cAMP) levels in the oocytes. cAMP is an important regulator of meiosis in oocytes, and at high concentrations, it sustains oocytes at the GV stage.
2. Approximately 48 h after PMSG injection, sacrifice the mouse using CO_2 sedation. Using a separate set of tools, cut the skin and peritoneum to reveal the abdominal organs, and then collect the ovaries, oviducts, and uterus underneath the fat pads [8].
3. Remove the ovarian versa and remove the ovary using curved forceps. Place the ovaries into a fresh 35 mm petri dish filled with pre-warmed M2 medium containing 0.2 mM IBMX (2.5 ml/5 ovaries).
4. Under a dissecting microscope, release the oocytes from the antral follicles by puncturing them with 27-gauge needles.

5. Only collect oocytes from the large antral follicles and transfer the collected GV oocytes to a 2-well dish filled with 1 ml pre-warmed M2 medium containing 0.2 mM IBMX.
6. Using a mouth pipette attached to a pulled capillary tube with a diameter slightly larger than the diameter of the oocyte, pipette the cumulus-oocyte-complexes collected from the follicles up and down to detach the cumulus cells.
7. Collect denuded GV oocytes into one 2-well dish filled with fresh pre-warmed M2 medium containing 0.2 mM IBMX and thoroughly remove the cumulus and wash the GV oocytes several times. The oocytes are now ready for the next experiment (Subheading 3.1.5).

Isolation of MII Oocyte

1. Prepare the 35 mm petri dish and 2-well dish filled with 2.5 ml or 1 ml of M2 medium, respectively, and warm the medium at 37 °C.
2. Approximately 16 h after hCG injection, sacrifice the mouse using CO_2 sedation. Using a separate set of tools, cut the skin and peritoneum to reveal the abdominal organs, and then collect the ovaries, oviducts, and uterus underneath the fat pads.
3. Transfer the oviducts into a 35 mm petri dish filled with pre-warmed M2 medium. Collect the oviducts from several mice in one dish.
4. Under a dissecting microscope, release the MII oocytes clumped into one mass of cumulus-oocyte-complexes by tearing the oviducts with 27-gauge needles.
5. Remove the expanded cumulus cells surrounding the MII oocytes by treatment with hyaluronidase (300 U/ml). Using a mouth pipette attached to a pulled capillary tube with a diameter slightly larger than the diameter of the oocyte, pipette up and down the cumulus-oocyte-complexes collected from the oviducts to detach the cumulus cells in M2 medium supplemented with hyaluronidase. Collect clean MII oocytes into one 2-well dish with fresh pre-warmed M2 medium after several thorough washings. The oocytes are now ready for the next experiment (Subheading 3.1.5).

3.1.3 Isolation of PN Stage Embryos

1. Prepare a 35 mm petri dish and a 2-well dish filled with 2.5 ml or 1 ml of M2 medium, respectively, and warm them at 37 °C.
2. Approximately 18–20 h after hCG injection and mating, sacrifice several mice using CO_2 sedation. Using a separate set of tools, cut the skin and peritoneum to reveal the abdominal organs and then collect the ovaries, oviducts, and uterus underneath the fat pads.

3. Transfer the oviducts into a 35 mm petri dish filled with 2.5 ml pre-warmed M2 medium. Collect the oviducts from several mice in one dish.
4. Under a dissecting microscope, release the PN embryos by tearing the oviducts with 27-gauge needles.
5. Remove the expanded cumulus cells surrounding PN embryos by treating M2 medium with hyaluronidase (300 U/ml). Using a mouth pipette attached to a pulled capillary tube with a diameter slightly larger than the diameter of PN embryos, pipette up and down the collected complexes from the oviducts to detach the expanded cumulus cells in M2 medium supplemented with hyaluronidase. Place clean PN embryos into fresh pre-warmed M2 medium after several thorough washings; the cells are now ready for the next step (Subheading 3.1.5).

3.1.4 Isolation of Embryos from the 2-Cell Up to Blastocyst Stages

1. Fill the 35 mm petri dish and a 2-well dish with 2.5 ml or 1 ml of M2 medium, respectively, and warm them at 37 °C.
2. To obtain embryos from the 2-cell to 8-cell stages, mouse oviducts were flushed. To obtain embryos from the morula and blastocyst stages, mouse uteri were flushed.
3. Dissect and collect the reproductive organs (oviducts or uterus). To obtain embryos from the 2-cell stage to the 8-cell stage, cut between the oviduct and ovary with scissors, reposition them with straight forceps, and then cut the uterus near the oviduct. To obtain morula and blastocyst stage embryos, cut between the uterine cervix and bladder with scissors. Pull the uterus upward to stretch the mesometrium and then use scissors to trim this membrane away close to the wall of the uterine horns. Then, cut the uterus below its junction with the oviduct. Make these dissections as clean as possible (less fat and debris) to make it easier to see the embryos after flushing them out of the oviducts and uterus.
4. Place the reproductive organs in a 35 mm petri dish filled with 2.5 ml pre-warmed M2 medium.
5. Test the syringe to be sure that it is free of air bubbles and that the M2 medium flows smoothly before inserting the needle.
6. Using a 27-gauge needle attached to a 1 ml syringe filled with pre-warmed M2 medium, insert the needle into one oviduct or each uterine horn and flush out the embryos with approximately 0.5 ml of M2. Flush again to release any remaining embryos.
7. Remove the flushed oviducts or uterus and visualize the embryos under the dissecting microscope. Collect the embryos using a mouth pipette and transfer them to fresh pre-warmed M2 medium.
8. Immediately clean the embryos and store them (Subheading 3.1.5).

3.1.5 Washing and Storage of Oocytes/Embryos

1. Prepare a 35 mm petri dish with three to six drops of 100 μl PBS containing 0.1% polyvinyl alcohol (PBS-PVA). Wash the isolated oocytes/embryos in PBS-PVA.
2. After several thorough washings of the oocytes/embryos with PBS-PVA, transfer a drop of PBS-PVA containing 10 or 20 oocytes/embryos into the very end of the 1.7 ml ultra-clear tube. Under the dissecting microscope, remove the drop of PBS-PVA solution using a mouth pipette to leave behind only the oocytes/embryos for snap freezing.
3. Immediately snap freeze tubes containing sets of 10 or 20 oocytes/embryos in liquid nitrogen and store at −70 °C prior to RNA isolation (*see* **Note 6**).

3.2 Isolation of mRNA from Mouse Oocytes/Embryos

Generally, 10 or 20 oocytes/embryos are used for mRNA extraction (*see* **Note 7**). Therefore, the traditional extraction method using TRIzol reagent is not applicable. Dynabeads® mRNA DIRECT™ Kit, designed for the simple and rapid isolation of pure, intact polyadenylated (poly A) mRNA directly from very small quantities of cells or tissues, is used because the oligo $(dT)_{25}$ residues are covalently linked to the surface of Dynabeads®, thus only mRNAs including poly A tails can be isolated from oocytes/embryos, while other RNA species will not be hybridized to the beads. This protocol can be scaled up or down according to the sample size.

Before starting,

1. Adjust the water bath or heat block to 70–73 °C.
2. If SDS has precipitated in the Lysis/Binding Buffer stored at 4 °C, warm it in a 37 °C water bath before use and the precipitate will dissolve.

3.2.1 Sample Preparation

1. For RNA extraction from oocytes/embryos, take the stored samples out of the −70 °C deep freezer and thaw slowly on ice.
2. Add 300 μl of Lysis/Binding Buffer directly to the frozen oocytes/embryos pellets, vortex for a few seconds, and then very briefly spin down.
3. Keep the tubes on ice.

3.2.2 Dynabeads® Oligo $(dT)_{25}$ Preparation

1. Resuspend Dynabeads® Oligo $(dT)_{25}$ by thoroughly vortexing before use.
2. Transfer 20 μl of beads to an RNase-free 1.5 ml tube and place the tube on a magnet for 1 min (*see* **Note 8**). The volume of beads can be scaled up or down depending on the number of oocytes/embryos.
3. Remove the supernatant while the tubes are still on the magnet.

3.2.3 Direct mRNA Isolation

1. Remove the tubes from the magnet and add 300 μl of sample lysate (Subheading 3.2.1) to the prewashed beads (Subheading 3.2.2).
2. Gently pipet the solution to resuspend the beads homogenously in the sample lysate. Be careful to not make excessive bubbles while pipetting. Do not vortex or spin down.
3. Leave the tubes for 3–5 min at RT to allow the poly A tail of the mRNA to hybridize to the oligo $(dT)_{25}$ on the beads.
4. Place the tubes on a magnet for 1 min and remove the supernatant on the magnet. If the solution is noticeably viscous, increase the time to place the tube on a magnet to approximately 10 min.
5. Wash the beads/mRNA complex with 500 μl of Washing Buffer A at RT. Place the tube on a magnet for 1 min and remove the supernatant on the magnet. Repeat this step two times.
6. Wash the beads/mRNA complex once with 300 μl of Washing Buffer B at RT. Place the tube on a magnet for 5 min and remove supernatant.
7. Add 13 μl of elution buffer and pipet the solution to resuspend beads/mRNA complex.
8. Incubate at 70–73 °C for 2 min to separate mRNAs from beads and immediately place the tube on a magnet for 3 min.
9. Transfer 13 μl of clear supernatants containing mRNAs to fresh RNase-free PCR tubes and leave the tubes on ice.
10. Use the extracted mRNAs immediately for cDNA synthesis (Subheading 3.3).

3.3 Reverse Transcription

Usually, before starting cDNA synthesis using RNA extracts from cells, the measurement of the RNA concentration is an essential step in order to use equal quantities of RNAs for each sample. However, because we are extracting mRNAs from the same number of oocytes/embryos, the concentration measurement is unnecessary.

1. Directly add 1 μl of oligo dT_{25} to 13 μl of extracted mRNA.
2. Incubate the mRNA/oligo dT_{25} complex at 70 °C for 10 min and then keep it on ice for at least 1 min.
3. Prepare the reaction mixture, as follows, for each reaction:

5× RT buffer	4.0 μl
M-MLV Reverse transcriptase	0.5 μl
RNasin® Ribonuclease Inhibitors	0.5 μl
10 mM dNTP mix	1.0 μl
Total	6.0 μl

4. Add 6 μl of reaction mixture to each tube containing 14 μl of the mRNA/oligo dT_{25} complex and mix briefly by pipetting.
5. Incubate at 42 °C for 60 min followed by heat activation for 2 min at 94 °C.
6. Store 20 μl of cDNAs at −20 °C until use for quantitative real-time RT-PCR (Subheading 3.4).

3.4 Quantitative Real-Time RT-PCR

Quantitative real-time RT-PCR analysis using the mouse oocytes/embryos relies on the iCycler iQ Detection System. The reaction mixture contains single oocyte- or single embryo-equivalent cDNA, 5 pmol forward and reverse gene-specific primers (*see* **Note 9**), and iQ SYBR Green Supermix. Any non-specific PCR products, including primer-dimers (*see* **Note 10**), contribute to an increase in fluorescence, which is detected by the quantitative real-time RT-PCR instruments. Because the designed primers of target genes have different annealing temperatures, it is necessary to optimize the annealing temperature (*X* °C) for each pair of primers before starting the quantitative real-time RT-PCR reaction. The templates are amplified through 40 cycles of denaturation (40 s, 95 °C) to optimize annealing (40 s, *X* °C), and extension (40 s, 72 °C). Upon the completion of PCR, fluorescence is monitored continuously as the samples are slowly heated from 60 to 95 °C at 0.5 °C intervals to obtain melting curves. The melting curves are then used to identify any nonspecific amplification products. The expression levels of each target mRNA in the oocytes/embryos are then normalized to those of β-actin (*Actb*), glyceraldehyde 3-phosphate dehydrogenase (*Gapdh*), *18S rRNA* [10], or spike-in mRNA [6, 9]. The relative expression levels of the target genes are evaluated using the comparative C_T method.

1. Turn on the computer and PCR machine and set up the program with the following thermal settings:

No. of cycle	Temperature	Time
1	95 °C	5 min
40	95 °C *X* °C[a] 72 °C	40 s 40 s 40 s
1[b]	60–95 °C (at 0.5 °C intervals)	30 s

[a]Optimized annealing temperature for each pair of primers
[b]Melting curve analysis

2. PCR mix per sample (*see* **Note 11**):

Single oocyte/embryo-equivalent cDNA	X µl[a]
Primer mix (Forward and Reverse, 5 pmol/µl)	1 µl
2× iQ SYBR Green Supermix	10 µl
Deionized H_2O	final volume to 20 µl

[a]If mRNA was isolated from 10 or 20 oocytes/embryos, the volume of single oocyte- or embryo-equivalent cDNA is 2 µl or 1 µl, respectively.

3. Prepare the reaction mixture containing cDNAs (single oocytes/embryo-equivalent cDNA), 2× iQ SYBR Green Supermix, and deionized H_2O without the primer mix.
4. Place 19 µl of the reaction mixture into the wells of a 96-well plate and then add 1 µl of the primer mix.
5. Place a sealing film over the 96-well plate to avoid evaporation. Briefly spin down the plate to remove air bubbles.
6. Place the plate into the thermal cycler and run the PCR program.
7. After PCR is complete, remove the tubes from the machine. PCR specificity is examined by 3% agarose gel using 10 µl of each reaction mixture (*see* **Note 12**).

3.5 Data Analysis by the $2^{-\Delta\Delta C_T}$ Method

For the $\Delta\Delta C_T$ calculation to be valid, the efficiency of the target gene amplification and the efficiency of the reference gene amplification (endogenous housekeeping genes or exogenous spike-in mRNA) must be approximately equal [11]. The mean, standard deviation (SD), and standard error (SE) are determined from the triplicate samples of the each oocyte/embryo sample group (*see* **Note 6**). In this way, the biological as well as the technical triplicates are performed.

1. After analysis, multiple peaks in the melting curve should be excluded for PCR specificity, whereas only one peak in the melting curve (Fig. 2a) is used for the relative quantitative analysis of the target gene. Because the nonspecific amplification leads to confounding quantitative real-time RT-PCR results (*see* **Note 12**), nonspecific amplification should always be identified by using melting curve analysis (*see* **Note 13**). Determine whether there is any bimodal dissociation curve (Fig. 2b) or abnormal amplification plot [12].
2. Calculate the threshold cycle (C_T) for each well using the instrument's software (Fig. 3). The C_T value represents the cycle number at which a fluorescent signal rises significantly above the background. Ensure that the threshold lines are the same across all PCR runs in the same analysis.

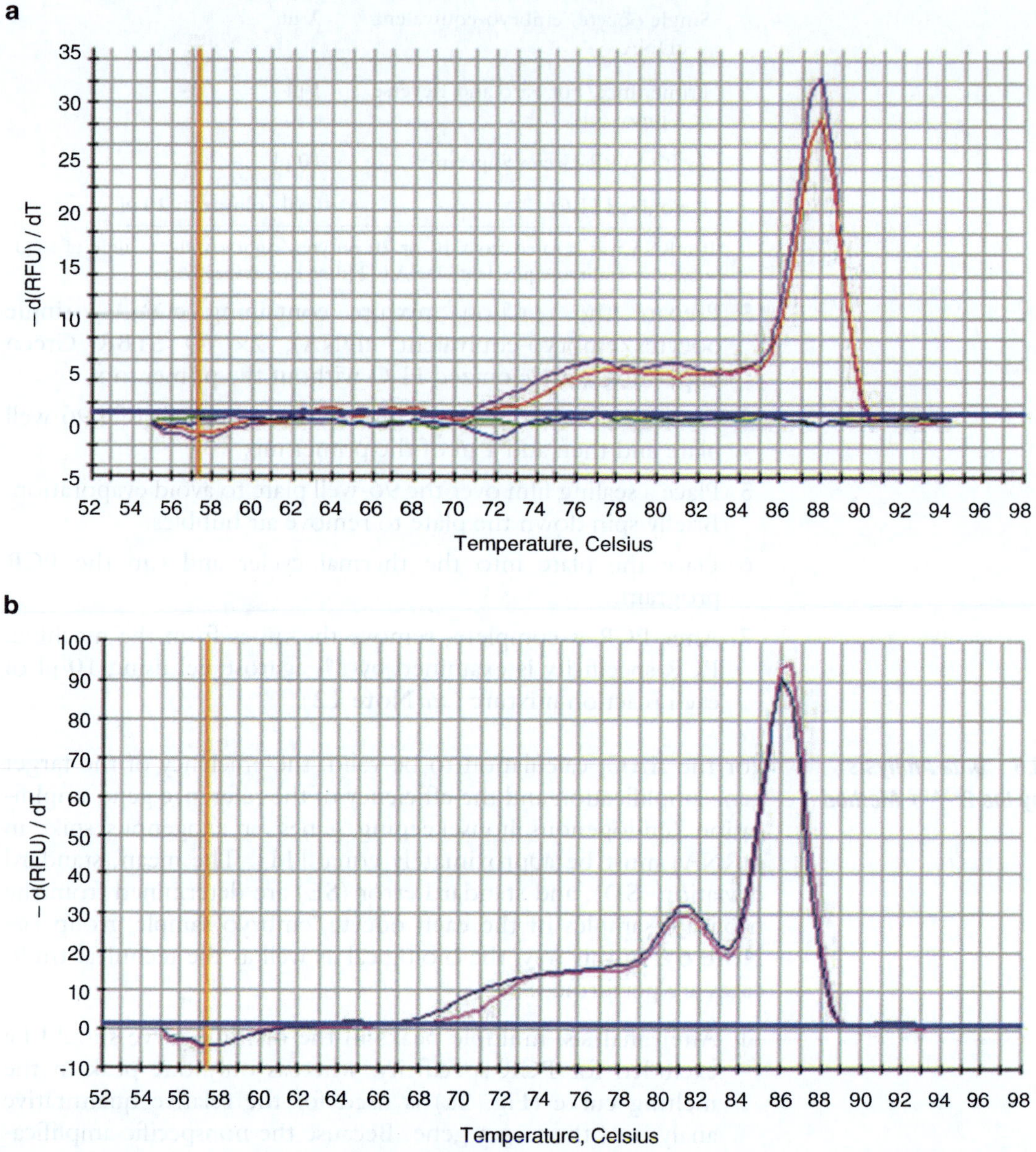

Fig. 2 Melting curve analysis of amplification from a SYBR® Green assay. After amplification, the instrument begins a melting program in which the reactions are cooled to anneal the specific primer and then slowly heated at 0.5 °C intervals while fluorescence is continuously monitored. (**a**) The violet and red melting curve lines are presented as a single, sharply defined melting curve with a narrow peak, which indicated that pure, specific gene amplification was conducted. In contrast, the green and blue melting curve lines are flat at the ground zero line, which indicated that gene amplification was not produced (see **Note 16**). (**b**) The blue and pink melting curve lines are presented as two clearly distinct peaks, a bimodal dissociation curve, which indicates that the gene amplification was not accomplished and was contaminated with other amplification or nonspecific products

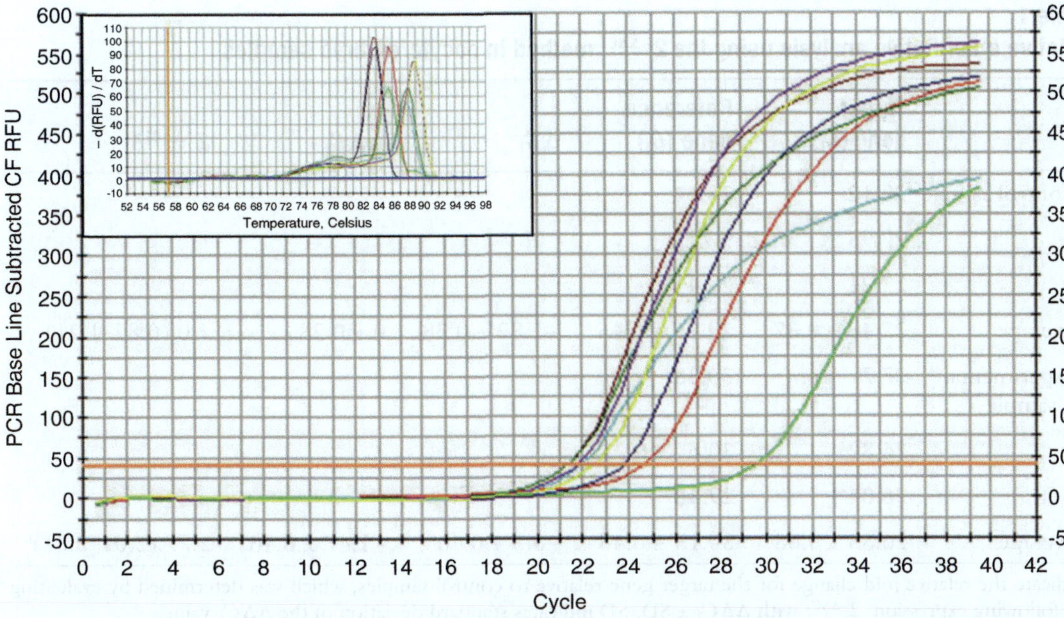

Fig. 3 Typical signal curves obtained from the quantitative real-time RT-PCR experiments with oocytes and embryos using the SYBR® Green assay. The threshold for the determination of the C_T values is indicated by the horizontal orange line. A smaller C_T value (green line at cycle 22) indicates the higher amounts of mRNAs in that sample, while a bigger C_T value (yellowish green line at cycle 30) indicates a smaller quantity of mRNAs in that sample. The upper inset graphs represent the normal melting curves, which confirms that the amplification of the target genes was accomplished without contamination with other products. RFU relative fluorescence units

3. To analyze the data, C_T values for all wells are exported to an Excel worksheet or Microsoft Word. The exported file contains columns with the sample well number, sample description, melting curve, and C_T values. The important parameter for quantitative analysis is C_T.

4. The C_T data are used to determine the quantities of mRNA of target and reference genes. Calculate the ΔC_T for control and experimental samples (Table 1). The C_T values of both control and experimental samples are normalized to the appropriate reference genes (*see* **Note 14**), such as *Actb*, *Gapdh*, *18S rRNA*, or spike-in mRNA (*see* **Note 15**). When biological and technical replicates are performed, calculate the average C_T value of each sample group. Average ΔC_T = average C_T (Target gene) − average C_T (Reference gene) [13]. The standard deviation of the difference is calculated from the standard deviations of the target gene and reference gene values.

5. Calculate the $\Delta\Delta C_T$ for the control sample and experimental sample (Table 1). $\Delta\Delta C_T$ = average ΔC_T − average ΔC_T (experimental sample) [13].

Table 1
Relative quantitative analysis using the $2^{-\Delta\Delta CT}$ method in oocyte/embryo samples.

	Target gene (C_T)	Reference gene (C_T)	ΔC_T	$\Delta\Delta C_T$	$2^{-\Delta\Delta CT}$ [a]
Control sample	27.12	28.64			
	27.95	9.71			
	27.21	29.41			
Average	27.43 ± 0.37	29.25 ± 0.45	−1.82 ± 0.28	0 ± 0.28	1.0 (0.82–1.21)
Experimental sample	27.7	30.38			
	26.82	30.6			
	26.04	29.48			
Average	26.85 ± 0.68	30.15 ± 0.48	−3.3 ± 0.46	−1.47 ± 0.46	2.77 (2.01–3.81)[b]

[a]Indicate the relative fold-change for the target gene relative to control samples, which was determined by evaluating the following expression: $2^{-\Delta\Delta CT}$ with $\Delta\Delta C_T$ ± SD. SD indicates standard deviation of the $\Delta\Delta C_T$ value.
[b]Represents statistical significance at $p < 0.05$ by a paired Student's t-test.

6. Calculate the fold-change for the target gene from experimental samples as $2^{-\Delta\Delta}C_T$ (Table 1) [13]. Because $\Delta\Delta C_T$ is equal to 0, 2^0 is equal to 1. The relative gene expression is usually set to 1 for control samples. If the fold-change of the target genes in experimental samples is greater than 1, it is a fold up-regulation. If the fold-change of the target gene in the experimental sample is less than 1, it is a fold down-regulation.

4 Notes

1. We used M2 and M16 media for oocyte/embryo collection and culture, respectively. These media are commonly used for the in vitro culture of oocytes and embryos for oocyte maturation and early embryo development [14]. M2 and M16 media contain glucose as well as pyruvate and lactate (Table 2). Embryos at the early developmental stage utilize pyruvate and lactate, but glucose is not efficiently used. M16 medium that contains sodium bicarbonate will quickly become alkaline upon exposure to air, whereas M2 medium, which contains both HEPES and bicarbonate, will not (Table 2). Therefore, M2 medium is suitable for the extended period of collection and handling for oocytes/embryos outside of the CO_2 incubator [15].
2. Environmental heat stress affects the embryonic gene expression before and after ZGA [16]. The collection of oocytes/

Table 2
Composition of the M2 and M16 media for oocytes/embryos manipulation

Component	M16 (g/l)	M2 (g/l)
$CaCl_2 \times 2H_2O$	0.251	0.251
$MgSO_4$ (anhyd)	0.165	0.165
KCl	0.356	0.356
KH_2PO_4	0.162	0.162
$NaHCO_3$	2.101	0.35
NaCl	5.532	5.532
d-Glucose	1	1
BSA (albumin, bovine fraction V)	4	4
Sodium lactate	2.95	2.95
Phenol red	0.01	0.01
Sodium pyruvate	0.036	0.036
HEPES	–	5.43

embryos is effective in spring and fall in terms of the quality and quantity of samples. During the summer period, with the environmental heat stress, the number of collected oocytes and embryos is decreased due to the reduced reproductive performance of the mouse [17]. In addition, exposing the oocytes to the environmental heat stress results in impaired maternal mRNA storage and/or transcriptional regulation. To collect as many good quality oocytes and embryos as possible, it is better to conduct the research during the spring, fall and winter seasons to avoid difficulties in collecting samples in the summer.

3. Female mice were injected with PMSG at 5–7 p.m. on Day 0 and hCG at 5–7 p.m. on Day 2. When mice were superovulated at this time schedule, we sacrificed the animals at 9–11 a.m. to collect PN, 2-cell, 8-cell and blastocyst stage embryos and at 9–11 p.m. for 4-cell and morula stage embryos (Fig. 1).
4. In general, clumsy handling results in the poor quality of oocytes and embryos. Therefore, do not sacrifice too many mice at once. This will depend on the skillfulness of the researcher, but 10 mice per one skillful person are typically adequate for collecting an appropriate number of oocytes.
5. To prevent the degradation of intracellular cAMP, i.e., to maintain the GV stage of oocytes during the sampling process, IBMX was added to the M2 medium immediately prior to use. Add 50 μl of 200× IBMX (40 mM) to the tube containing 10 ml of the M2 medium and mix well.

6. Collect the oocytes/embryos from at least three biological replicates to increase the power of the data for statistical analyses.
7. For mRNA isolation, at least five oocytes/embryos are required to minimize the pipetting errors. Thus, we typically used 10 or 20 oocytes/embryos in one tube, and we recommend collecting five oocytes/embryos in one tube when the samples are very precious.
8. The components of the Dynabeads® mRNA DIRECT™ Kit should be adjusted for mRNA isolation according to the number of oocytes/embryos.

	10–20 oocytes/ embryos	50 oocytes/ embryos	100 oocytes/ embryos
Beads (μl)	20	30	40
Lysis/binding buffer (μl)	300	500	700
Washing buffer A (μl)	500	700	900
Washing buffer B (μl)	300	500	700

9. Make the gene-specific PCR primers using a standard set of criteria for primer design
   (a) Primer T_m = 58–61 °C (The difference of T_m of both forward and reverse primers must not be greater than 2 °C),
   (b) Primer lengths of 19–21 bp,
   (c) Guanine cytosine (GC) content between 45% and 50%,
   (d) Unique short PCR product lengths between 100 and 200 bp,
   (e) The 3′-end of coding sequence (CDS) region is better for targeting in the primer design because it is generally more unique than the 5′-end of CDS and closer to the RT start site (poly A tails).
10. When the transcript level of target genes is very low, we often observe the increased assembly of primer-dimers despite well-designed target-specific primers. To avoid this problem, increase the amount of the template, i.e., cDNA of the sample, to eliminate the formation of primer-dimers.
11. Reduce technical errors in the PCR reaction by minimizing the number of pipetting steps. According to our standardized protocol, mix single oocyte/embryo-equivalent cDNA with SYBR Green reagents (master mix), aliquot a standard volume (19 μl) of master mix into each reaction well, and then load a standard volume (1 μl) of specific primer mix (*see* Subheadings 3.4, **steps 2–4**).

12. To avoid misleading results from the quantitative real-time RT-PCR reaction, make sure to perform a hot-start PCR with an annealing temperature greater than 60 °C. If the nonspecific amplification is persistent, design a new set of different primers for that specific target gene.
13. Although confirming the bands by gel electrophoresis or checking the melting curves provides a good indication of the target gene-specific amplification, we use the combination of both gel inspection and melting curve analysis to ensure specific PCR products are made.
14. Reference genes are selected based on the literatures [18, 19]. Reference genes, such as *Actb*, *18S rRNA*, *B2m*, *Gapdh*, *Hmbs*, *Hprt1*, *Ppib*, and *Rplp0*, are routinely used in gene expression studies. The selection of suitable reference genes is of importance for the accurate interpretation of data generated by quantitative real-time RT-PCR using oocytes/embryos.
15. We use *H1foo* as a reference gene for mouse oocytes and *Actb* for mouse embryos. When comparing the gene expression between oocytes and embryos, the expression levels of target genes are then normalized to the level of an exogenous spike-in mRNA. We use homemade green fluorescent protein (*GFP*) mRNA for spiking.
16. When we obtain a completely negative sign of PCR amplification in a melting curve or signal curve data, as shown as the minimally fluctuating green and blue lines near the ground zero line (thick blue straight line in Fig. 2a), we assess the function of the primer sets by working with positive control samples that are known to express those specific genes. If the primers amplify the specific gene in the positive samples, then the negative PCR amplification can be interpreted as "no expression" of that specific gene in that specific sample. If the primers did not work in the positive control sample, then a new PCR primer set must be designed for the next experiment.

Acknowledgments

This research was supported by the Priority Research Centers Program through the National Research Foundation of Korea (NRF) funded by the Ministry of Education (2009-0093821) and the Basic Science Research Program (NRF-2015R1D1A1A 01056595).

References

1. Yoon SJ, Chung HM, Cha KY, Kim NH, Lee KA (2005) Identification of differential gene expression in germinal vesicle vs. metaphase II mouse oocytes by using annealing control primers. Fertil Steril 83(Suppl 1):1293–1296
2. Gandolfi TA, Gandolfi F (2001) The maternal legacy to the embryo: cytoplasmic components and their effects on early development. Theriogenology 55:1255–1276
3. Su YQ, Sugiura K, Woo Y, Wigglesworth K, Kamdar S, Affourtit J, Eppig JJ (2007) Selective degradation of transcripts during meiotic maturation of mouse oocytes. Dev Biol 302:104–117
4. Kim KH, Lee KA (2014) Maternal effect genes: findings and effects on mouse embryo development. Clin Exp Reprod Med 41:47–61
5. Langley AR, Smith JC, Stemple DL, Harvey SA (2014) New insights into the maternal to zygotic transition. Development 141:3834–3841
6. Park MW, Kim KH, Kim EY, Lee SY, Ko JJ, Lee KA (2015) Associations among Sebox and other MEGs and its effects on early embryogenesis. PLoS One 10:e0115050
7. Kim EA, Kim KH, Lee HS, Lee SY, Kim EY, Seo YM, Bae J, Lee KA (2011) Downstream genes regulated by Bcl2 l10 RNAi in the mouse oocytes. Dev Reprod 15:61–69
8. Hogan B, Beddington R, Costantini F, Lacy E (eds) (1994) Manipulating the mouse embryo. Cold Spring Harbor Laboratory Press, New York, NY
9. Kim KH, Kim EY, Lee KA (2008) SEBOX is essential for early embryogenesis at the two-cell stage in the mouse. Biol Reprod 79:1192–1201
10. Willems E, Mateizel I, Kemp C, Cauffman G, Sermon K, Leyns L (2006) Selection of reference genes in mouse embryos and in differentiating human and mouse ES cells. Int J Dev Biol 50:627–635
11. Udvardi MK, Czechowski T, Scheible WR (2008) Eleven golden rules of quantitative RT-PCR. Plant Cell 20:1736–1737
12. Yuan JS, Reed A, Chen F, Stewart CN Jr (2006) Statistical analysis of real-time PCR data. BMC Bioinformatics 7:85
13. Livak KJ, Schmittgen TD (2001) Analysis of relative gene expression data using real-time quantitative PCR and the 2(-Delta Delta C(T)) Method. Methods 25:402–408
14. Whittingham DG (1971) Culture of mouse ova. J Reprod Fertil Suppl 14:7–21
15. Quinn P, Barros C, Whittingham DG (1982) Preservation of hamster oocytes to assay the fertilizing capacity of human spermatozoa. J Reprod Fertil 66:161–168
16. Gendelman M, Roth Z (2012) Seasonal effect on germinal vesicle-stage bovine oocytes is further expressed by alterations in transcript levels in the developing embryos associated with reduced developmental competence. Biol Reprod 86:1–9
17. Gendelman M, Aroyo A, Yavin S, Roth Z (2010) Seasonal effects on gene expression, cleavage timing, and developmental competence of bovine preimplantation embryos. Reproduction 140:73–82
18. Stephens AS, Stephens SR, Morrison NA (2011) Internal control genes for quantitative RT-PCR expression analysis in mouse osteoblasts, osteoclasts and macrophages. BMC Res Notes 4:410
19. Fu J, Bian L, Zhao L, Dong Z, Gao X, Luan H, Sun Y, Song H (2010) Identification of genes for normalization of quantitative real-time PCR data in ovarian tissues. Acta Biochim Biophys Sin 42:568–574

Chapter 5

Detection of miRNA in Mammalian Oocytes and Embryos

Malavika K. Adur, Benjamin J. Hale, and Jason W. Ross

Abstract

MicroRNAs (miRNAs) play a crucial role in the regulation of many post-transcriptional processes in reproductive cells. Regulation of maternal mRNA translation and activation of zygotic mRNA are essential to successful embryonic development. Moreover, the precise development of embryonic cell and/or tissue lineages requires temporal and spatial control of gene expression, mRNA abundance, and translation into proteins, which is in part regulated via miRNA. Here, we describe some key protocols that can be utilized to detect and quantify miRNA in in vitro produced oocytes and embryos.

Key words Oocyte, Embryo, Pig, miRNA detection

1 Introduction

During the period of oocyte meiotic maturation after the breakdown of the germinal vesicle (GVBD), or nuclear envelope, transcriptional activity is almost nonexistent and transcription of new messenger RNA (mRNA) does not reoccur until early embryo development and activation of the embryonic genome, which occurs around the two- to eight-cell stage of development depending on the species [1]. Following GVBD, the maturing oocyte must then rely on post-transcriptional regulation to bring about the dynamic changes necessary before fertilization [2, 3]. Among the numerous small noncoding RNA (ncRNA) molecules in the mammalian genome, microRNAs (miRNAs) play a pivotal role in the regulation of many post-transcriptional physiological processes in reproductive tissues and cells [4]. MiRNAs are integrated in regulatory networks that govern a wide variety of processes ranging from the maintenance of tissue homeostasis to the specification of cell fate during development. The correct specification and development of embryonic and extra-embryonic lineages requires precise temporal and spatial control of gene expression and mRNA abundance, which is in part regulated via miRNA. The generation of embryos

Kiho Lee (ed.), *Zygotic Genome Activation: Methods and Protocols*, Methods in Molecular Biology, vol. 1605,
DOI 10.1007/978-1-4939-6988-3_5,

lacking mature miRNAs through the deletion of proteins necessary for their processing has revealed that these regulators play an essential role in early embryo development in numerous species.

1.1 MicroRNA Biogenesis

Biogenesis of miRNA (Fig. 1) is initiated by transcription of primary miRNA transcripts (pri-miRNA) by RNA polymerase II. These primary transcripts are then processed by Drosha/DGCR8 complex and transported from the nucleus to cytoplasm. Finally, Dicer processes the miRNA precursors into mature miRNAs [5]. A key function of miRNAs is to repress expression of their target genes through sequence complementation, which reduces the abundance of the target mRNAs and/or inhibits their translation [6].

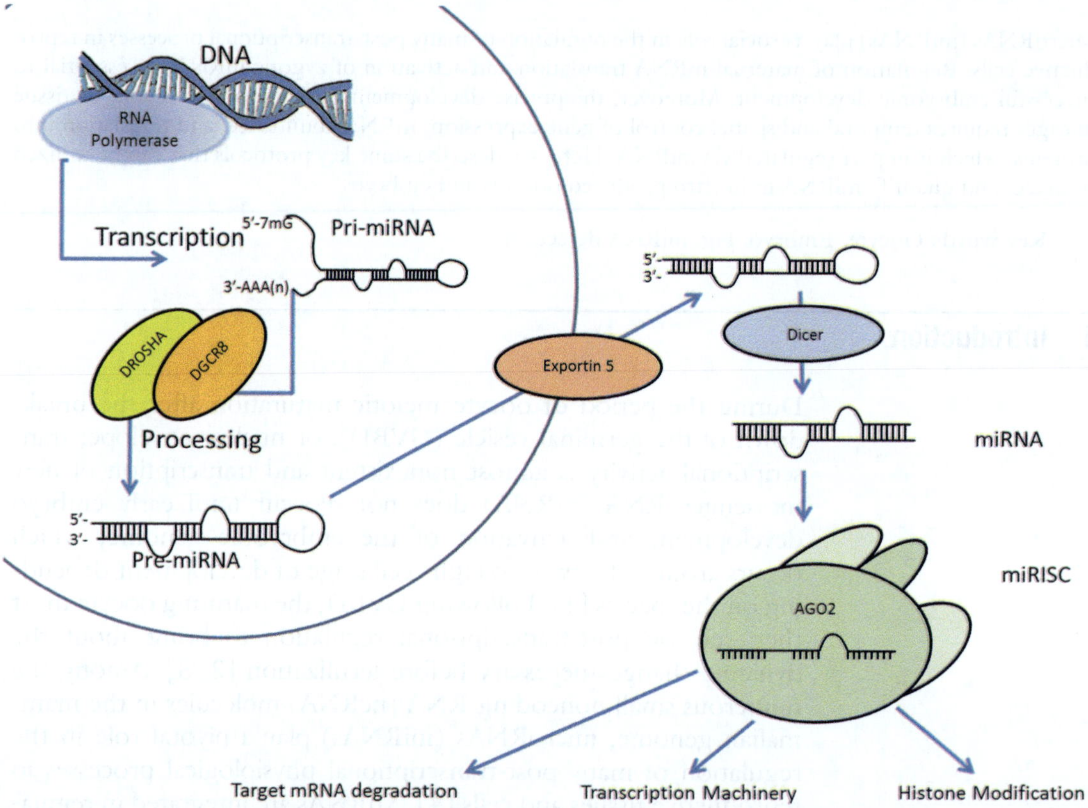

Fig. 1 MicroRNA biogenesis is initiated via the activity of RNA polymerase II resulting in synthesis of a primary miRNA transcript (pri-miRNA) that is both capped and polyadenylated. The appropriate spatial complementation of specific nucleotides within the transcript results in the formation of hairpin secondary structures that are recognized by the RNA-processing complex consisting of DROSHA and DGCR8. The enzymatic activity of this protein complex results in cleavage and removal of the hairpin structure (now considered a pre-miRNA) from the primary transcript. Exportin 5 facilitates the transport of pre-miRNA from the nucleus into the cytoplasm of the cell where it is recognized by Dicer, and the loop is cleaved leaving a short duplex mature miRNA molecule. Upon dissociation, either strand from the duplex miRNA structure can be utilized by the RNA-induced silencing complex to contribute to post-transcriptional gene regulation through impacting mRNA stability and/or translation efficiency in addition to contributing to chromatin modifications to control gene expression. (The Male Role in Pregnancy Loss and Embryo Implantation Failure, Small RNAs: Their possible roles in reproductive failure, Hale BJ, Keating AF, Yang, CX, Ross JW (2015) 868:49–79)

1.2 MicroRNA Mechanism of Action

Target recognition of miRNA primarily occurs through binding of the first eight residues on the 5′ end of the miRNA (also known as the "seed region") to a complementary sequence in the 3′UTR of mRNA targeted for post-transcriptional gene regulation (PTGR) [7, 8]. Significant binding of the seed region was initially thought to be a requirement for miRNA-regulated PTGR, although less than perfect complementation can lead to translational inhibition [6]. Interestingly, unique examples exist in which even weak complementation between the 3′ end of the miRNA and the 3′UTR of the target gene can confer PTGR through mRNA degradation [9]. Some miRNA cause both transcript degradation and translation inhibition due to their interaction with mRNA, while other miRNA:mRNA interactions result only in the latter. Transcriptional degradation and translational repression are widely involved in miRNA function and are of direct importance to mammalian developmental biology [10].

1.3 Biological Importance of MicroRNA

Maternal inheritance of miRNAs is thought to be crucial for early mammalian development. In the mouse, while there is a significant global loss of maternal miRNAs between the one- and two-cell stages of development, de novo synthesis of miRNAs commences at the two-cell stage [11]. Following fertilization, successful embryonic development requires broad transcriptional arrest and mRNA clearance to deplete maternally stored mRNA transcripts in coordination with the activation of zygotic genome and subsequent mRNA and protein production [5].

The biological importance of miRNA function is due to its role in conferring robustness to the cell's mRNA and protein profile [12]. With respect to embryonic development, tissue reorganization, and cellular differentiation, miRNA functions are not only strategically involved with cell fate, but are clearly required components for specific cell lineage destinations [12–14]. Alteration of only a few miRNA during embryonic development may dictate large changes in the transcript and protein abundance that occur during developmental stage progression [15], since the average miRNA is estimated to have recognition sites for approximately 100 target mRNAs [7], and furthermore, individual miRNAs may confer PTGR on a few to more than 800 mRNAs [16]. A more comprehensive review of miRNA and other RNA classes, such as small interfering RNA, and PIWI interacting RNA in reproductive tissues has already been published [5].

1.4 Importance of MicroRNA Function for Oocyte and Early Embryonic Development

Successful embryonic development requires broad transcriptional arrest and mRNA clearance to deplete maternally derived proteins and eliminate stored mRNA transcripts in coordination with the activation of zygotic mRNA and protein production. Maternal mRNA depletion and regulation is, in part, controlled via the 3′UTR of the expressed transcripts [17, 18]. Thus, it is not surprising that differential miRNA expression has been demonstrated during oocyte maturation and early embryonic development in *Xenopus laevis* [19], *Drosophila* [20, 21], zebrafish [10, 22], and mice [11]. Furthermore, DICER, an enzyme mechanically responsible for a key step in miRNA production, is required for early embryonic development in mice [11, 23, 24] and zebrafish [22]. DICER's requirement for successful oocyte development suggests that specific mature miRNA are responsible for the PTGR that is necessary for the production of a functional oocyte capable of producing a developmentally competent embryo. Mice with a ZP3-driven conditional DGCR8 (an enzyme responsible for canonical miRNA production) knockout are still fertile, although they have reduced fecundity with litter size less than half of controls suggesting that miRNA do contribute to developmental competency [25, 26].

Precise and quantitative estimation of miRNA profiles and/or specific miRNA expression is the key to better understanding their function and correlation to a given condition. Each detection technique has certain benefits and limitations [27]. Recent fascinating and fast progress in bioinformatic, high-throughput sequencing, and other biochemical approaches have fueled rapid growth in our appreciation of the tremendous number, diversity, and biological importance of these ncRNAs in gene regulation at both transcriptional and post-transcriptional levels [5]. The protocols mentioned in this chapter refer to strategies and methods utilized to detect and quantify miRNA in in vitro produced oocytes and embryos.

2 Materials

2.1 Oocyte and Sperm Wash Media

2.1.1 Oocyte Wash Media: HEPES-Buffered Tyrode's Lactate Media (TL-HEPES-PVA)

To make 1 L, add 6.6622 g NaCl, 0.2386 g KCl, 0.2940 g $CaCl_2.2H_2O$, 0.0408 g Na_2HPO_4, 0.1017 g $MgCl_2.6H_2O$, 1.8680 mL Na-Lactate (60% syrup), 2.3830 g HEPES, 0.0220 g Na-Pyruvate, 2.1864 g Sorbitol, 0.1680 g $NaHCO_3$, 0.1 g PVA, 50 μg/mL Gentamicin to Nanopure water. The pH should range between 7.2 and 7.4 and osmolarity between 285–295 mOsm. Filter and store at 4 °C. Warm it to 37 °C before actual use.

2.1.2 Sperm Wash Media: Dulbecco's Phosphate-Buffered Saline with Bovine Serum Albumin

To make 100 mL, add 10 μg/mL Gentamicin and 0.1 g BSA to 1× DPBS. Filter and keep warm at 38.5 °C until use.

2.2 In Vitro Maturation Media for Oocytes

2.2.1 Modified TCM-199 (mTCM-199) Stock

To make 200 mL, add 1.9 g M-199 powder, 0.44 g $NaHCO_3$, 0.2 g PVA, 0.02 g Na-Pyruvate, 0.11 g D-Glucose, 0.0002 g Gentamicin to Nanopure water. The pH should be 7.4 and osmolarity between 295–310 mOsm. Filter and store it at 4 °C. This can be used for up to three weeks.

2.2.2 mTCM-199 Working Media

Made fresh each time it is needed. To make 20 mL, add 0.0014 g L-Cysteine, 20 µL Follicle Stimulating Hormone (FSH), 20 µL Luteinizing Hormone (LH), 20 µL Epidermal Growth Factor (EGF) to mTCM-199 stock made earlier. (Refer to Subheading 4 (*see* **Note 1**) for details on FSH, LH and EGF stock preparation). Filter it and make 500 µL drops in a dish, covered with oil. Incubate for at least three hours at 38.5 °C, 5% CO_2 in air before adding in the oocytes for maturation.

2.3 In Vitro Fertilization Media

2.3.1 Modified Tris-Buffered Media (mTBM) Stock

To make 100 mL, add 0.6611 g NaCl, 0.0224 g KCl, 0.1102 g $CaCl_2.2H_2O$, 0.2423 g Tris (crystalized free base), 0.1982 g D-Glucose, 0.0550 g Na-Pyruvate to Nanopure water. Mix well, but do not adjust the pH. Filter it and store at 4 °C for up to one week.

2.3.2 mTBM Working Media

Made fresh on the day oocyte maturation begins. Pipette 15 mL of mTBM stock made earlier into a 50 mL conical tube and add 0.01 g caffeine (final concentration 2 mM) and 0.03 g BSA (final concentration 0.2% w/v). Mix it well and filter.

2.4 Utilization of a Percoll Gradient for Sperm Preparation

2.4.1 1× PBS

To make 50 mL, add 5 mL of 10× DPBS to 45 mL Nanopure water.

2.4.2 100% Percoll

To make 50 mL, add 5 mL 10× DPBS to 45 mL Percoll.

2.4.3 60% Percoll

To make 50 mL, add 30 mL of 100% Percoll from Subheading 2.4.2 to 20 mL 1× DPBS. Filter it and store at 4 °C. Warm it to room temperature before use.

2.5 In Vitro Culture Media

2.5.1 Porcine Zygote Medium-3 (PZM-3) Stock

To make 500 mL, add 3.155 g NaCl, 0.3728 g KCl, 0.0238 g KH_2PO_4, 0.0493 g $MgSO_4$, 1.053 g $NaHCO_3$, 10 µg/mL Gentamicin to Nanopure water. Do not adjust the pH, but osmolarity should range between 280–285 mOsm. Filter it and store at 4 °C.

2.5.2 PZM-3 Working Media

To make 100 mL, add 0.0022 g Na-Pyruvate, 0.0617 g Ca-Lactate, 0.0146 g L-Glutamine, 0.0546 g Hypotaurine, 2 mL BME (amino acid 50×), 1 mL MEM (nonessential amino acid 100×), 0.3 g fatty acid free BSA to 97 mL of PZM-3 stock made in Subheading 2.5.1. The pH should range between 7.2–7.4 and the osmolarity between 280–290 mOsm.

2.6 Hyaluronidase Stock

Use TL-HEPES to make a 0.1% solution. Aliquot and store at −20 °C.

2.7 Miscellaneous Items for Embryo Production

Vortex, Table-top centrifuge, 15 mL and 50 mL conical tubes, 1.5 mL microcentrifuge tubes, 60 × 15 mm plastic petri dishes, 4-well culture dishes, mineral oil (light density), slide warmer, CO_2 incubator, stereo microscope, inverted compound microscope, counting chamber slide, glass slides and coverslip glass, glass Pasteur pipets.

2.8 mirVana miRNA Isolation Kit (Ambion)

30 mL miRNA Wash solution I (add 21 mL 100% ethanol before use), 50 mL Wash solution 2/3 (add 40 mL 100% ethanol before use), 100 mL Lysis/Binding buffer, 10 mL miRNA Homogenate Additive, 100 mL Acid-Phenol:Chloroform, 1.4 mL Gel-Loading Buffer, 5 mL Elution Buffer, collection tubes, filter cartridges, 1.5 mL microcentrifuge tubes.

2.9 Buffers for Northern Blot

2.9.1 10× TBE Buffer for Northern Blot

Add 109 g Tris Base, 55 g Boric acid, and 40 mL 0.5 M EDTA to 850 mL nuclease-free water. Adjust the final volume to 1 L.

2.9.2 Electrophoretic Transfer Buffer (20×)

Add 35.29 g Trisodium Citrate and 22.71 g Sodium phosphate, dibasic (Na_2HPO_4) to 850 mL nuclease-free water. Adjust final volume to 1 L. Adjust the pH to 8.3, autoclave, and store for up to one to two years at room temperature.

2.9.3 Prehybridization/Hybridization Mix

To make 50 mL, add 25 mL Formamide, 12.5 mL SSC for Northern (20×), 1.5 mL Denhardt's solution for Northern (100×), 2 mL Herring testis carrier DNA, 0.5 mL SDS (10%) to 8.5 mL nuclease-free water. Prepare fresh before each use.

2.9.4 SSC for Northern (20×)

To make 1 L, add 175.3 g NaCl, 88.3 g Trisodium Citrate to 850 mL nuclease-free water. Adjust the final volume to 1 L. Adjust the pH to 7.0 with NaOH or HCl. Autoclave the solution and store it at room temperature. It will usually last for one year or longer.

2.10 Denaturing 15% Acrylamide Gels

To make 15 mL, add 7.2 g Urea, 1.5 mL 10× TBE, and 5.6 mL of 40% Acrylamide in nuclease-free water. Mix it well and then add 75 μL of 10% ammonium persulfate and 15 μL Tetramethylethylenediamine (TEMED).

2.11 TaqMan™ Gene Expression Cells-to-CT™ Kit (Applied Biosystems)

DNase I, 20× RT Enzyme Mix, Lysis solution, 2× RT Buffer, TaqMan® Gene Expression Master Mix, Stop solution. Other general laboratory equipment will be needed such as 96-well plates, PCR tubes, Thermal Cycling equipment, and nuclease-free PCR grade water.

2.12 TaqMan® MicroRNA Reverse Transcription Kit (Applied Biosystems)

10× RT buffer, 100 mM dNTP mix (with dTTP), RNase inhibitor (20 U/μL), and MultiScribe™ RT enzyme (50 U/μL).

2.13 MinElute PCR Purification Kit (Qiagen)

30 mL Buffer PB, 2 × 6 mL Buffer PE (concentrate), Buffer EB, MinElute spin columns, pH indicator, collection tubes, loading dye.

2.14 Cell Differentiation and Development RT² miRNA PCR Array (SABiosciences)

It contains formatted array plates and optical adhesive film. Additionally, RT² miRNA First Strand kit and RT² SYBR Green qPCR Mastermix are also needed.

2.15 SOLiD Small RNA Expression Kit (Ambion)

Adaptor Mix A, Adaptor Mix B, Control RNA, nuclease-free water, Hybridization solution, 2× Ligation buffer, Ligation Enzyme mix, 10× RT Buffer, ArrayScript™ Reverse Transcriptase, RNase H (*E. coli*), 10× PCR Buffer I, AmpliTaq® DNA Polymerase, 2.5 mM dNTP Mix, SOLiD PCR Primers 1-10, SYBR gold dye. General laboratory equipment such as NanoDrop ND-1000 spectrophotometer, Agilent 2100 Bioanalyzer with Small RNA Chip Kit, Thermal Cycler and Polyacrylamide gel electrophoresis equipment, RNase-free PCR and 1.5 mL microcentrifuge tubes, Ambion spin columns.

3 Methods

3.1 Porcine Embryo Production (Adapted from Refs. 28–30)

3.1.1 Oocyte Collection and Maturation

1. Sow ovaries are obtained from an abattoir and transported to the laboratory in a thermos maintained at approximately 35 °C.
2. Follicular fluid from 3 to 5 mm (in diameter) antral follicles is aspirated using an 18-gauge needle attached to a 10 mL disposable syringe. Keep the bevel of the needle facing upward, insert needle at the base of the follicle while pulling on the syringe piston. This will allow minimal loss of follicular fluid.
3. The cumulus-oocyte complexes (COCs) with multiple layers of intact cumulus cells and uniform ooplasm are selected for GV-stage sample collection or in vitro maturation (IVM).
4. IVM is conducted in TCM-199 media at 38.5 °C and 5% CO_2 in air for 42 to 44 h.
5. Following IVM, the cumulus cells of matured COCs are separated from oocytes by vortexing COCs in 0.1% hyaluronidase in HEPES-buffered Tyrode medium containing 0.1% PVA.

6. MII oocytes are identified by the presence of an extruded polar body.
7. Following separation, GV and MII oocytes are collected and stored as needed for subsequent procedures.

3.1.2 In Vitro Fertilization

1. Oocytes with a polar body after IVM are rinsed in mTBM working medium and transferred into equilibrated 50 μL droplets of mTBM working medium, with approximately 35 oocytes added per drop.
2. Fresh boar semen is rinsed twice using DPBS plus 1 mg/mL of BSA, then diluted to the desired concentration of sperm per milliliter using mTBM working medium.
3. Fifty microliters of the sperm sample are added into the droplets with oocytes.
4. After 4–6 h of incubation, presumptive zygotes are rinsed and cultured in 500 μL of PZM-3 in 4-well Nunclon dishes (Nunc) at 38.5 °C and 5% CO_2 in air.
5. Following IVF, two-cell embryos, four- to eight-cell embryos, and blastocysts can be collected from the culture at 30, 60, and 114 h post-insemination, respectively, for further processing.

3.2 Isolating Small RNA from Small Volumes (Adapted from Refs. 28, 31, 32)

(mirVana™ miRNA Isolation Kit; Ambion)

The ability to analyze miRNAs in small populations of pure cells has significantly advanced our understanding of the role of miRNAs. Total RNA enriched for small RNAs can be isolated from embryos using the *mir*Vana™ miRNA Isolation Kit. Experiment specifics (such as starting template volume, concentration, etc.) can be calculated as needed for the pertinent protocol, but certain aspects need to be taken into consideration while processing the samples (Refer Subheading 4 (*see* **Notes 2–5**) for details on laboratory practices while handling samples for miRNA isolation and downstream processing).

1. Add 600 μL Lysis/Binding buffer to each sample. Vortex samples for 30 s to completely lyse embryos.
2. Add 1/10th volume of miRNA homogenate additive to each sample. Mix well by vortexing and incubate on ice for 10 min.
3. Add 600 μL Acid-Phenol:Chloroform to each sample and vortex for 45 s, followed by a 10 min centrifugation at 10,000 RCF at room temperature.
4. The supernatant containing the RNA should then be removed carefully, without disturbing the lower phase, and transferred to a fresh tube.
5. For RNA precipitation, add one third volume of room temperature 100% ethanol to each aqueous phase sample obtained from **steps 3** and **4**. Mix well by vortexing.

6. The sample is then applied to a filter cartridge (placed in a collection tube) and centrifuged for ~15 s at 10,000 RCF. Collect the filtrate and measure the total volume for each sample. *At this point the filter contains the RNA fraction, depleted of small RNAs, which are now in the filtrate in the collection tube.*
7. For each sample, add two thirds volume room temperature 100% ethanol to the filtrate and mix well.
8. Place a fresh filter cartridge in a fresh 1.5 mL collection tube for each sample. Apply the sample (filtrate + ethanol mix) to the respective filter cartridge.
9. Centrifuge for ~15 s at 10,000 RCF. Discard the flow-through and transfer the filter to a fresh collection tube. (Refer Subheading 4 (*see* **Note 6**) for details on collecting clean elution fractions).
10. Add 700 μL Wash solution I to each filter cartridge and centrifuge for 10 s. Discard flow-through and place the filter in a fresh 1.5 mL collection tube.
11. Add 500 μL Wash solution 2/3 to each filter cartridge and centrifuge for 10 s. Discard flow-through and insert filter in fresh collection tube. Repeat this step once.
12. After the three wash steps, place the filter in a fresh 1.5 mL collection tube and centrifuge for 1 min to remove residual fluid.
13. Transfer each filter to a fresh collection tube. Add 100 μL of pre-warmed Elution buffer or nuclease-free water to the center of the filter and close the cap. Centrifuge at maximum speed for 30 s to recover RNA.
14. Label the tubes appropriately and store at −20 °C or below for long-term storage.
15. Sample concentration and quality can be determined by spectrophotometry using the NanoDrop-1000 spectrophotometer.

3.3 Northern Blotting to Detect Small RNAs (Adapted from Refs. 33, 34)

Traditional Northern blotting is very useful for detecting large RNAs (over ~100 bases). However, denaturing Acrylamide gels are much better for precise detection of small RNAs.

1. Prepare 1× TBE by diluting the 10× TBE (Refer to Subheading 2) in nuclease-free water. Adjust final volume to 1 L.
2. Prepare 15% acrylamide gel by dissolving urea, 10× TBE and 40% acrylamide in nuclease-free water. Then add 10% ammonium persulfate and TEMED (as mentioned in Subheading 2). Mix briefly and pour the gel immediately, allow it to set (Refer to Subheading 4 (*see* **Note 7**) for tips on gel setting).
3. To prepare the samples, mix 1–2 μg of RNA sample with an equal volume of gel-loading buffer. Heat samples at 95 °C for 4 min.

4. Load the samples on a denaturing 15% polyacrylamide gel and conduct electrophoresis at 30–45 mA. Stop the electrophoresis when the bromophenol blue dye has migrated to the bottom of the gel.

   From this step on, there are two options that may be followed to visualize the results.

   For a quick and direct visualization of bands, follow the next two steps.

5. Soak the gel in an ethidium bromide solution for 5 min. Wash the gel in 1× TBE for 5 min.
6. Visualize the RNA using a UV illuminator. Small RNA of high quality should have distinct bands for tRNA, 5S, rRNA, and 5.8S rRNA.

   Alternatively, transfer and hybridization may be conducted to visualize results.

7. Remove the gel from the gel box and separate the plates. Wet two sheets of Whatman 3MM paper with 20× SSC and place them on the top of the gel.
8. Soak a sponge from the electroblot apparatus in electrophoretic transfer buffer until wet. Transfer the gel with the paper onto the sponge, such that the paper is on the top of the sponge.
9. Float the nylon transfer membrane in water until wet, and then immerse it in 20× SSC. Place the nylon membrane on the top of the gel and remove any bubbles.
10. Wet two more sheets of Whatman 3MM paper with 20× SSC, and place them on the top of the membrane.
11. Add the second sponge on the top of the Whatman paper sheets and clamp the cassette closed. (Refer to Subheading 4 (*see* **Note 8**) for further "sandwich" assembly details).
12. Place the assembled gel cassette into the electroblot chamber and fill the chamber with 1× electrophoretic transfer buffer, precooled to 4 °C. Carry out the transfer at 250 mA for 2 h at 4 °C and then at 350 mA for a further 2 h at 4 °C. Disassemble the apparatus.
13. Remove and save the nylon membrane. Carefully remove any gel fragments from the membrane and mark the orientation of the membrane with a ballpoint ink pen. Wrap the membrane in plastic wrap and place it on a sheet of Whatman 3MM paper.
14. To cross-link the RNA to the nylon transfer membrane, use a 254 nm UV transilluminator, exposing the membrane with the RNA surface facing down against the glass surface; or bake the membrane in a vacuum oven for 1 h at 80 °C.
15. Place the membrane in a heat-seal bag or cylindrical glass hybridization bottle and cover with prehybridization/hybridization mix. Prehybridize the membrane with shaking or rotation for at least 2 h at 42 °C.

16. Replace the buffer with an appropriate volume of fresh prehybridization/hybridization mix containing radiolabeled probe at 106 cpm/mL, and hybridize overnight at 42 °C.
17. Remove and safely dispose of the radioactive hybridization buffer.
18. Prepare three wash solutions: 2× SSC/1% SDS, 1× SSC/1% SDS, and 0.1× SSC.
19. Perform the following series of 15 min washes on the membrane, each at 42 °C with agitation: 2× SSC/0.1% SDS, 1× SSC/0.1% SDS, 0.1× SSC.
20. Air-dry the membrane, cover it with plastic wrap, and expose it to a phosphor screen for phosphorimaging or to X-ray film for autoradiography. Exposure for several hours to overnight may be necessary. Image on a phosphorimager or develop the X-ray film.

3.4 MicroRNA Quantitative Reverse Transcription-Polymerase Chain Reaction (qRT-PCR)

It is in this step that the cDNA is amplified through primer-driven DNA synthesis. The relative abundance of DNA is quantified through the FAM dye-quencher system. As DNA is duplicated from the template strand, the DNA polymerase comes in contact with the miRNA-specific probe. This causes the quencher molecule on the probe to no longer be in contact with the FAM dye, releasing it and allowing it to fluoresce. Relative fluorescence is directly correlated to relative abundance of amplified DNA. For oocyte and embryo samples, quantitative RT-PCR may be conducted in a manner that enables comparisons to be made on the amplification of RNA template equivalent to one oocyte or embryo.

3.4.1 Template Preparation (Adapted from Refs. 28, 30, 35)

(TaqMan Gene Expression Cells-to-CT Kit; Ambion)

1. Collect pools of cumulus cell-free GV oocytes, MII oocytes, two-cell embryos, and four-to eight-cell embryos to represent each stage of oocyte and embryo development (as needed for the experiment). Collect exactly 25 of each stage in 5 μL PBS in a separate RNase-free PCR tube for each stage of development.
2. Add 4.95 μL of Lysis solution and 0.05 μL DNase to each pool. Mix gently and incubate at room temperature for 5 min.
3. Add room temperature stop solution (0.5 μL) and mix by gently pipetting few times. Incubate for an additional 2 min at room temperature and then place samples on ice.
4. Add 2 μL of nuclease-free water to reach a total volume of 12.5 μL. Hereon, 1 μL of lysate may be considered equivalent to two oocytes or embryos.

3.4.2 cDNA Reverse Transcription (Adapted from Refs. 28, 36)

(TaqMan® MicroRNA RT kit; Applied Biosystems)

1. Thaw the samples and kit components on ice.

2. Prepare the RT Mastermix on ice according to the number of samples to be tested. For each sample to be tested, combine (in an RNase-free PCR tube) 0.15 μL of 100 mM dNTPs, 1 μL of MultiScribe™ Reverse Transcriptase, 1.5 μL of 10× Reverse Transcription buffer, 0.19 μL of RNase inhibitor, and 4.16 μL of nuclease-free water (total volume 7 μL). Mix gently.
3. Dilute the RT primers to a 5× working stock using 0.1× TE buffer.
4. The RT reaction is 15 μL consisting of 7 μL master mix, 3 μL primers, and 1 μL sample, made up to final volume with nuclease-free water.
5. Thermal Cycler conditions for reverse transcription are 16 °C for 30 min, 42 °C for 30 min, and 85 °C for 5 min.
6. Store the products at −20 °C or lower for further use.

3.4.3 Quantitative Polymerase Chain Reaction (Adapted from Refs. 28, 29, 37, 38)

(TaqMan MicroRNA Assays; Applied Biosystems)

1. Thaw all kit components on ice. Mix gently before use.
2. All reactions should be performed in triplicate (or at least duplicates) to determine an average threshold cycle (C_T) value. Calculate volumes accordingly.
3. For a single reaction, add 1 μL TaqMan® Small RNA Assay (20×), 10 μL Taqman® Universal PCR Mastermix II (2×), and 7.67 μL of nuclease-free water, to a fresh RNase-free PCR tube.
4. From the RT reaction prepared earlier, 1.33 μL of cDNA is used for each PCR amplification reaction, to make a total of 20 μL reaction volume. Mix gently and centrifuge.
5. Transfer the mixtures to appropriate wells of a 96-well plate. Seal the plate with appropriate optical film cover. Centrifuge plate briefly.
6. Real-time Thermal Cycler parameters for PCR are 95 °C for 10 min, followed by 50 cycles of 95 °C for 15 s and 60 °C for 60 s.
7. Relative quantification of the expression level may be evaluated with the comparative C_T method.
8. Following normalization to endogenous control, the $\Delta\Delta C_T$ is calculated using the single greatest sample ΔC_T value (i.e., the sample with the lowest expression) to subtract from all other sample ΔC_T values.
9. Assuming that each cycle difference is equivalent to a twofold difference, relative fold-change for each sample is calculated by applying the equation $2^{-\Delta\Delta CT}$.

3.5 MicroRNA Microarray (Adapted from Ref. 39)

Total cellular RNA from pools of oocytes and/or embryos can be subjected to qRT-PCR probing for mature miRNAs that have been correlated with cellular differentiation and development. The Cell Differentiation & Development RT² miRNA PCR Array

(SABiosciences) can be used for this procedure. This technique allows us to evaluate the differential expression of pathway focused miRNA that are relevant to cell fate and lineage decisions, thus potentially identifying miRNAs with specific functions in cells or tissues.

3.5.1 cDNA Reverse Transcription

(RT² miRNA First Strand Kit; SABiosciences)

1. Use 50–400 ng of small RNA sample in each reaction. The amount may need to be decided depending on type of array to be used (Refer User Manual for recommended starting material concentration).
2. Combine 1 μL of miRNA RT primer & ERC mix, 2 μL of 5× miRNA RT Buffer 2, 1 μL of miRNA RT enzyme mix, 1 μL miRNA nucleotide mix, appropriate volume of small RNA sample and nuclease-free water to a final volume of 10 μL. Mix contents gently and briefly centrifuge.
3. Incubate at 37 °C for 2 h. Then heat to 95 °C for 5 min to degrade the RNA and inactivate the reverse transcriptase.
4. Place on ice for at least one minute and add 90 μL of nuclease-free water to each cDNA reaction. Mix gently.

3.5.2 Quantitative Polymerase Chain Reaction

(RT² miRNA PCR Arrays; SABiosciences)

1. In a multichannel reservoir, combine 2× RT² SYBR Green qPCR Mastermix, appropriately diluted cDNA reaction, and nuclease-free water in proportions appropriate to the chosen array.
2. For loading the PCR arrays, carefully remove the array plates from their sealed bags, and add the appropriate volume of the Mastermix cocktail to each well of the array plate. Seal with optical film.
3. Thermal Cycler parameters for PCR are 95 °C for 10 min, followed by 40 cycles of 95 °C for 15 s, 60 °C for 30 s, and 72 °C for 30 s.
4. Specific miRNA abundance can be classified according to the manufacturer's recommendations and miRNAs that exhibit a C_T of >35 can be categorized as nondetectable, while detectable miRNAs can be categorized as displaying low expression (C_T = 33–34.9), moderate expression (C_T = 30–32.9), or high expression (C_T = 25.5–29.9).
5. Data may be analyzed using the RT² Profiler PCR Array Data Analysis web-based tool (SABiosciences; http://www.sabiosciences.com/pcrarraydataanalysis.php).

3.6 Sequencing by Oligonucleotide Ligation and Detection (SOLiD) Sequencing (Adapted from Refs. 29, 40)

(SOLiD Small RNA Expression Kit; Ambion)

This is a robust method for whole genome analysis of expression patterns that leave out the bias of microarray pre-formatted assays. The total RNA can be converted into a library in a single day using a single kit, which not only simplifies the process but also reduces variability.

3.6.1 Template Preparation

1. Small RNA are isolated using the *mir*Vana miRNA Isolation Kit (as described earlier in Subheading 3.2) and used to create barcoded small RNA libraries.
2. On ice, combine 2 μL Adaptor Mix A (Refer to Subheading 4 (*see* **Note 9**) for details on how to choose Adaptor mix), 3 μL Hybridization solution, RNA sample, and nuclease-free water to make a final volume of 8 μL.
3. Incubate at 16 °C for 16 h to yield templates for SOLiD sequencing from the 5′ end of the sense strand.
4. Prepare the Reverse transcription Mastermix by combining 13 μL of nuclease-free water, 4 μL of 10× RT Buffer, 2 μL of 2.5 mM dNTP mix, 1 μL of ArrayScript™ Reverse Transcriptase to make a total of 20 μL for each sample. Add this RT Mastermix to each sample and mix gently.
5. Incubate at 42 °C for 30 min to synthesize cDNA.
6. Transfer 10 μL of cDNA to a fresh tube and add 1 μL RNase H. Vortex gently and centrifuge briefly. Incubate at 37 °C for 30 min. RNase H treatment is used to digest the RNA from RNA/cDNA duplexes and to reduce the concentration of unligated adaptors and adaptor by-products.

3.6.2 PCR Amplification and Product Evaluation

1. To provide sufficient quantity of cDNA for SOLiD sequencing, each library can be amplified using the supplied barcoded primer sets and 18–22 cycles of PCR according to the recommended cycling parameters.
2. In a fresh tube combine 38.9 μL nuclease-free water, 5 μL 10× PCR Buffer 1, 1 μL SOLiD PCR Primers, 4 μL of 2.5 mM dNTP mix, 0.6 μL of AmpliTaq® DNA Polymerase to make a total volume of 49.5 μL.
3. Add 0.5 μL RNase H-treated cDNA to each aliquot of PCR Mastermix.
4. Thermal Cycler parameters for PCR are 95 °C for 5 min, followed by 18–22 cycles of 95 °C for 30 s, 62 °C for 30 s, and 72 °C for 30 s. These are followed by a final extension step at 72 °C for 7 min.
5. Evaluate the PCR product by running 5–10 μL on a 6% polyacrylamide gel.
6. Replicated PCRs from each sample can be combined into a single tube and cleaned using the MinElute PCR Purification Kit (Qiagen) to remove unincorporated primers, enzymes, and salt.
7. Samples can be separated on a 6% polyacrylamide gel and stained with SYBR gold dye.
8. The 105- to 150-bp region for each library can be excised and purified, by elution in 0.3 M NaCl, and then precipitated with 100% ethanol and resuspended in 20 μL of nuclease-free water.

9. The quality and quantity of cDNA libraries can be determined using a 2100 Agilent Bioanalyzer and a NanoDrop-1000 Spectrophotometer.
10. Small RNA cDNA libraries can be subjected to massively parallel deep sequencing using the SOLiD sequencing system following the vendor's recommended protocol for small RNA sequencing.

3.7 Fluorescence In Situ Hybridization (Adapted from Refs. 28, 41, 42)

1. At the time of collection, preserve ovarian tissue in 4% paraformaldehyde for 24 h and then transfer to 70% Ethanol. Ovarian tissue sections of 5 μm thickness are then mounted on slides for further analysis.
2. Alternatively, embryos can be fixed by immersion in 4% paraformaldehyde for 24 h, washed in PBS, dehydrated in 15% sucrose, and embedded in OCT. Sections of 12 μm thickness can then be cut and mounted on slides and stored at −20 °C.
3. Each section is subjected to CitriSolv® twice for 5 min, rehydrated in two changes of 100% ethanol for 3 min, followed by 95% ethanol for 1 min and finally 80% ethanol for 1 min. Then rinse in distilled water.
4. Slides are then kept immersed in heated citrate buffer (95 °C) for 30 min and then cooled to room temperature.
5. Once at room temperature slides are blocked for 30 min with 5% BSA.
6. Slides are then placed in hybridization solution for 1 h at 65 °C.
7. The probe is then added and slides are incubated in high humidity overnight at 65 °C.
8. On day 2, slides are washed in standard saline citrate (SSC) solution and PBST at room temperature as follows—50% hybridization solution/2× SSC at 60 °C for 10 min twice, 2× SSC for 10 min three times, 0.2× SSC/25% PBST at room temperature for 10 min three times, 0.2× SSC/50% PBST at room temperature for 10 min three times, 0.2× SSC/75% PBST at room temperature for 10 min three times, PBST at room temperature for 10 min three times.
9. Antifade DAPI is added and a cover slip is placed over each section. Seal edges with clear nail polish.
10. Visualize stained sections using a compound microscope that detects fluorescence.

3.8 MIR Inhibition During In Vitro Maturation of Oocytes Using PNA (Adapted from Ref. 28)

Peptide nucleic acids (PNA) are artificially constructed oligonucleotides that bind complementary RNA with a strong affinity and specificity. Moreover, they do not require transfection reagents since they are conjugated to certain peptides that allow them easy access into the cell. These PNAs can hence be used as efficient miRNA inhibitors, thus being able to evaluate the role of the miRNA in a particular developmental pathway.

1. A scrambled PNA with no predicted targets may be used as a negative control.
2. PNA oligonucleotides are diluted in maturation media at a stock concentration of a 100 nM/μL and then added to maturation media on the day of COC collection to acquire a final concentration of 2.0 and 0.2 nM. Working concentrations should be determined empirically for individual labs and specific experiments. Utilization of fluorescently labeled PNA linked oligonucleotides can also be utilized to confirm the ability of the molecule to be translocated into the cell (Fig. 2).
3. A control group of oocytes should be matured without PNA that is used to evaluate the potential toxicity of the PNA.
4. GV oocytes and cumulus cells are collected at the time of COC collection and MII oocytes and cumulus cells are collected at 42–44 h of maturation.
5. Parthenogenetic activation of MII oocytes is performed with 50 oocytes from each treatment to determine developmental competence up to 60 h.
6. MII arrested oocytes are washed in a high calcium activation media (Mannitol 0.28 M, $CaCl_2$ 1.0 mM, $MgCl_2$ 0.1 mM, HEPES 0.5 mM and BSA 1 mg/mL), and then placed between two electrodes covered with activation media and activated by two consecutive 30 μs pulses at 1.2 kV/cm.
7. Following activation, zygotes are washed and cultured in PZM-3 at 38.5 °C in 5% CO_2.
8. At 60 h embryos are evaluated for development and the number of embryos with four or more uniform blastomeres is recorded.

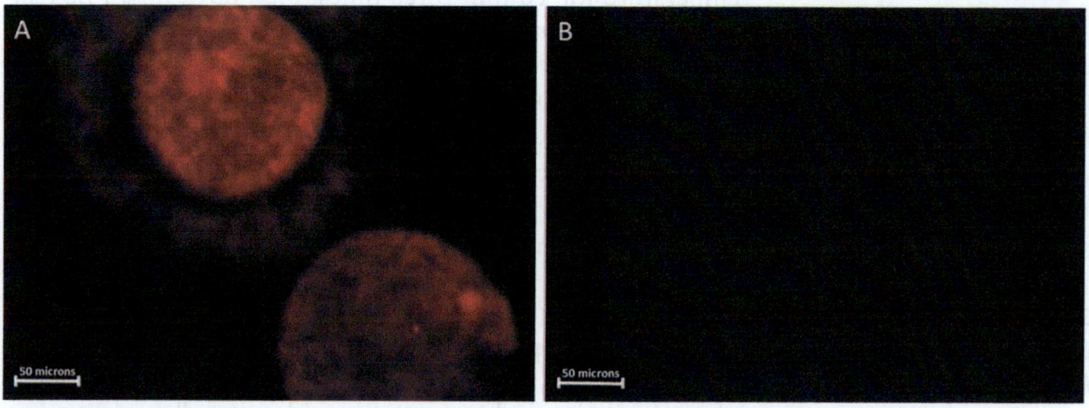

Fig. 2 Translocation of miRNA inhibitor into pig oocytes during in vitro maturation. (**a**) Antisense MIR21 FAM labeled PNA added to maturation media to demonstrate translocation of PNA during in vitro maturation into cytoplasm of cells in the cumulus oocyte complex (*top left*) and denuded oocyte (*bottom right*). (**b**) Control oocytes lacking detectable autofluorescence

4 Notes

1. Recipes for FSH, LH, and EGF preparation for IVM media:
   (a) *To prepare FSH stock:*
   - Add 50 mg FSH in 10 mL TCM-199 to obtain a 5000 μg/mL concentration.
   - Take 1 mL from solution prepared in previous step, and add it to 9 mL of TCM-199, to obtain a 500 μg/mL concentration. Use this solution for the mTCM-199 working media.

   (b) *To prepare LH stock:*
   - Add 30 mg LH in 6 mL TCM-199 to obtain a 5000 μg/mL concentration.
   - Take 1 mL from solution prepared in previous step, and add it to 9 mL TCM-199, to obtain a 500 μg/mL concentration. Use this solution for the mTCM-199 working media.

   (c) *To prepare EGF stock:*
   - Add 1 mg of EGF in 10 mL TCM-199 to obtain a 100 μg/mL concentration.
   - Take 1 mL from solution prepared in previous step, and add it to 9 mL TCM-199 to obtain a 10 μg/mL concentration. Use this solution for the mTCM-199 working media.
2. Detection of miRNA is not a trivial task in small sample sizes. Hence, it is even more important to follow good laboratory and pipetting practices when working with small RNA samples. Always wear a clean laboratory coat and clean gloves. Change gloves intermittently. Spray your work station and gloves with 70% ethanol before starting work. An RNase decontamination solution should be used to wipe surfaces, such as RNaseZap® (Applied Biosystems).
3. RNA molecules are highly degradable. Frozen RNA samples should always be thawed on ice or close to 4 °C.
4. Controls: Each experiment should have positive and negative controls for each variable being tested such as treatment effect, time effect, enzyme effect, template effect, etc.
5. Maintaining the same number of oocytes or embryos of the same stage per lysis reaction decreases the variability in starting RNA abundance.
6. For the extraction techniques using kits, transferring the filter cartridge to a fresh collection tube after each centrifugation step will improve your chances of getting a clean eluted sample.

7. Leave behind a little acrylamide gel solution in the conical tube (it was prepared in) and set it at room temperature next to the poured gel in the gel-box apparatus. Once the gel solution in the conical tube sets/solidifies, one may safely assume that the gel poured in the apparatus is also set and ready to be used.
8. The layers of the sandwich are as follows, starting with the negative (−) pole: sponge, two sheets of Whatman 3MM paper, gel, nylon membrane, two sheets of Whatman 3MM paper, sponge. The RNA will migrate to the positive (+) pole, i.e., from the gel to the nylon membrane.
9. For SOLiD sequencing, Adaptor Mix A or B can be used depending on which end (5′ or 3′, respectively) needs to be the starting point of sequencing.

References

1. Hale BJ, Keating AF, Yang CX, Ross JW (2015) Small RNAs: their possible roles in reproductive failure. In: Bronson R (ed) The male role in pregnancy loss and embryo implantation failure. Springer, Switzerland
2. Eichenlaub-Ritter U, Peschke M (2002) Expression in in vivo and in vitro growing and maturing oocytes: focus on regulation of expression at the translational level. Hum Reprod Update 8:21–41
3. Wrenzycki C, Herrmann D, Niemann H (2007) Messenger RNA in oocytes and embryos in relation to embryo viability. Theriogenology 68S:S77–S83
4. Hossain M, Salilew-Wondim D, Schellander K, Tesfaye D (2012) The role of microRNAs in mammalian oocytes and embryos. Anim Reprod Sci 134:36–44
5. Hale BJ, Yang CX, Ross JW (2014) Small RNA regulation of reproductive function. Mol Reprod Dev 81:148–159
6. Bartel DP (2004) MicroRNAs: genomics, biogenesis, mechanism and function. Cell 116: 281–297
7. Brennecke J, Stark A, Russell RB, Cohen SM (2005) Principles of microRNA-target recognition. PLoS Biol 3(3):e85
8. Doench JG, Sharp PA (2004) Specificity of microRNA target selection in translational repression. Genes Dev 18(5):504–511
9. Bagga S, Bracht J, Hunter S, Massirer K, Holtz J, Eachus R, Pasquinelli AE (2005) Regulation by let-7 and lin-4 miRNAs results in target mRNA degradation. Cell 122(4):553–563
10. Giraldez AJ, Mishima Y, Rihel J, Grocock RJ, Van Dongen S, Inoue K, Enright AJ, Schier AF (2006) Zebrafish MiR-430 promotes deadenylation and clearance of maternal mRNAs. Science 312(5770):75–79
11. Tang F, Kaneda M, O'Carroll D, Hajkova P, Barton S, Sun YA, Lee C, Tarakhovsky A, Lao K, Surani MA (2007) Maternal microRNAs are essential for mouse zygotic development. Genes Dev 21:644–648
12. Stark A, Brennecke J, Bushati N, Russell RB, Cohen SM (2005) Animal MicroRNAs confer robustness to gene expression and have a significant impact on 3′UTR evolution. Cell 123(6):1133–1146
13. Babak T, Zhang W, Morris Q, Blencowe BJ, Hughes TR (2004) Probing microRNAs with microarrays: tissue specificity and functional inference. RNA 10(11):1813–1819
14. Sood P, Krek A, Zavolan M, Macino G, Rajewsky N (2006) Cell-type-specific signatures of microRNAs on target mRNA expression. Proc Natl Acad Sci USA 103(8):2746–2751
15. Schier AF (2007) The maternal-zygotic transition: death and birth of RNAs. Science 316(5823):406–407
16. Rajewsky N (2006) microRNA target predictions in animals. Nat Genet 38(Suppl):S8–S13
17. Brevini TA, Cillo F, Antonini S, Tosetti V, Gandolfi F (2007) Temporal and spatial control of gene expression in early embryos of farm animals. Reprod Fertil Dev 19(1):35–42
18. Tadros W, Lipshitz HD (2005) Setting the stage for development: mRNA translation and stability during oocyte maturation and egg activation in Drosophila. Dev Dyn 232(3):593–608
19. Watanabe T, Takeda A, Mise K, Okuno T, Suzuki T, Minami N, Imai H (2005) Stage-specific expression of microRNAs during

Xenopus development. FEBS Lett 579(2): 318–324

20. Aboobaker AA, Tomancak P, Patel N, Rubin GM, Lai EC (2005) Drosophila microRNAs exhibit diverse spatial expression patterns during embryonic development. Proc Natl Acad Sci USA 102(50):18017–18022
21. Biemar F, Zinzen R, Ronshaugen M, Sementchenko V, Manak JR, Levine MS (2005) Spatial regulation of microRNA gene expression in the Drosophila embryo. Proc Natl Acad Sci USA 102(44):15907–15911
22. Wienholds E, Kloosterman WP, Miska E, Alvarez-Saavedra E, Berezikov E, de Bruijn E, Horvitz HR, Kauppinen S, Plasterk RH (2005) MicroRNA expression in zebrafish embryonic development. Science 309(5732): 310–311
23. Bernstein E, Kim SY, Carmell MA, Murchison EP, Alcorn H, Li MZ, Mills AA, Elledge SJ, Anderson KV, Hannon GJ (2003) Dicer is essential for mouse development. Nat Genet 35(3):215–217
24. Murchison EP, Stein P, Xuan Z, Pan H, Zhang MQ, Schultz RM, Hannon GJ (2007) Critical roles for Dicer in the female germline. Genes Dev 21(6):682–693
25. Ma J, Flemr M, Stein P, Berninger P, Malik R, Zavolan M, Svoboda P, Schultz RM (2010) MicroRNA activity is suppressed in mouse oocytes. Curr Biol 20(3):265–270
26. Suh N, Baehner L, Moltzahn F, Melton C, Shenoy A, Chen J, Blelloch R (2010) MicroRNA function is globally suppressed in mouse oocytes and early embryos. Curr Biol 20(3):271–277
27. Koshiol J, Wang E, Zhao Y, Marincola F, Landi MT (2010) Strengths and limitations of laboratory procedures for microRNA detection. Cancer Epidemiol Biomarkers Prev 19(4):907–911
28. Wright EC, Hale BJ, Yang CX, Njoka JG, Ross JW (2016) MicroRNA-21 and PDCD4 expression during in vitro oocyte maturation in pigs. Reprod Biol Endocrinol 14:21
29. Yang CX, Du ZQ, Wright EC, Rothschild MF, Prather RS, Ross JW (2012) Small RNA profile of the cumulus oocyte complex and early embryos in the pig. Biol Reprod 87(5):117
30. Yang CX, Wright EC, Ross JW (2012) Expression of RNA binding proteins DND1 and FXR1 in the porcine ovary and during oocyte maturation and early embryo development. Mol Reprod Dev 79:541–552
31. Li M, Xia Y, Gu Y, Zhang K, Lang Q, Chen L, Guan J, Luo Z, Chen H, Li Y, Li Q, Li X, Jiang A, Shuai S, Wang J, Zhu Q, Zhou X, Gao X, Li X (2010) MicroRNAome of porcine pre- and postnatal development. PLoS One 5:7
32. *mir*Vana™ miRNA Isolation Kit; Ambion. https://tools.thermofisher.com/content/sfs/manuals/cms_055423.pdf
33. Rio D (2011) Northern blots for small RNAs and microRNAs. Adapted from RNA: a laboratory manual by Donald C Rio, Manuel Ares Jr, Gregory J Hannon, Timothy W Nilsen. CSHL Press, Cold Spring Harbor, NY
34. Reddy A, Zheng Y, Jagadeeswaran G, Macmil S, Graham W, Roe B, Desilva U, Zhang W, Sunkar R (2009) Cloning, characterization and expression analysis of porcine microRNAs. BMC Genomics 10:65
35. TaqMan Gene Expression Cells-to-CT Kit; Ambion. https://tools.thermofisher.com/content/sfs/manuals/cms_056225.pdf
36. TaqMan® Small RNA Assays; Applied Biosystems. https://tools.thermofisher.com/content/sfs/manuals/cms_042167.pdf
37. Lin Y, Zeng Y, Zhang F, Xue L, Huang Z, Li W, Guo M (2013) Characterization of microRNA expression profiles and the discovery of novel microRNAs involved in cancer during human embryonic development. PLoS One 8:8
38. TaqMan MicroRNA Assays; Applied Biosystems. https://genome.med.harvard.edu/documents/qpcr/microRNATaqManAssayProtocol.pdf
39. RT² miRNA PCR Arrays; SABiosciences. http://www.sabiosciences.com/Manual/mirnapcrarray.pdf
40. SOLiD Small RNA Expression Kit; Ambion. http://tools.thermofisher.com/content/sfs/manuals/cms_054973.pdf
41. Deo M, Yu J, Chung K, Tippens M, Turner D (2006) Detection of mammalian microRNA expression by in situ hybridization with RNA oligonucleotides. Dev Dyn 235:2538–2548
42. Wright E (2012) MicroRNA expression and function during porcine oocyte maturation and early embryonic development. Dissertation, Iowa State University

Chapter 6

Detection of Bidirectional Promoter-Derived lncRNAs from Small-Scale Samples Using Pre-Amplification-Free Directional RNA-seq Method

Nobuhiko Hamazaki, Kinichi Nakashima, Katsuhiko Hayashi, and Takuya Imamura

Abstract

Development of high-throughput sequencing technologies has uncovered the immensity of the long noncoding RNA (lncRNA) world. Divergently transcribed lncRNAs from bidirectional gene promoters, called promoter-associated noncoding RNAs (pancRNAs), account for ~20% of the total number of lncRNAs, and this major fraction is involved in many biological processes, such as development and cancer formation. Recently, we have found that the pancRNAs activate their partner genes, as represented by the fact that *pancIl17d*, a pancRNA that is transcribed from the antisense strand of the promoter region of *Interleukin 17d (Il17d)* at the onset of zygotic gene activation (ZGA), is essential for mouse preimplantation development through *Il17d* upregulation. The discovery of the expression of a specific set of pancRNAs during ZGA was achieved by using a method that generates directional RNA-seq libraries from small-scale samples. Although there are several methods available for small-scale samples, most of them require a pre-amplification procedure that frequently generates some amplification biases toward a subset of transcripts. We provide here a highly sensitive and reproducible method based on the preparation of directional RNA-seq libraries from as little as 100 mouse oocytes or embryos without pre-amplification for the quantification of lncRNAs as well as mRNAs.

Key words Directional RNA-seq, Long noncoding RNA, Mouse preimplantation embryos

1 Introduction

In mammals, transcription in the zygote begins after fertilization. This transcriptional wave is called zygotic gene activation (ZGA) and is believed to be an essential step to form a zygote-type transcriptional network. In mouse, ZGA generally begins at the 2-cell stage [1, 2]. In actuality, selected genes, not all genes, are transcribed from the zygote genome [3]. Thus, sequence-specific machineries should regulate the gene activation pattern. One key issue is how such sequence-specific gene activation is achieved in the process of pluripotency acquisition in early mouse embryos.

Kiho Lee (ed.), *Zygotic Genome Activation: Methods and Protocols*, Methods in Molecular Biology, vol. 1605,
DOI 10.1007/978-1-4939-6988-3_6, © Springer Science+Business Media LLC 2017

Long noncoding RNAs (lncRNAs) constitute one group of factors that play roles in such a specific gene upregulation [4]. Among such lncRNAs, promoter-associated ncRNAs (pancRNAs), which are polyA+ long (>200 nt) antisense ncRNAs transcribed from promoter regions of coding genes, account for 20% of the total number of lncRNAs (Fig. 1) [5–7]. Thousands of pancRNAs are expressed in mouse, rat, and human in tissue-specific manners and activate their partner genes [6, 8–10]. Recently, we have found that pancRNAs also play essential roles in early mouse development through a similar mechanism. For example, *Interleukin 17d (Il17d)* is activated by its pancRNA, and is important for the cell's pluripotency acquisition and maintenance (Fig. 2) [11]. We named this pancRNA *pancIl17d*. The *pancIl17d* is upregulated by the 2-cell stage, followed by *Il17d* upregulation from the 4-cell stage, and knockdown of *pancIl17d* results in the inhibition of the corresponding promoter DNA demethylation triggered by the TET3 enzyme, leading to embryonic lethality by the blastocyst stage, along with increased apoptosis. Sequence-specific DNA methylation can be also observed upon the knockdown of other pancRNAs [11], suggesting that this novel class of lncRNAs may specify the DNA methylome according to the embryonic stage in order for preimplantation development to proceed appropriately.

The discovery of the expression and functional annotation of a specific set of pancRNAs during ZGA was achieved by using the method that we devised for generating directional RNA-seq libraries from small-scale samples. Indeed, conventional directional RNA-seq techniques require more than 100 ng of total RNAs [12, 13]. However, it is difficult to apply these techniques to the analysis of early stages of mammalian embryos since 100 ng of total RNA corresponds to the amount obtained from thousands of oocytes or

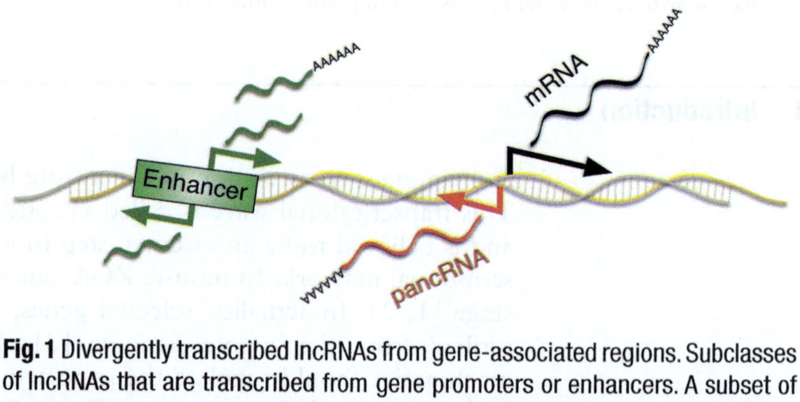

Fig. 1 Divergently transcribed lncRNAs from gene-associated regions. Subclasses of lncRNAs that are transcribed from gene promoters or enhancers. A subset of divergently transcribed lncRNAs, promoter-associated noncoding RNAs (pancRNAs), are transcribed from gene promoter regions in the opposite direction to the paired coding RNA, On the other hand, another subset of divergent lncRNAs that are derived from enhancer regions are called enhancer RNAs (eRNAs). Both classes of lncRNAs are thought to contribute to gene activation

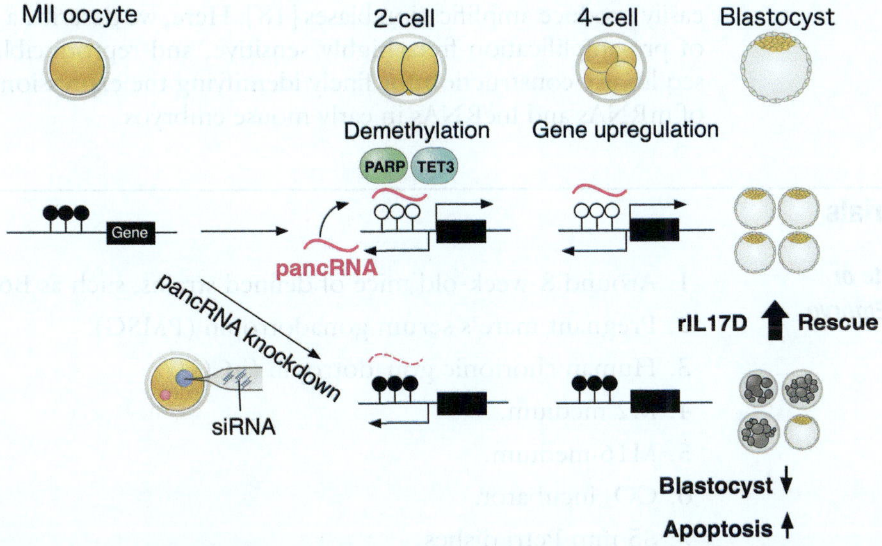

Fig. 2 A schematic model for pancRNA-mediated gene activation in early mouse development. At the 2-cell stage, *pancIl17d* expression, together with TET3 and PARP, leads to establishment of the hypomethylated status at the *Il17d* promoter in a sequence-specific manner, and thus to *Il17d* mRNA expression starting from the 4-cell stage. When these steps are compromised, apoptosis is increased at the morula stage and embryos die by the blastocyst stage

embryos. In fact, 10,000 mouse oocytes or embryos were reported to have been used for obtaining a transcriptome in a previous study [14]. To overcome this difficulty, methods for obtaining the transcriptome information even from a single cell are rapidly being developed, including SC3-seq, smart-seq2, and Quartz-seq [15–17]. Since the directionality of RNA is an important property for its function in the cells [7–9], it would be ideal if RNA directionality could be recognized in small-scale sample datasets. However, current RNA-seq data for small-scale samples frequently lose directional information. In this context, our directional RNA-seq method is useful because it enables reproducible data acquisition from samples as small as 100 oocytes, for example. It is also useful to apply this method for the functional annotation of a set of genes/lncRNAs. In order to annotate the phenotypic function of genes/lncRNAs expressed in early embryos, embryo manipulation techniques including nucleic acid microinjection have been used as powerful tools. Since our method can be also applied to just a hundred of such manipulated embryos for directional RNA-seq data acquisition [11], utilization of this method can facilitate the accumulation of knowledge on genome-wide level effects of a gene/lncRNA in addition to the phenotypic annotation. Another advantage of our method is that it is based on a pre-amplification-free procedure [11], while most of the RNA-seq methods for small-scale samples require pre-amplification steps by PCR or in vitro transcription, which may easily produce amplification biases [18]. Here, we describe a method of pre-amplification-free, highly sensitive, and reproducible RNA-seq library construction for finely identifying the expression pattern of mRNAs and lncRNAs in early mouse embryos.

2 Materials

2.1 Oocyte or Fertilized Embryo Collection

1. Around 8-week-old mice of defined strains, such as B6C3F1.
2. Pregnant mare's serum gonadotropin (PMSG).
3. Human chorionic gonadotropin (hCG).
4. M2 medium.
5. M16 medium.
6. CO_2 incubator.
7. 35 mm Petri dishes.
8. Mouth pipet constituted with mouthpiece and proximal tubing.
9. 1% hyaluronidase (Sigma) dissolved in M2 medium.
10. Paraffin oil.
11. Surgical forceps.
12. Surgical scissors.
13. 100 mm Petri dishes.
14. Glass capillary for mouth pipette.

2.2 polyA+ RNA Purification

1. RNase-free water.
2. Dynabeads mRNA DIRECT™ Micro Kit (ThermoFisher).
3. Magnetic rack.

2.3 Directional RNA-Seq Library Construction

1. NEBNext Ultra Directional RNA Library Prep Kit for Illumina (NEB).
2. NEBNext Singleplex (NEB) or NEBNext Multiplex (NEB #E7335, #E7500, #E7600) Oligos for Illumina.
3. Agencourt AMPure XP (Beckman Coulter).
4. 5 μg/μL Actinomycin D stock solution.

2.4 Library Quality Check

1. Agilent High Sensitivity DNA Kit (Agilent).

2.5 Library Quantification

1. KAPA library quantification kit (KAPA Biosystems).

3 Methods

***Important**

Before collecting samples, you should consider the type of RNA in which you are interested. Around 10–50 million reads/sample is generally sufficient for mRNA analysis. On the other hand, it is strongly recommended to perform sequencing of samples to obtain more than 50 million reads/sample for lncRNA analysis since lncRNAs generally show low expression levels [5].

3.1 Superovulation of Mice

Day 1

1. Administer 5–7.5 U of PMSG to 8-week female mice by intraperitoneal injection.

Day 3

1. 46–48 h after injection of PMSG, inject the mice with 5 U of hCG. To collect fertilized embryos, mate the hCG-treated female mice with male mice.

3.2 Preparation of Mouth Pipettes for Embryo Transfer

1. To make a mouth pipette for embryo transfer and manipulation, melt a glass capillary in a flame and pull it immediately.
2. Cut the pulled glass capillary using a diamond pencil or scissors.
3. The diameter of the pipette tip should be around 90–100 μm.

3.3 Preparation of In Vitro Embryo Culture Medium

1. Make 50 μL drops of M16 on the 35 mm dishes and cover them with mineral oil or liquid paraffin to avoid evaporation.
2. For the pH equilibration of drops, place these dishes in a 37 °C CO_2 incubator overnight.

3.4 Collection of Oocytes or Zygotes from Super Ovulated Mice

Day 4

1. After 16 h of hCG treatment, sacrifice superovulated mice and collect mature oocytes or zygotes with cumulus cells from the ampulla of oviduct tubes (*see* **Note 1**).
2. Dilute the stock 1% hyaluronidase tenfold with M2 medium (final conc.: 0.1% hyaluronidase).
3. Place the oocytes or zygotes with cumulus cells in the hyaluronidase solution, incubate them in a 37 °C incubator for 3–5 min, and remove the cumulus cells by pipetting.
4. Transfer the oocytes or zygotes to a drop of M2 successively at least four times.
5. In the case of collecting unfertilized oocytes, transfer these washed oocytes to RNase-free tubes (*see* **Notes 2** and **3**) and store at −80 °C until use or directly proceed to polyA+ RNA purification steps.

6. In the case of collecting fertilized 2-cell embryos, transfer zygotes into pre-equilibrated M16 medium and culture at 37 °C until the first cleavage.

Day 5

1. When the embryos grow to the 2-cell stage, transfer the embryos to RNase-free tubes and store them at −80 °C until use, or you can directly proceed to the polyA+ RNA purification step (*see* **Notes 2** and **3**).

3.5 polyA+ RNA Purification from Embryos

See Fig. 3 for an overview of lysis/polyA selection steps.

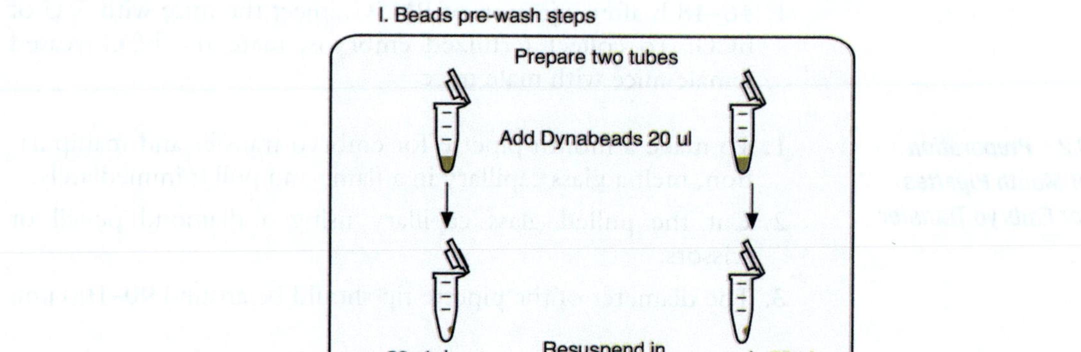

Fig. 3 An overview of polyA+ RNA extraction steps. Oocytes or embryos are lysed with 100 μL of lysis/binding buffer. Lysed samples are transferred to prewashed oligo dT beads, followed by two cycles of wash and elution steps. Eluted polyA+ RNA can be stored at −80 °C or directly subjected to library construction steps

3.5.1 Prewash of dT Beads

1. Bring a Dynabeads® mRNA DIRECT™ Micro Kit to room temperature before use (*see* **Note 4**).
2. Resuspend the Dynabeads® Oligo $(dT)_{25}$ by vortexing for >30 s, or by tilting and rotating for 5 min before use.
3. Prepare two RNase-free tubes/sample.
4. Transfer the Dynabeads® Oligo $(dT)_{25}$ (20 μL/sample) to each RNase-free tube.
5. Add an equivalent volume of Lysis/Binding Buffer (20 μL/sample) to each tube, then mix.
6. Place the tubes on a magnetic rack for 1 min (*see* **Note 5**), then remove and discard the supernatant.
7. Remove the tubes from the magnetic rack, then resuspend the beads in one of the tubes in 20 μL of Lysis/Binding Buffer, and resuspend the beads in the other tube in 70 μL of Lysis/Binding Buffer.

3.5.2 mRNA Isolation from Frozen Embryo

Keep the sample frozen by floating the tubes in liquid nitrogen if you are working with frozen embryos until lysis, then perform a rapid lysis in Lysis/Binding buffer as soon as possible. This is critical for obtaining undegraded mRNA. Avoid thawing of frozen material before lysis.

1. Incubate 1 mL of H_2O at 80 °C using a heat block.
2. Add 100 μL Lysis/Binding Buffer immediately to the fresh/frozen embryos.
3. Perform repeated passage of the solution through a pipette tip to obtain complete lysis. The lysate can be frozen (−80 °C) and stored for later use.
4. Transfer the clear lysate to the tube containing 20 μL of prewashed Dynabeads® Oligo $(dT)_{25}$ prepared in the previous prewash steps.
5. Pipet up and down more than ten times to mix.
6. Place the tube on a sample mixer or roller at room temperature for 5 min for annealing polyA+ RNA to the Dynabeads®.
7. Place the sample tube on a magnetic rack for 1 min, then discard the supernatant.
8. Remove the sample tube from the magnetic rack and resuspend the Dynabeads®-mRNA complex in 100 μL Washing Buffer A by careful pipetting.
9. Place the sample tube on the magnetic rack for 1 min, then discard the supernatant.
10. Repeat **steps 8** and **9** once.
11. Resuspend the Dynabeads®-mRNA complex in 100 μL Washing Buffer B.

12. Place the sample tube on the magnetic rack for 1 min, then discard the supernatant.
13. Repeat **steps 11** and **12** once.
14. Add 50 μL preheated (80 °C) nuclease-free water and incubate at 80 °C for 2 min.

3.5.3 mRNA Re-Purification from Eluted RNA Samples

1. Place the tube on the magnetic rack and immediately transfer the supernatant to a new tube containing prewashed Dynabeads® Oligo $(dT)_{25}$ suspended in 70 μL of Lysis/Binding Buffer.
2. Pipet the mixture up and down ten times, then incubate at room temperature for 5 min. The mixture should appear homogeneous.
3. Place the tube on the magnetic rack. After the solution clears, remove and discard the supernatant without disturbing the pellet.
4. Remove the tube from the magnetic rack. Add 100 μL Washing Buffer A to each tube, then pipet up and down ten times. The mixture should appear homogeneous.
5. Place the tube on the magnetic rack. After the solution becomes clear, remove and discard the supernatant without disturbing the pellet.
6. Remove the tube from the magnetic rack. Add 100 μL Washing Buffer B to each tube, then pipet up and down ten times. The mixture should appear homogeneous.
7. Place the tube on the magnetic rack. After the solution becomes clear, remove and discard the supernatant without disturbing the pellet.
8. Remove the tube from the magnetic rack.
9. Add 25 μL of preheated (80 °C) nuclease-free water to each tube, pipet ten times to resuspend the beads in the nuclease-free water, then let it sit at room temperature for 30 s.
10. Add 25 μL of Lysis/Binding Buffer to each tube, then pipet up and down ten times.
11. Incubate at room temperature for 5 min.
12. Place the tube on the magnetic rack. After the solution becomes clear, remove and discard the supernatant without disturbing the pellet.
13. Repeat **steps 4–7** to wash the RNA.
14. Remove the tube from the magnetic rack, then add 5–10 μL of the warmed (80 °C) nuclease-free water to each tube.
15. Place the tube on the magnetic rack. After the solution becomes clear, transfer the supernatant containing the mRNA to a new tube without disturbing the pellet.
16. *Stop point*: you can store the eluted polyA+ RNA at −80 °C for a few weeks.

3.6 Library Construction Using NEBNext Ultra Directional RNA Library Prep Kit

See Fig. 4 for an overview of library construction.

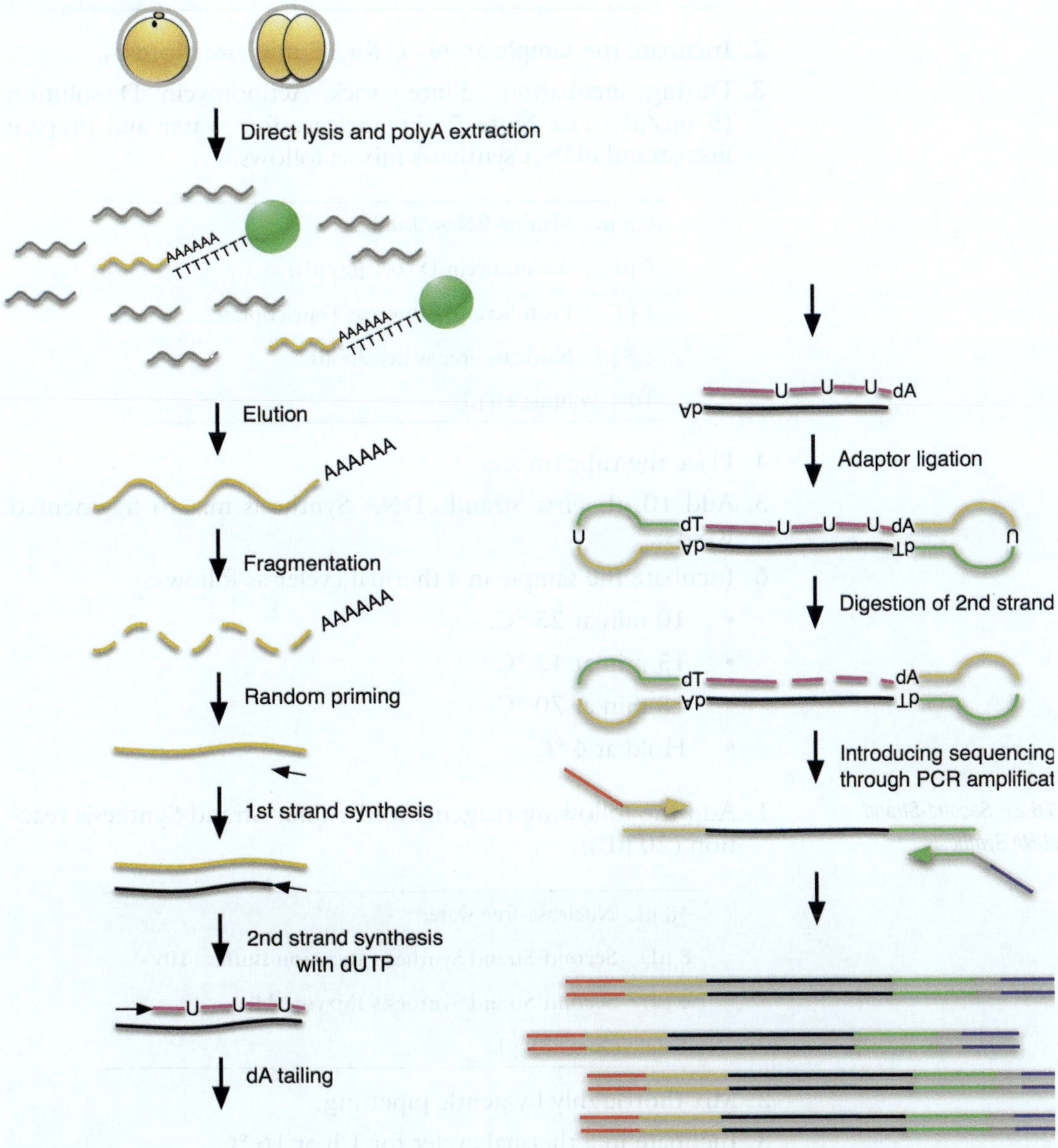

Fig. 4 A schematic view of our directional RNA-seq from small-scale samples. Oocytes or embryos are directly put into lysis and polyA extraction buffer. polyA+ RNAs extracted by magnetic oligo dT beads are eluted and fragmented. These fragmented RNAs are subjected to reverse transcription with random primers to generate first-strand cDNAs. Using these single-strand cDNAs as templates, dUTP-containing second-strand cDNAs are synthesized. Circulizing adaptors are ligated by TA ligation reaction. In the reaction of USER enzyme, this circularized adaptor becomes linear and the dUTP-containing second-strand becomes fragmented. As a result, only first-strand cDNAs are amplified by the subsequent PCR reaction. During this PCR reaction, sequencing tags are added. These constructed libraries can immediately be subjected to high-throughput sequencing

3.6.1 RNA Fragmentation and Priming Starting from Purified mRNA

1. Mix the following materials:

5 μL	Purified mRNA/ribosomal-depleted RNA (10–100 ng)
4 μL	NEBNext First Strand Synthesis Reaction Buffer (5×)
1 μL	Random Primers

2. Incubate the sample at 94 °C for 15 min (*see* **Note 6**).
3. During incubation, dilute stock Actinomycin D solution (5 μg/μL) (*see* **Note 7**) in nuclease-free water and prepare first-strand cDNA synthesis mix as follows:

0.5 μL	Murine RNase Inhibitor
5 μL	Actinomycin D (0.1 μg/μL)
1 μL	ProtoScript II Reverse Transcriptase
3.5 μL	Nuclease-free water 3.5 μL
Total volume 10 μL	

4. Place the tube on ice.
5. Add 10 μL First-Strand cDNA Synthesis mix to fragmented RNAs.
6. Incubate the sample in a thermal cycler as follows:
   - 10 min at 25 °C.
   - 15 min at 42 °C.
   - 15 min at 70 °C.
   - Hold at 4 °C.

3.6.2 Second-Strand cDNA Synthesis

1. Add the following reagents to the First-Strand Synthesis reaction (20 μL):

48 μL	Nuclease-free water
8 μL	Second-Strand Synthesis Reaction Buffer (10×)
4 μL	Second-Strand Synthesis Enzyme Mix
Total volume 80 μL	

2. Mix thoroughly by gentle pipetting.
3. Incubate in a thermal cycler for 1 h at 16 °C.

3.6.3 Purification of Double-Strand DNA Using AMPure XP Beads

1. Make 500 μL/sample 80% ethanol.
2. Vortex AMPure XP beads to resuspend.
3. Add 144 μL (1.8×) of resuspended AMPure XP beads to the second-strand synthesis reaction (~80 μL). Mix well on a vortex mixer or by pipetting up and down at least ten times.

4. Incubate at room temperature for 5 min.
5. Place the tube on an appropriate magnetic rack to separate beads from the supernatant. After the solution is clear (about 2 min), carefully remove and discard the supernatant.
6. Add 200 μL of freshly prepared 80% ethanol to the tube while in the magnetic rack. Incubate at room temperature for 30 s, and then carefully remove and discard the supernatant.
7. Repeat **step 6** once for a total of two washing steps.
8. Air dry the beads for 2 min while the tube is on the magnetic rack with lid open.
9. Elute the DNA target from the beads into 60 μL nuclease-free water. Mix well on a vortex mixer or by pipetting up and down.
10. Place tubes on the magnetic rack until the solution becomes clear.
11. Remove 55.5 μL of the supernatant and transfer to a clean nuclease-free PCR tube.
12. *Stop point*: you can store the eluted DNA at −20 °C.

3.6.4 End Repair/dA-Tailing of cDNA Library

Day 6

1. To the purified double-stranded cDNA (55.5 μL), add the following components:

6.5 μL	NEBNext End Repair Reaction Buffer (10×)
3 μL	NEBNext End Prep Enzyme Mix
Total volume 65 μL	

2. Incubate the sample in a thermal cycler as follows:
   - 30 min at 20 °C.
   - 30 min at 65 °C.
   - Hold at 4 °C.
3. Proceed immediately to Adaptor Ligation.

3.6.5 Perform Adaptor Ligation

Dilute the NEBNext Adaptor for Illumina (15 μM) to 1.5 μM with nuclease-free water at a tenfold dilution (*see* **Notes 8** and **9**).

1. To the dA-Tailed cDNA (65 μL), add the following components:

15 μL	Blunt/TA Ligase Master Mix
1 μL	Tenfold Diluted NEBNext Adaptor
2.5 μL	Nuclease-free water
Total volume 83.5 μL	

2. Incubate at 20 °C for 15 min in a thermal cycler.

3.6.6 Purification and Size-Selection of the Ligated cDNA Using AMPure XP Beads

1. To the ligation reaction mixture (83.5 μL), add 16.5 μL nuclease-free water to bring the reaction volume to 100 μL.
2. Add 100 μL (1.0×) resuspended AMPure XP beads and mix well on a vortex mixer or by pipetting up and down at least ten times.
3. Incubate at room temperature for 5 min.
4. Quickly spin down and place the tube on an appropriate magnetic rack to separate beads from the supernatant. After the solution becomes clear, discard the supernatant, which contains unwanted fragments.
5. Add 200 μL of freshly prepared 80% ethanol to the tube while in the magnetic rack. Incubate at room temperature for 30 s, and then carefully remove and discard the supernatant.
6. Repeat **step 5** once.
7. Briefly spin down, and put the tube back in the magnetic rack.
8. Completely remove the residual ethanol, and air dry beads for 2 min while the tube is on the magnetic rack with the lid open.
9. Elute DNA target from the beads with 50 μL nuclease-free water. Mix well on a vortex mixer or by pipetting up and down, and put the tube in the magnetic rack until the solution is clear.
10. Transfer the 50 μL supernatant to a clean PCR tube. Discard beads.
11. To the 50 μL supernatant, add 50 μL (1.0×) of the resuspended AMPure XP beads and mix well on a vortex mixer or by pipetting up and down at least ten times.
12. Incubate it at room temperature for 5 min.
13. Quickly spin down and place the tube on an appropriate magnetic rack to separate beads from the supernatant. After the solution becomes clear, discard the supernatant, which contains unwanted fragments.
14. Add 200 μL of freshly prepared 80% ethanol to the tube while in the magnetic rack. Incubate at room temperature for 30 s, and then carefully remove and discard the supernatant.
15. Repeat **step 14** once.
16. Briefly spin down, and put the tube back in the magnetic rack.
17. Completely remove the residual ethanol, and air dry beads for 2 min while the tube is in the magnetic rack with the lid open.
18. Elute DNA target from the beads with 20 μL nuclease-free water. Mix well on a vortex mixer or by pipetting up and down, and put the tube in the magnetic rack until the solution becomes clear.
19. Without disturbing the bead pellet, transfer 17 μL of the supernatant to a clean PCR tube and proceed to PCR enrichment.

3.6.7 USER Excision and PCR Library Enrichment

1. To the cDNA (17 μL) add the following components:

3 μL	NEBNext USER Enzyme
25 μL	2xNEBNext High-Fidelity PCR Master Mix
2.5 μL	Universal PCR Primer (10 μM)
2.5 μL	Index (X) Primer (10 μM) (*see* **Note 10**)
Total volume 50 μL	

PCR reaction

1. 37 °C, 15 min.
2. 98 °C, 30 s.
3. 98 °C, 10 s.
4. 65 °C, 75 s.

12 cycles of **steps 3** and **4**

5. 65 °C, 5 min.
6. 4 °C, hold.

3.6.8 Purification of the PCR Reaction Products Using AMPure XP Beads

1. Add 50 μL (1.0×) of vortexed AMPure XP beads to the PCR reaction.
2. Mix well on a vortex mixer or by pipetting up and down at least ten times.
3. Incubate at room temperature for 5 min.
4. Quickly spin down and place the tube in a magnetic rack. After the solution becomes clear, carefully remove and discard the supernatant.
5. Add 200 μL of freshly prepared 80% ethanol to the tube and wait for 30 s, and then remove the supernatant.
6. Repeat **step 5** once.
7. Air dry the beads for 2 min.
8. Elute the DNA from the beads by resuspending beads into 50 μL nuclease-free water. Mix well by pipetting up and down, quickly spin down and place the tube in the magnetic rack until the solution is clear.
9. Transfer 50 μL of the supernatant to a clean PCR tube.
10. Repeat **steps 1–7** again.
11. Elute the DNA into 22 μL nuclease-free water, and store at −20 °C.

3.7 Quality Check Before High-Throughput Sequencing

3.7.1 Visualizing cDNA Fragment Size Using BioAnalyzer

In order to visualize the size distribution of cDNA fragments, we usually use Bioanalyzer 2100 (Agilent) according to the manufacturer's instruction. Occasionally, additional peaks at 80 bp and/or 128 bp are observed (Fig. 5a). These peaks are derived from PCR residuals: a peak at 80 bp = PCR primers and a peak at 128 bp = primer dimers. These can be removed by additional AMPure purification cycles. Add water to the library to adjust the sample volume to 50 μL. Then, add 50 μL AMPure XP to the library and perform an additional purification step. If the size selection is successfully completed, the plot should appear like that in Fig. 5b. If unintended larger peaks are observed, such peaks come from over-amplification of the libraries. In such a case, it is necessary to reduce the PCR cycles for the optimization of library preparation.

3.7.2 Library Quantification by qPCR

It is essential to precisely quantify the amount of DNA in libraries for obtaining an appropriate amount of RNA-seq reads. BioAnalyzer, Nanodrop or Qubit can measure the amount of DNA in the library. These systems, however, also measure the DNA fragments lacking 5′ or 3′ tags. In this context, quantitative PCR (qPCR) is one of the best methods to quantify the fully tagged DNA in libraries. We use a KAPA library quantification kit (KAPA Bioscience) to measure the fully tagged DNA concentration in the library according to the manufacturer's instructions. For HiSeq2000/2500 High output mode, 1 μL of 1.2 nM library is a minimum amount for sequencing (depending on the systems and versions). Representative amounts of libraries quantified by qPCR are shown in Table 1. Even though the library contains only the smallest amount of DNA (20 μL of 2.71 nM), the amount corresponds to ~45 times as much as the quantity to be applied to the sequencing in one lane of a flowcell. If the library is indexed by four index primers, the required amount of

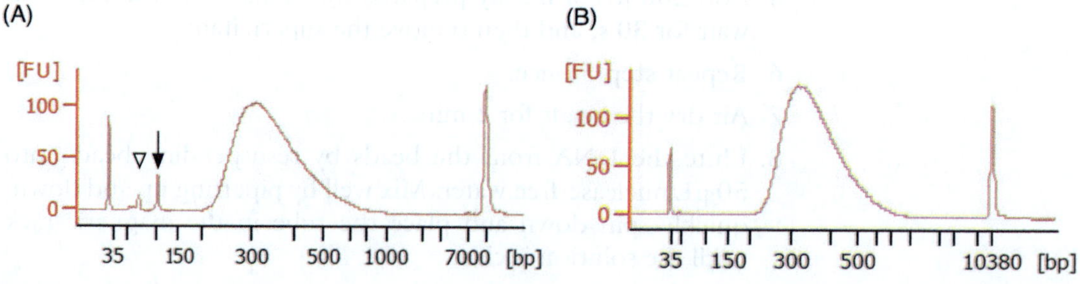

Fig. 5 Representative images of constructed libraries visualized using BioAnalyzer 2100. Horizontal axis indicates the fragment size. Vertical axis indicates the amount of DNA. Narrow peaks appearing on the left and right sides are markers for electrophoresis. (**a**) An example of a library containing PCR residuals. The profile has a major peak at 200–500 bp. The white arrowhead indicates the peak that appeared at the 80 bp position and that corresponds to primers used in the PCR reaction. The arrow indicates the peak that appeared at the 128 bp position and that corresponds to primer dimers generated by the PCR reaction. (**b**) An example of an ideal library after the purification

Table 1
Library yields from ES cells and MII oocytes and estimated minimum cell number for RNA-seq

Sample	Cell no.	Average fragment length	Concentration (nM)	Lanes	Lanes for four indexes	Estimated minimum cell number
ESCs	10000	360	17.6	293.3	1173.3	8.5
ESCs	10000	363	15.6	259.4	1037.4	9.6
ESCs	13200	365	10.1	167.8	671.2	19.7
MII oocytes	110	358	5.5	92.3	369.0	0.3
MII oocytes	108	365	3.5	58.9	235.6	0.5
MII oocytes	99	340	2.7	45.1	180.5	0.5

DNA is even smaller: the library contains 180 times as much as the quantity to be applied to the sequencing in one lane of a flowcell. Taking this information into consideration, the minimum number of required cells to generate a directional RNA-seq library is theoretically around 10 and 1 for ES cells and oocyte, respectively. Therefore, the method described here would be applicable to prepare a sequencing library from a rare cell population or clinically limited samples, such as punched-out tissues.

3.7.3 Capillary Sequencing

It is recommended to confirm the contents of the library before performing high-throughput sequencing to avoid potential risks, such as contamination by ribosomal-RNA-derived sequences or contamination by RNA derived from nontarget cells. We usually confirm the contents of the library by Sanger capillary sequencing (~30 clones). In our experience, less than 3% of sequenced clones contained rRNA-derived sequences. BLAST analysis (http://blast.ncbi.nlm.nih.gov/Blast.cgi) finds the genes/lncRNA detected in one's library. This helps one to know the level of the possible contamination as a quality check of one's library preparations before proceeding to high-throughput sequencing.

3.8 Quality Check After High-Throughput Sequencing

After high-throughput sequencing, quality of RNA-seq data must be checked regarding several aspects. Here, we briefly show some critical points.

3.8.1 Reproducibility Among Replicates

A high degree of variability among technical replicates can be reflected by noise, which compromises many statistical analyses, such as differential expression analysis. Most RNA-seq methods for small-scale samples require pre-amplification steps by PCR or in vitro transcription, which can cause low reproducibility [18]. In contrast, our RNA-seq method does not require such additional amplification steps (Fig. 4). This allows us to keep the variability low. In addition, in our method, the cycle number of PCR is just 12, the same number as in standard RNA-seq protocols starting from 100 ng total RNAs (10,000–100,000 cells). Our method shows a high value of Pearson correlation coefficient between biological replicates (Fig. 6a), indicating that the reproducibility of our RNA-seq library preparations is high enough to perform robust analysis.

3.8.2 Marker Gene Expression

We usually check the expression pattern of known marker genes in order to check whether RNA-seq data can recapitulate the expected expression pattern of such marker genes (Fig. 6b). This enables you to check whether the RNA-seq libraries come from the cells that you expected to collect.

3.8.3 Directionality of Sequenced Reads

Directional RNA-seq methods enable one to know which strand is the template of the transcription in relation to the gene of interest. When you explore an "antisense" transcript, you can calculate the strand-specificity of reads by using RSeQC package [19]. In our experience, most of the libraries show excellent strandedness (more than 98% of reads were aligned in the same orientation against annotated genes), leading to the detection of antisense RNAs, including pancRNAs, stringently and comprehensively (Figs. 7 and 8).

4 Notes

1. Usually, 20–35 oocytes or embryos were recovered from one B6C3F1 female mouse.
2. Reduce carryover of medium as much as possible.
3. We recommend starting from 100 embryos as an initial trial.
4. Confirm that the Lysis/Binding Buffer has not precipitated. If any precipitation is observed, warm to room temperature and shake to dissolve.
5. Wait until the solution becomes clear.
6. Optimization of fragmentation conditions may be required. In case, change incubation time.
7. Actinomycin D is more stable in a high concentration at −20 °C, such as 5 μg/mL. Diluted Actinomycin D should not be used more than once.
8. The adaptor is provided in NEBNext Singleplex (NEB #E7350) or NEBNext Multiplex (NEB #E7335, #E7500) Oligos for Illumina.
9. Dilution of adaptor is necessary; otherwise, many primer-dimers are formed at the PCR step and result in a decrease in the yield of the library.

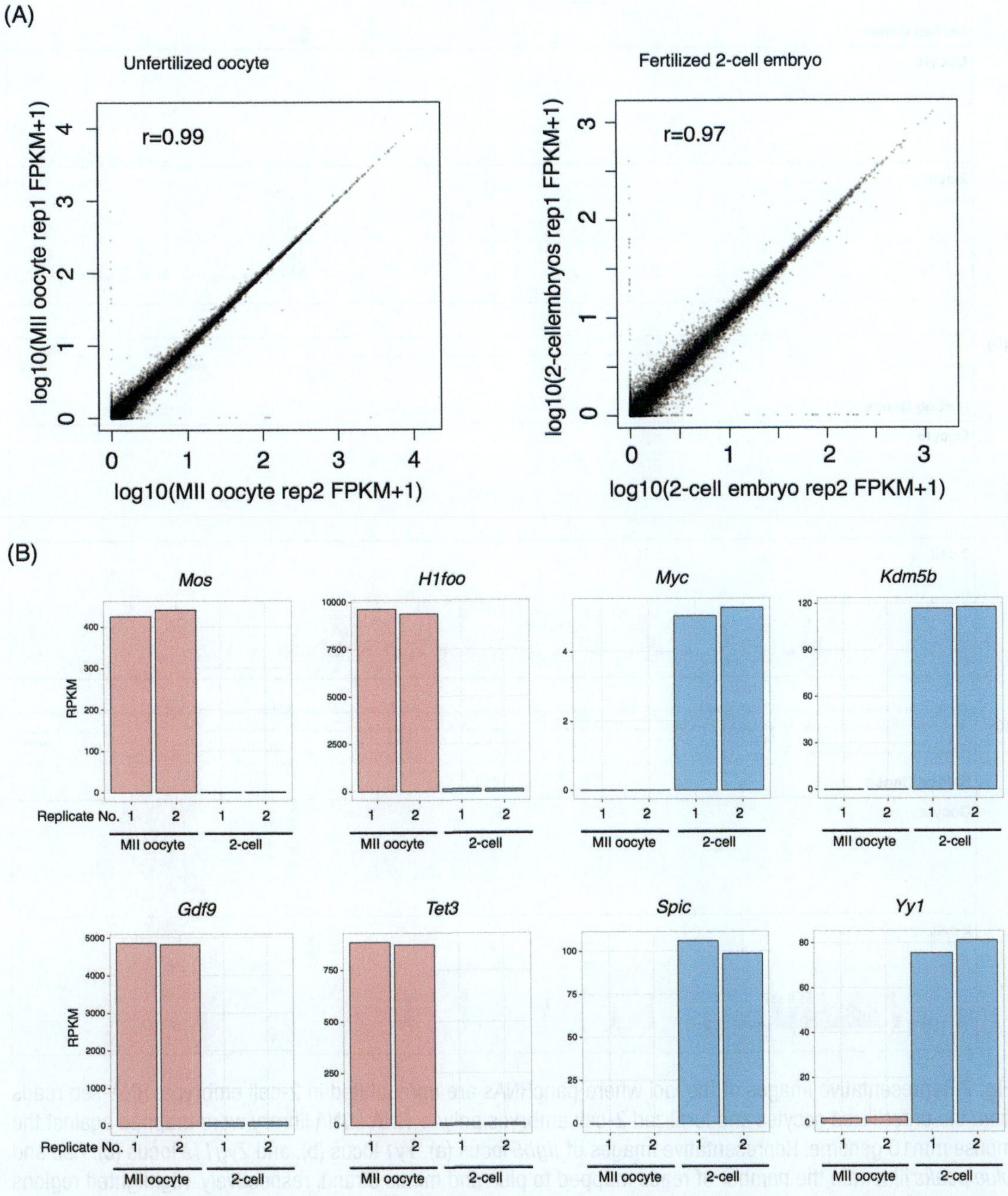

Fig. 6 Qualities of the directional RNA-seq. (**a**) Scatter plots of gene expression levels calculated from RNA-seq data, of all genes in two biological replicates. Expression levels of RefSeq genes in unfertilized oocytes and fertilized 2-cell embryos are shown in left and right panels, respectively. Values depicted on the upper-left side indicate the Pearson correlation coefficient between biological replicates. (**b**) Bar graphs representing the expression levels of genes known to be highly expressed in oocytes or 2-cell embryos. Note that the RNA-seq data recapitulate the expression pattern of these genes

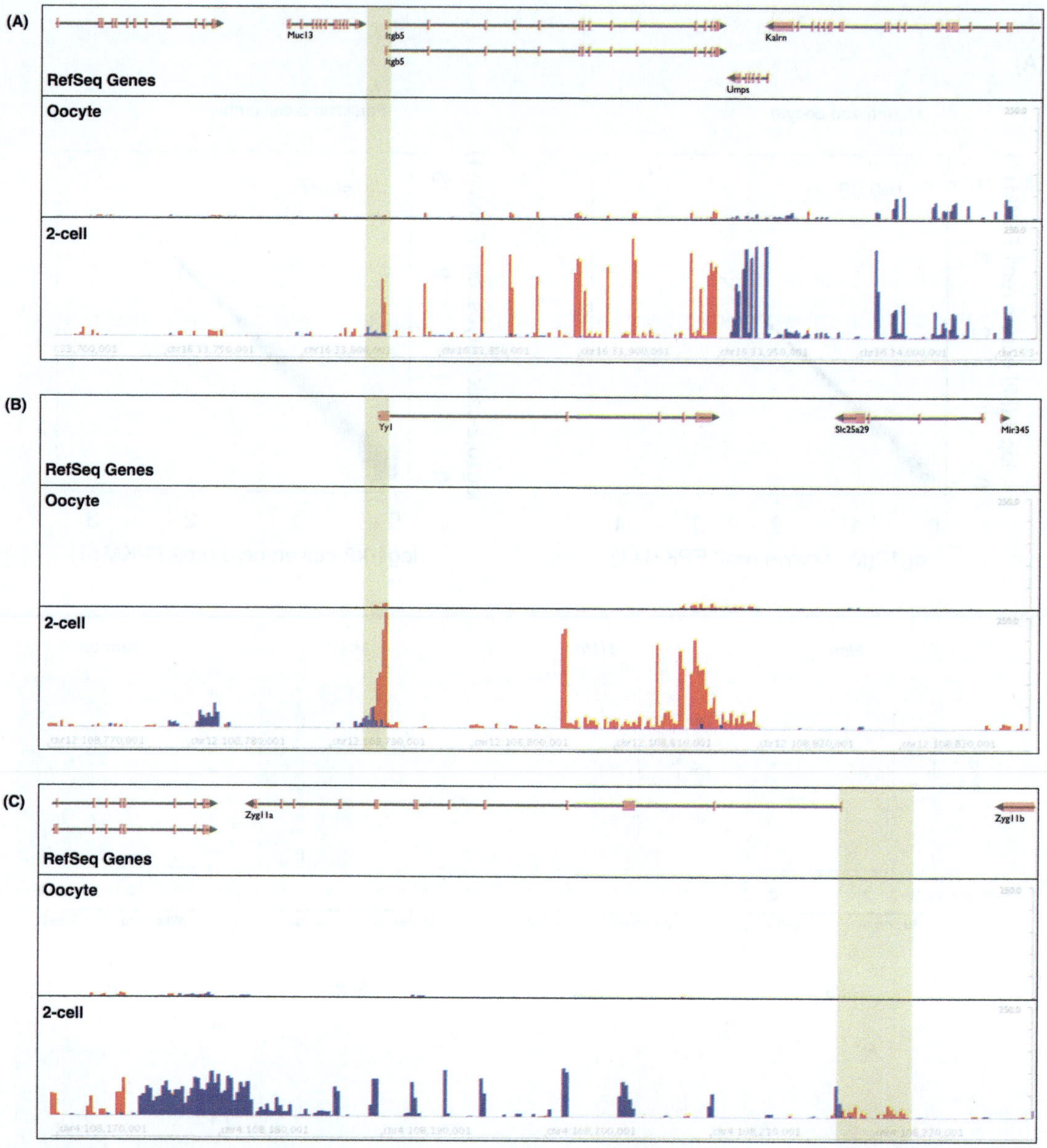

Fig. 7 Representative images of the loci where pancRNAs are upregulated in 2-cell embryos. RNA-seq reads from the unfertilized oocytes and fertilized 2-cell embryos polyA+ RNA cDNA library were mapped against the mouse mm10 genome. Representative images of *Itgb5* locus (**a**), *Yy1* locus (**b**), and *Zyg11a* locus (**c**). *Red* and *blue peaks* illustrate the number of reads mapped to plus and minus strand, respectively. Highlighted regions indicate the expression of putative pancRNAs. Note that these pancRNAs are upregulated together with their partner genes

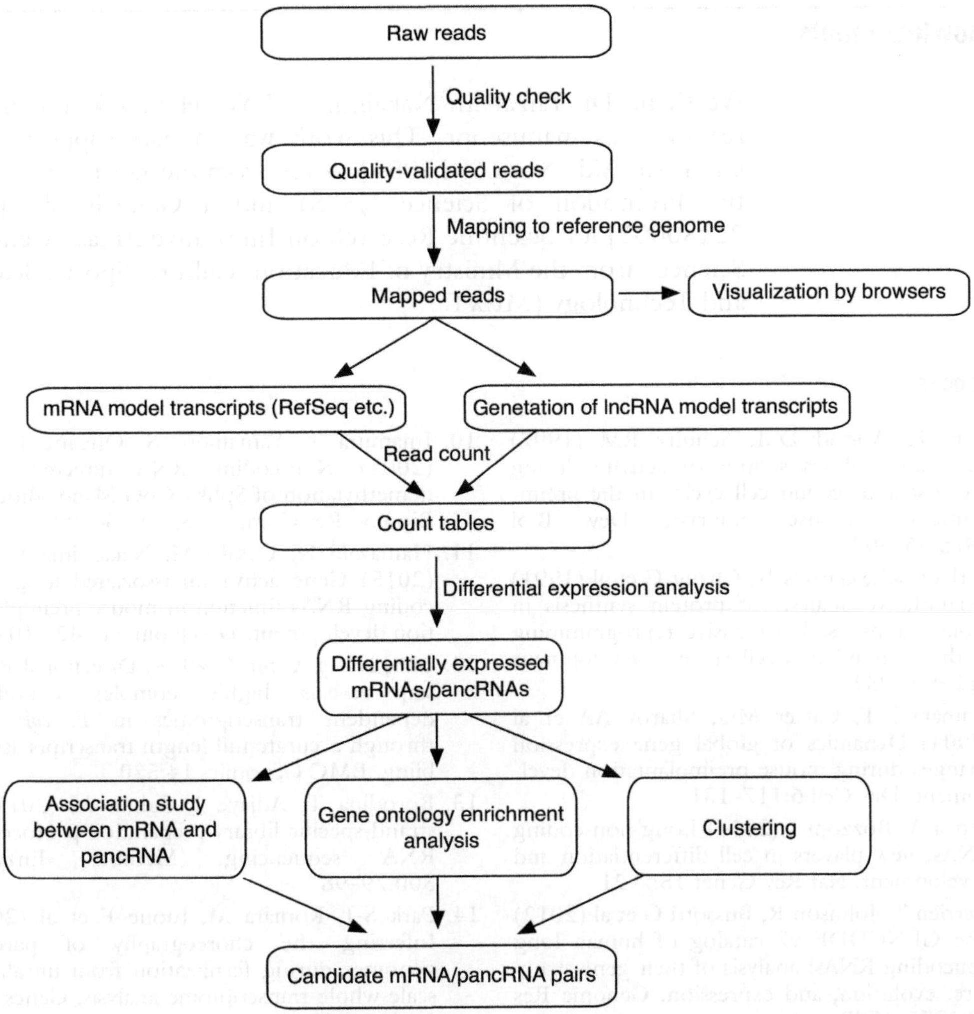

Fig. 8 A Work flow for detecting candidates pancRNAs. Raw reads are filtered by their quality values and mapped to a reference genome. Mapped reads can be visualized and checked on a genome browser, such as Integrative Genomics Viewer (IGV). Count tables are generated by counting reads of features, such as mRNA transcripts and promoter regions. In order to obtain mRNAs or pancRNAs that showed statistically significant differences according to condition or environmental changes, open source applications, such as Cufflinks [20], edgeR [21], and RSEM [22], are available. After the acquisition of a list of differentially expressed mRNAs or pancRNAs, it can be utilized for further analyses, such as analysis of association between mRNA/pancRNA pairs, analysis of gene ontology enrichment or clustering analysis to annotate potential functions of candidate mRNA/pancRNA pairs

10. The proper combination of indexing ensures high-quality sequencing. Therefore, choose a proper combination pattern for your sequencing according to the latest manufacturer's instruction, such as https://www.neb.com/protocols/2015/01/26/please-see-manual-neb-e7335-for-protocols.

Acknowledgments

We thank Dr. Elizabeth Nakajima and Yui Hamazaki for proofreading this manuscript. This work was in part supported by Grants-in-Aid [No. 16K15054] to T.I. from the Japan Society for the Promotion of Science (JSPS) and a Grant-in-Aid [No. 221S0002] for Scientific Research on Innovative Areas "Genome Science" from the Ministry of Education, Culture, Sports, Science and Technology (MEXT).

References

1. Aoki F, Worrad DM, Schultz RM (1997) Regulation of transcriptional activity during the first and second cell cycles in the preimplantation mouse embryo. Dev Biol 181:296–307
2. Latham KE, Garrels JI, Chang C et al (1991) Quantitative analysis of protein synthesis in mouse embryos. I. Extensive reprogramming at the one-and two-cell stages. Development 112:921–932
3. Hamatani T, Carter MG, Sharov AA et al (2004) Dynamics of global gene expression changes during mouse preimplantation development. Dev Cell 6:117–131
4. Fatica A, Bozzoni I (2014) Long non-coding RNAs: new players in cell differentiation and development. Nat Rev Genet 15:7–21
5. Derrien T, Johnson R, Bussotti G et al (2012) The GENCODE v7 catalog of human long noncoding RNAs: analysis of their gene structure, evolution, and expression. Genome Res 22:1775–1789
6. Uesaka M, Nishimura O, Go Y et al (2014) Bidirectional promoters are the major source of gene activation-associated non-coding RNAs in mammals. BMC Genomics 15:35
7. Luo S, Lu JY, Liu L et al (2016) Divergent lncRNAs regulate gene expression and lineage differentiation in pluripotent cells. Cell Stem Cell 18(5):637–652
8. Yamamoto N, Agata K, Nakashima K et al (2016) Bidirectional promoters link cAMP signaling with irreversible differentiation through promoter-associated non-coding RNA (pancRNA) expression in PC12 cells. Nucleic Acids Res 44(11):5105–5122
9. Tomikawa J, Shimokawa H, Uesaka M et al (2011) Single-stranded noncoding RNAs mediate local epigenetic alterations at gene promoters in rat cell lines. J Biol Chem 286:34788–34799
10. Imamura T, Yamamoto S, Ohgane J et al (2004) Non-coding RNA directed DNA demethylation of Sphk1 CpG island. Biochem Biophys Res Commun 322:593–600
11. Hamazaki N, Uesaka M, Nakashima K et al (2015) Gene activation-associated long noncoding RNAs function in mouse preimplantation development. Development 142:910–920
12. Li S, Dong X, Su Z (2013) Directional RNA-seq reveals highly complex condition-dependent transcriptomes in *E. coli* K12 through accurate full-length transcripts assembling. BMC Genomics 14:520
13. Borodina T, Adjaye J, Sultan M (2011) A strand-specific library preparation protocol for RNA sequencing. Methods Enzymol 500:79–98
14. Park S-J, Komata M, Inoue F et al (2013) Inferring the choreography of parental genomes during fertilization from ultralarge-scale whole-transcriptome analysis. Genes Dev 27:2736–2748
15. Picelli S, Björklund ÅK, Faridani OR et al (2013) Smart-seq2 for sensitive full-length transcriptome profiling in single cells. Nat Methods 10:1096–1098
16. Sasagawa Y, Nikaido I, Hayashi T et al (2013) Quartz-Seq: a highly reproducible and sensitive single-cell RNA-Seq reveals non-genetic gene expression heterogeneity. Genome Biol 14:R31
17. Nakamura T, Yabuta Y, Okamoto I et al (2015) SC3-seq: a method for highly parallel and quantitative measurement of single-cell gene expression. Nucleic Acids Res 43:e60
18. Bhargava V, Head SR, Ordoukhanian P et al (2014) Technical variations in low-input RNA-seq methodologies. Sci Rep 4:3678
19. Wang L, Wang S, Li W (2012) RSeQC: quality control of RNA-seq experiments. Bioinformatics 28:2184–2185

20. Trapnell C, Roberts A, Goff L et al (2012) Differential gene and transcript expression analysis of RNA-seq experiments with TopHat and Cufflinks. Nat Protoc 7:562–578
21. Robinson MD, McCarthy DJ, Smyth GK (2010) edgeR: a Bioconductor package for differential expression analysis of digital gene expression data. Bioinformatics 26:139–140
22. Li B, Dewey CN (2011) RSEM: accurate transcript quantification from RNA-Seq data with or without a reference genome. BMC Bioinformatics 12:323

Chapter 7

Detection and Characterization of Small Noncoding RNAs in Mouse Gametes and Embryos Prior to Zygotic Genome Activation

Jesús García-López, Eduardo Larriba, and Jesús del Mazo

Abstract

Small noncoding RNAs (ncRNAs) are regulatory elements of gene expression in all cell types and tissues. An ever-increasing number of studies have implicated ncRNAs in differentiation and developmental processes. In mammals, as a consequence of fertilization, the content of ncRNAs in the zygote is mostly the result of the maternal material included on oocytes and the potential sperm-borne paternal contributions. The genetic identity program of any individual is the reprogramming of each parental contribution to the zygotic genome activation. In mouse, this activation occurs at 2-cell stage. In this program of early development the small ncRNAs can play important roles. Here, we describe protocols for collection of oocytes, spermatozoa, and zygotes in mouse, followed by RNA purification to analyze the different types of small ncRNA by next-generation sequencing approaches (NGS). Bioinformatics protocols also describe the methodology able to characterize microRNAs (miRNAs) as the most well-known and widespread regulatory small ncRNA. The comparative analysis allows identifying the changes and background previous to zygotic genome activation.

Key words Small noncoding RNAs, Oocyte, Spermatozoa, Zygote, miRNAs, NGS

1 Introduction

The correct embryo development requires deep but selective changes in the pattern of gene expression from the mostly based maternal contribution in the oocyte to the transcriptome of the totipotent zygote. After fertilization, this transition involves the replacement of specific genetic information coded abundantly in the maternal RNAs from oocyte, and potentially as sperm-borne, with zygotic genome expression. Consequently, an assessment of the transcriptome landscape as a result of zygotic activation, occurring at 2-cell stage in mouse [1], requires a comparative analysis of the previous transcript background. That is, the whole mRNAs

Kiho Lee (ed.), *Zygotic Genome Activation: Methods and Protocols*, Methods in Molecular Biology, vol. 1605,
DOI 10.1007/978-1-4939-6988-3_7, © Springer Science+Business Media LLC 2017

and ncRNAs contribution of both gametes to the zygote and the zygote itself previously to the cleavage and zygotic activation.

Small noncoding RNAs (ncRNAs) are regulatory components of gene expression and genome structure in all cell types and tissues. The role of ncRNAs in the maternal to zygotic transition has been remarked as key regulatory elements in the control of maternal transcript elimination and new gene expression regulation [2]. On the other hand, Yuan et al. [3] recently reported the relative importance of some sperm-borne ncRNAs during embryo development, but not in the fertilization neither in the zygotic activation.

Both specific mRNA and ncRNA present in the zygote have to be functionally conserved in the early preimplantation embryo development. Among the different types of ncRNAs: microRNAs (miRNAs), piwi-interacting RNAs (piRNAs), endogenous siRNAs (endo-siRNAs), the miRNAs are the best characterized ncRNAs in the biogenesis and function [4]. Maternal miRNAs are essential in the zygotic development and the earliest stages of mouse embryonic development [5]. However, some other regulated mRNA and miRNAs are selectively eliminated before or as a consequence of zygotic activation. In fact, some miRNAs are actively participating in the selective "erase" of maternal mRNAs [6]. Maternal, or putative paternal mRNAs are clearance by MZT-specific miRNAs (maternal-to-zygotic transition). A clear example is miR-430 in zebrafish (miR-294 orthologous in mouse) involved in mRNA degradation by deadenylation of maternal transcripts [7].

The NGS approach allows the analysis and characterization of different types of small ncRNAs. Although other types of small ncRNA, which are present in both gametes and zygotes [8, 9], could act in the elimination of mRNA [10] facilitating the reprogramming of zygote activation, here we focus on the study of miRNAs through bioinformatics approaches.

In this chapter, we describe the cytological, molecular, and bioinformatics protocols to analyze qualitatively and quantitatively the expression profiles of small ncRNAs, focusing on miRNAs, in spermatozoa, oocytes, and zygotes. We provide the methodology for isolating both male and female gametes and zygotes, RNA isolation from these cells, and finally, the bioinformatics analysis of miRNAs from data obtained by NGS technology.

Different stages of bioinformatics pipeline present in these protocols required the use of the terminal (Bash, Bourne again shell), these stages are detailed in format copy and paste to facilitate the use of terminal. In addition, we include the use of different free software programs that have detailed documentation and tutorials. Analysis of miRNA encompassing different steps, including: evaluation of sRNA-Seq reads, trimming and cleaning sequencing reads, and a specific framework for identification and quantification a miRNA and isomiRs. Finally, miRNA regulatory elements and targets genes were predicted using different web-tools (Fig. 1).

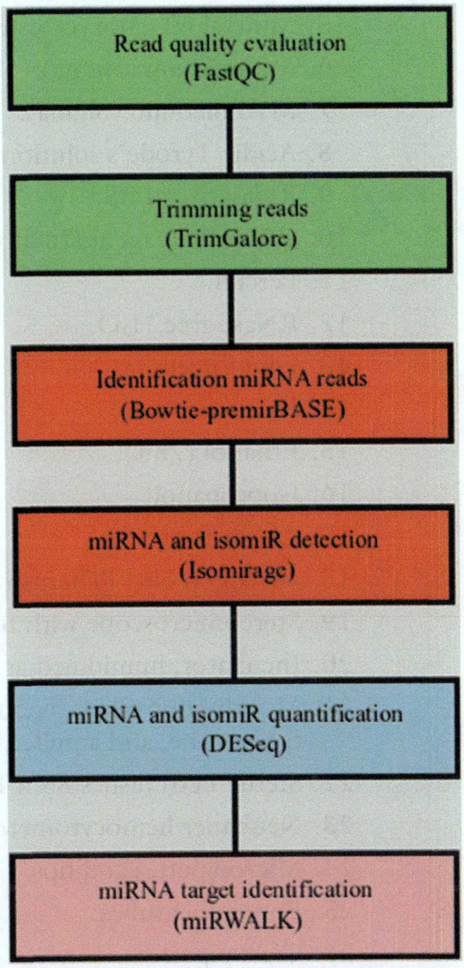

Fig. 1 An overall pipeline designed to process small ncRNA data after NGS. Workflow is colored based in the different stages of bioinformatics protocol. In green, quality control of sequencing libraries and trimming. In red, identification of miRNA and isomiR detection and characterization. In blue, normalization of miRNAs and isomiRs and differential expression analysis. In pink, miRNA target identification. Software used in each phase is indicated in brackets

2 Materials

2.1 Isolating Gametes and Zygotes and RNA Purification

1. Mice: 4–5-week-old C57BL6 females and adult DBA/6J males bred under specific pathogen-free (SPF), temperature (22 ± 1 °C), humidity-controlled (50–55%), 12 h light/dark cycles with ad libitum access to food and water.
2. Pregnant mare serum (PMSG).
3. Human chorionic gonadotropin (HCG).
4. Hyaluronidase solution (0.3 mg/ml) (Sigma).

5. Mineral oil embryo-tested light (Sigma).
6. M2 medium (Sigma).
7. M16 medium (Sigma).
8. Acidic Tyrode's solution (Sigma).
9. RNase-free PBS.
10. TRIzol® Reagent (Invitrogen).
11. Percoll.
12. RNase-free H_2O.
13. Fine forceps.
14. Fine scissors.
15. Ethanol (75%).
16. Isopropanol.
17. Chloroform.
18. Glycogen blue (Thermo Fisher Scientific).
19. Stereomicroscope with transmitted light.
20. Incubator, humidified at 37 °C, 5% CO_2 in air.
21. Mouth-controlled pipette (consisting in a mouth piece, a tygon tube, and a pulled Pasteur pipette).
22. Sterile Petri dishes 35 × 10 mm (Nunc).
23. Neubauer hemocytometer.
24. Micropipettes and tips.
25. Eppendorf tubes.
26. Glass pipettes.
27. Tube shaker.
28. NanoDrop Spectrophotometer (ND-1000).
29. Microcentrifuge (Eppendorf 5424 or equivalent).
30. Centrifuge (Eppendorf 5403 or equivalent).
31. Heat block.
32. Bioanalyzer 2100 (Agilent Technologies).

2.2 Bioinformatics Analysis from NGS Data

2.2.1 Hardware

We recommend a High Performance Computing (HPC) especially when planning to analyze a large number of NGS sequencing files. Alternatively, this protocol is designed for a personal computer with the following minimum configuration: 4 core processor with 8 threads and 16 GB of RAM memory.

2.2.2 Software

We suggest organizing both software and databases in different folders (e.g., "Software/" and "Databases/").

This protocol was designed to be as user-friendly as possible in a personal computer. It also includes the strategies and commands to enable the analysis of different libraries at once using bash scripts.

All software/packages in this protocol contain a complete manual available online.

1. Operating System: Unix-based operating system such as Ubuntu or Mac OS (*see* **Note 1**).
2. Databases files: miRBase: http://www.mirbase.org/ (Version 21).
3. Isomirage: http://cru.genomics.iit.it/Isomirage/ [11].
4. R and Bioconductor (DESeq package): https://cran.r-project.org/bin/linux/ubuntu/ and https://www.bioconductor.org/install/ (*see* **Note 2**).
5. Bowtie: https://sourceforge.net/projects/bowtie-bio/files/bowtie/1.1.2/ (for Ubuntu users, type in the terminal: sudo apt-get install bowtie).
6. SAM tools: https://sourceforge.net/projects/samtools/files/ (for Ubuntu users, type in the terminal: sudo apt-get install samtools).
7. FastQC and Trim Galore: http://www.bioinformatics.babraham.ac.uk/projects/ (for Ubuntu users, type in the terminal: sudo apt-get install fastqc).
8. FASTX-toolkit: http://hannonlab.cshl.edu/fastx_toolkit/download.html (for Ubuntu users, type in the terminal: sudo apt-get install fastx-toolkit).
9. miRWalk2.0: http://zmf.umm.uni-heidelberg.de/apps/zmf/mirwalk2/.

3 Methods

3.1 Setting Up the Oocytes and Zygotes Collection

3.1.1 Female Hormone Induction

1. To obtain large amount of oocytes and zygotes, we induce hormonal superovulation of females. For most mouse strains, 5 IU of Pregnant mare's serum gonadotropin (PMSG) injected intraperitoneally (i.p.) induce the maturation of oocytes mimicking follicle-stimulating hormone (FSH).
2. Then, 48 h later, human chorionic gonadotropin (hCG) is administrated i.p. to induce the rupture of the matured follicles mimicking the effect of luteinizing hormone (LH). Generally, 5 IU are sufficient for the majority of the strains. The age and weight of the female usually affects the number of oocytes/zygotes recovered. The most appropriate age for inducing superovulation is different for each strain, but commonly varies between 4 and 5 weeks. Also consider that underweight females yield a low number of oocytes/zygotes after superovulation.
3. Finally, for zygote harvesting purposes it is necessary to set the mating after hCG i.p. injection. The following day after mating

the female will be checked for a copulation plug and the oocytes/zygotes will be harvested between 8 and 10 h after midnight of hCG administration day.

3.1.2 Making Glass Pipettes

1. Melt a Pasteur pipette by rotating it in a burner until the glass become soft.
2. Take away the glass from the flame and pull both ends to produce a tube with a thinner diameter. Do not pull the glass while is still in the flame.
3. Cut the extreme of the glass pipette with a diamond pencil.

3.1.3 Preparing Microdrops Culture Plates

1. Dispense 100 μL drops of M2 medium on the bottom of 35 mm plastic dish.
2. Cover the drops with mineral oil to avoid desiccation.
3. Place the microdrop culture dish in the incubator at 37 °C and 5% CO_2.

3.1.4 Collecting Oocytes and Zygotes

1. Open the abdominal cavity of female mice and pull the uterus, oviduct, and ovary away from the body.
2. First, cut between the oviduct and ovary, and then cut again between uterus and the oviduct close to the oviduct.
3. Transfer the oviduct to a 35 mm petri dish containing M2 medium.
4. Oocytes and zygotes are located in the ampulla, which resembles an enlarged bag on the upper part of the oviduct. Using one pair of forceps to tear the ampulla and the oocytes/zygotes will come out themselves.
5. At that point, oocytes/zygotes are surrounded by cumulus cells. To remove them, add hyaluronidase solution in the petri dish containing the M2 medium and wait for 1–3 min. Cumulus cells will start to disaggregate from oocytes/zygotes facilitating their individual collection.
6. Transfer the oocytes/zygotes to fresh M2 medium and clean up them pipetting up and down several times to remove any trace of hyaluronidase.
7. Finally, transfer the desired number of oocytes/zygotes into a new petri dish containing 1× PBS solution.
8. Use a pipette to transfer the oocytes/zygotes to a new petri dish containing Tyrode's solution to remove the zona pellucida. The oocytes/zygotes have to be monitored constantly under a stereomicroscope during the treatment to prevent prolonged exposure to acidic pH. Once the zona pellucida is removed, transfer them to a new petri dish with different drops containing 1× PBS to clean them from Tyrode's solution treatment. Then the oocytes and zygotes are ready to be stored.

9. Collect the oocytes/zygotes on a PCR Eppendorf tube in a final volume of 4 μL using a P2 micropipette.

3.2 Isolating Epididymal Spermatozoa

1. After euthanasia, adult mouse males were abdominal dissected and epididymis are collected with fine forceps and scissors and placed in M2 medium in a petri dish (the number of animals depends on the amount of RNA, about 20 mice).
2. Fat, veins, and cauda epididymis are discarded. Caputs containing mature spermatozoa are collected.
3. Each 4–5 caputs are immersed in drops of 200 μl of M2 media covered with mineral oil.
4. By 30-G needle fine cuts in the epididymis facilitate the release of motile spermatozoa to the media.
5. Drops of M2 with caputs are incubated at 37 °C during 10–15 min to allow swim-out and dispersion of spermatozoa into the media.
6. The media with suspension of spermatozoa from groups of 3–4 drops are collected in Eppendorf tubes at 37 °C until use.
7. Spermatozoa are isolated by Percoll gradient. In 15 ml conical tube, in a horizontal position as much as possible, place carefully successive layers of 3 ml of 90% Percoll in PBS followed by 45% and finally 1 ml of spermatozoa suspension.
8. Tubes are centrifuged at 800 × *g* during 20 min.
9. Spermatozoa are as pellet that is washed two times with PBS and centrifuged each wash.
10. To eliminate potential somatic cell contamination by osmotic shock, the pellet can be resuspended in RNAse-free water centrifuge for 3 min at 6000 × *g* discarding supernatant.
11. Resuspend in PBS. Take 1 μL of suspension, diluted 400× and count concentration by Neubauer chamber.
12. Resuspend in 100 μL of Trizol and frozen at −80 °C until use for RNA isolation.

3.3 RNA Isolation from Low Amount of Cells

1. To isolate RNA for samples with small amount of cells (oocytes and zygotes) (1×10^4–1×10^5 cells) add 250 μL of TRIzol, as well as for 5×10^7 spermatozoa. For samples >1×10^5 add <1×10^6 we recommend to add 500 μL of TRIzol.
2. Homogenize the sample pipetting up down several times and incubate the sample for 5 min at room temperature.
3. Add 66 μL of chloroform and shake the tube during 30 s. and incubate at room temperature for 2–5 min.
4. Centrifuge the sample at 12,000 × *g* for 15 min at 4 °C.
5. Place the aqueous phase into a new Eppendorf tube.

6. Add 5 μg of glycogen blue to make visible the RNA pellet in the next steps.
7. Add 166 μL of isopropanol and incubate for 2 h at −20 °C.
8. Centrifuge at 12,000 × *g* for 10 min at 4 °C.
9. Remove the supernatant leaving just the bluish pellet containing precipitated RNA.
10. Wash the RNA pellet, with 250 μL of 75% ethanol (or 80% if an enrichment of small RNAs is desired). Vortex the sample to mix.
11. Centrifuge the sample at 7500 × *g* for 5 min at 4 °C, and discard the supernatant.
12. Vacuum or air dry the RNA pellet for 5–10 min. Do not dry the pellet by vacuum centrifuge.
13. Resuspend the RNA pellet in RNase-free water and incubate in a water bath or heat block set at 55–60 °C for 15 min.
14. In nanoDrop Spectrophotometer, use absorbance of RNA at 260 and 280 nm to determine total RNA concentration.
15. Finally, determine the quality of RNA by 2100 Bioanalyzer using supplier procedures (for spermatozoa *see* **Note 3** and Fig. 2).

3.4 Next-Generation Sequencing (NGS) and Quality Control and Trimming of Files from NGS

Sequencing RNA using a small RNASeq protocol in a Massive Parallel Sequencing platform (NGS) (*see* **Note 4**).

This protocol was designed to work with FASTQ format files (*.fastq), which are commonly generated by Illumina sequencers. This format can also be easily obtained from different Massive Parallel Sequencing platforms (NGS). We recommend to download or copy all the *.fastq files to the same folder (e.g., into a folder called Reads/inside Documents/).

3.4.1 Quality Control

For quality control we used FastQC software (*see* Subheading 2.2.2, **item 7**). A web-browser will open showing the quality report by opening the report .html file. The graph: "Per base sequence quality" shows a box plot of quality scores in relation to the global sequences. FastQC website details some examples of different quality reports.

1. Run FastQC inside your Reads/folder:

```
cd Documents/Reads/
fastqc -t 7 input_file.fastq
```

   where -t is the number of threads that will be used to run FastQC (seven in last command) and input_file is the .fastq file generated by the NGS platform.

2. For running FastQC on multiple files located in the same directory follow:

```
fastqc -t 7 input_file_1.fastq input_
file_2.fastq input_file_N.fastq
```

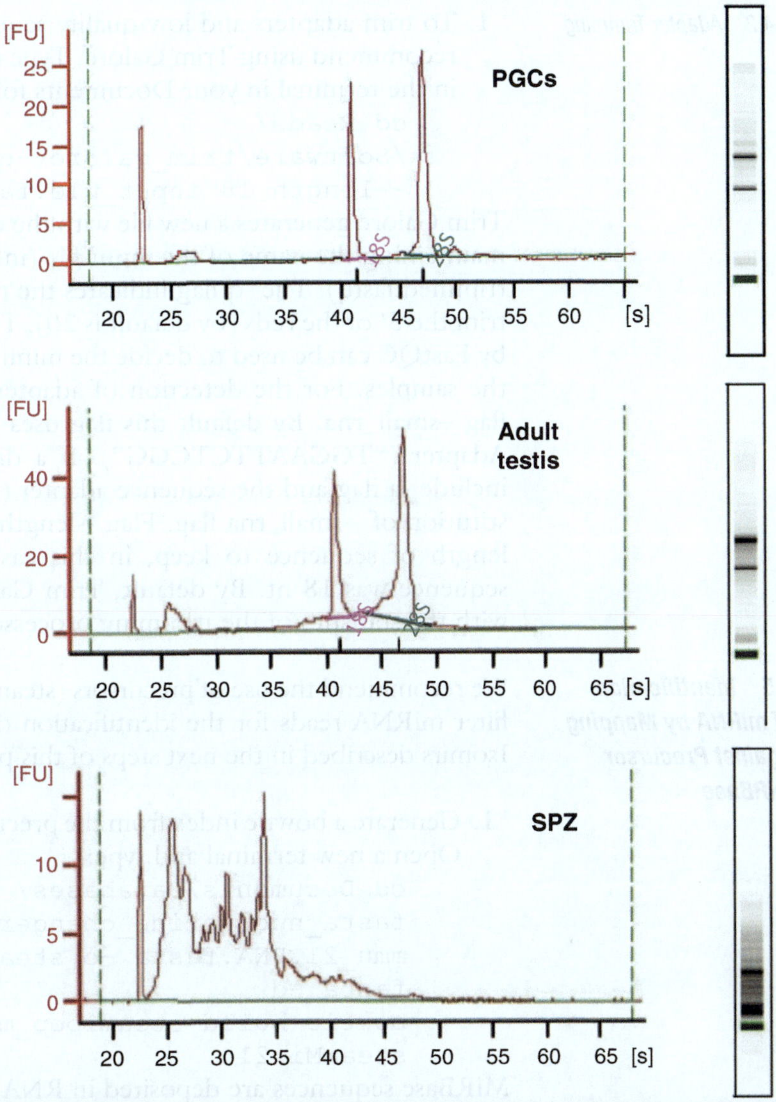

Fig. 2 Electropherograms of total RNAs from male germ cell and tissue assessed by Bioanalyzer. PGCs correspond to Primordial Germ Cells from male embryos of 13.5 postcoitum days; SPZ isolated spermatozoa following the protocols detailed in this chapter; Adult testis homogenate. Note the high proportion of small RNAs and practically absence of 18S and 28S RNAs in SPZ sample

where -t is the number of threads to run FastQC and input_file_1, 2, N are the different .fastq files to analyze.

3. For running FastQC interactively, type in the terminal:

```
fastqc -t (number of threads)
```

A window of FastQC will open go to "File," then "Open" and select your *. fastq file and run the analysis.

3.4.2 Adapter Trimming

1. To trim adapters and low-quality sequences from raw data we recommend using Trim Galore. Type the following commands in the terminal in your Documents folder:

```
cd Reads/
/Software/trim_galore -q 28 --small_rna
--length 18 input_file.fastq
```

Trim Galore generates a new file with the extension .trimmed.fastq maintaining the name of the input file (in this example, input_file.trimmed.fastq). The -q flag indicates the minimum Phred score to trim the 3′ of the reds (by default is 20). The information provided by FastQC can be used to decide the minimum Phred score to trim the samples. For the detection of adapters in the sample use the flag –small_rna. By default this flag uses an Illumina Small RNA Adapter ("TGGAATTCTCGG"). If a different adapter is used, include -a flag and the sequence adapter (in DNA format) in substitution of --small_rna flag. Flag --length indicates the minimum length of sequence to keep, in this case the minimum length sequence was 18 nt. By default, Trim Galore generates a .txt file with the statistics of the trimming processes.

3.5 Identification of miRNA by Mapping Against Precursor miRBase

We recommend the use of precursors (steamloop_mmu_21.fasta) to filter miRNA reads for the identification of canonical miRNA and Isomirs described in the next steps of this protocol (*see* **Note 5**).

1. Generate a bowtie index from the precursor miRNA sequences. Open a new terminal and type:

```
cd Documents/Databases/
fasta_nucleotide_changer -i steamloop_
mmu_21.RNA.fasta -o steamloop_mmu_21.
fasta -d
bowtie-build steamloop_mmu_21.fasta
steamMir21
```

MiRBase sequences are deposited in RNA nucleotide code, so the first step is the transformation of RNA to DNA using FASTX-toolkit (first command). Bowtie-build builds a Bowtie index from a set of DNA sequences, generating six files with extension: .ebwt. Check them typing in the terminal "ls." To index any file in fasta format only change the first argument for the fasta file and the second argument for name of database (*see* **Note 5**).

2. The next step is to map all trimmed sncRNA sequences to miRBase precursors to filter sequences that were not complementary to reported miRNA. For this purpose, we use the bowtie aligner. Type in your terminal:

```
cd ../
mkdir MapPrecursors/
mkdir SAMPrecursors/
cd Reads/
```

bowtie ../Databases/steamMir21 input_file.trimmed.fastq -p 4 -v 3 -a --strata --best --tryhard --chunkmbs 256 -S --al ../MapPrecursors/input_file.steam21.fastq>../SAMPrecursors/input_file.steam21.sam

To run bowtie on multiple fasta sequences in the same directory:

for i in 'ls *.trimmed.fastq'; do bowtie ../Databases/steamMir21 -f input_file.trimmed.fasta -p 4 -v 3 -a --strata --best --tryhard --chunkmbs 256 -S --al ../MapPrecursors/$i.steam21.fastq > ../SAMPrecursors/$i.steam21.sam; done

Here, -p indicates the number of threads to run bowtie; -v is the bowtie alignment mode followed by the number of mismatches, in this case mismatches3 mismatches are allowed (to perform a stringent analysis, put 0 to 1 mismatches, change -v 3 by -v 0); last flags are allow bowtie to obtain the best alignments for each read (*see* bowtie manual, Subheading 2.2.2, **item 5**). Finally, to control the output of bowtie use the flag --al (directory where the mapped sequences are redirected in this case to MapPrecursors) and -S (to return the mapping file in .sam format instead of .map format). Symbol > is to redirect the output to a specific folder (in this case SAMPrecursors/). SAM format is easy to transform to BAM format.

3.6 miRNA and isomIR Identification Using Isomirage

1. Download and install the Isomirage. Go to Isomirage [11] web page, download section (second tab), and download "Custom libraries mouse" (is a zipped file) in the Database folder and "Executable" (is a jar file) in the Software folder (*see* **Note 6**).
2. Unzip the Custom libraries; make a bowtie database with bowtie-build and run the alignment. To set up the database of Isomirage type in the terminal, in the documents folder:
   ```
   cd Database/
   unzip mmu21t.zip
   ```
3. Decompression of zip generate two files (you can type "ls" in the terminal to list them), one is a genome file that is needed for running Isomirage and the second is a fasta file with sequences of different miRNA forms. The next step is to generate a bowtie index database and map the fastq reads to such database for Isomirage analysis. In the Database folder type in the terminal:
   ```
   Bowtie-build mmu21t.fasta mmu21t
   cd ..
   mkdir IsomirageAnalysis/
   cd MapPrecursors/
   bowtie ../Databases/steamMir21 in-
   put_file.trimmed.steam21.fastq -p 4 -v 0
   --best --tryhard --chunkmbs 256 > ../
   IsomirageAnalysis/input_file.steam21.map
   ```

4. For running multiple fasta sequences in the same directory:

```
for i in 'ls *. trimmed.steam21.fastq';
do bowtie ../Databases/steamMir21 $i -p
  4 -v 0 --best --tryhard --chunkmbs 256 >
  ../IsomirageAnalysis/$i.steam21.map; done
```

5. Description of the bowtie flags is detailed in Subheading 3.6, **step 2**. After the mapping process, a map file for each fastq file will be generated. When bowtie finishes, a report in the terminal indicating the total sequences, mapped sequences, not mapped sequences, and number of alignments will be reported.
6. Isomirage runs in Java, and has a user-friendly interface. To run Isomirage, type in the terminal in the Documents/folder:

```
cd Software/
java -jar Isomirage.jar
```

To start the analysis, go to Files in the menu, and select a .map file (in the Documents/IsomirageAnalysis folder) using the option *Open bowtie map*. It is possible to include different files of mapping with the option "*Add more map files*" from the Files menu. Finally, select a genome file for the analysis in File menu and "*Choose genome.*" In this case is a mmu21t.genome file in the Documents/Database/. Start the analysis using "*Run*" option in the analysis section. During analysis, Isomirage indicates the different phases of the analysis in the progress window.

7. Isomirage generates different output files (in this case located at Documents/IsomirageAnalysis folder) in .txt format (it is possible to open the files using any spreadsheet software). Two types of reports are generated based in the first column: one called _data.txt (name of the miRNA, trimmed variants and sequence) and the other called _summary.txt (containing names of the miRNA and trimmed variants) and three types of normalization methods, raw counts, Robust Multi-array Average (RMA) counts and Reads Per Kilobase Million (RPKM) normalized counts. Isomirage always produces six output files independently of the number of samples analyzed. If different samples are processed at the same time, Isomirage output is consistent in the six files in which, each column represents a value for each file. Header of column is renamed with the name of the map file.
8. Generate a list of raw counts only containing the canonical miRNA sequences using: input_file.trimmed.steam21_linked.txt_raw_summary.txt using the next bash scripts, by typing in the terminal:

```
grep "miRNA" input_file.trimmed.steam21_
linked.txt_raw_summary.txt | awk -F'_'
'BEGIN{print "miRNA""\t""Counts"} {print
$1""$3}' | sed 's/t//g' > input_file.
trimmed.steam21_raw_summary.CANONICAL.txt
```

9. Obtain the list of miRNA and isomiR variants by

```
awk -F '\t' '$2 >=1 {print $1"\"$2}'
input_file.trimmed.steam21_linked.
txt_raw_summary.txt| awk -F'_'
'{print $1"\t"$2"\t"$3"\t"$4}' > in-
put_file.trimmed.steam21_raw_summary.
FILTERCOUNTS.csv
```

This script transforms the output of Isomirage to tabular format: miRNA names, and modifications of sequences. To change the threshold of counts to consider any sequence, modify the value of >=1. For example for 5 counts >=5.

10. To classify the different isomiRs, apply

```
awk -F '\t' '$2 >=1 {print $1}' in-
put_file.trimmed.steam21_linked.txt_
raw_summary.txt t | awk -F'_' '{print
$2"_"$3}'| sort | uniq -c -d > in-
put_file.trimmed.steam21_raw_summary.
NUMBEROFMODIFICATIONS.csv
```

3.7 Differential Expression Analysis Using DESeq Package

For differential expression analysis we recommend processing all samples at the same time. When processing different samples at once using Isomirage, inputs_raw_summary.txt file is ready to run DESeq on it. We performed the analysis with two files, input_file1 and input_file2. We use for differential expression analysis the file: input_files.trimmed.steam21.map_linked.txt_raw_summary.txt

1. For differential expression analysis of two libraries analyzed with Isomirage, in the folder containing the input_file.trimmed.steam21.map_linked.txt_raw_summary.txt, type in the terminal:

```
R
```

Alternatively, run R in Rstudio (*see* **Note 2**).

2. In the R terminal or in the Rstudio script screen:

```
library("DESeq")
```

Call DESeq and the associated packages.

```
countsTable <- read.delim("input_file.
trimmed.steam21.map_raw_summary.txt")
rs <- rowSums(countsTable)
use <- (rs > 1)
countTableFilt <- countsTable[use,]
```

These codes are referred to input counts table in R (in this case the input table is: input_file.trimmed.steam21.map_raw_summary.txt), and perform the filter threshold as one count.

3. Raw counts from the different sequencing libraries are normalized adjusting for different library sizes among the treatments (*see* **Note 7**):

```
conds <- factor(c("input_file1","input_
file1"))
```

```
cds <- newCountDataSet(countTableFilt,
conds)
cds <- estimateSizeFactors(cds)
write.table(counts(cds,normalized=TRUE
), file="input_file.trimmed.steam21.map_
NORCOUNTS_summary.txt", sep="\t")
```

Normalized counts are generated in txt file called: input_file.trimmed.steam21.map_NORCOUNTS_summary.txt, that can be open using any spreadsheet software.

4. For differential expression analysis type:

```
cds<-estimateDispersions(cds,method="blind",
fitType="local", sharingMode="fit-only")
res <- nbinomTest(cds, "input_file1",
"input_file2")
write.table(res, file="input_file1vsinput_
file2.txt", sep="\t")
```

In these cases do not use the *p*-values or *q*-values to filter the genes because no replicates are present. The differential expression is contained in:

input_file1vsinput_file2.txt.

3.8 Uploading the NGS Data to GEO-NCBI

We recommend the submission of sncRNA-seq data to the Gene Expression Omnibus (GEO) genomics data repository. Submission of sncRNA transcriptomic data in GEO facilitates a GEO accession number(s), which is necessary for the most publication processes. For submission of samples it is necessary to collect all the information concerning the experiment and individual for complete Metadata worksheet, a normalization data (such as generated in Subheading 3.8) needed for Matrix worksheet and reads generated by sequencing platform in fastq format.

A nice and user-friendly manual of submission is available in the NCBI portal:

http://www.ncbi.nlm.nih.gov/geo/info/seq.html

4 Notes

1. Ubuntu operating system can be downloaded free from: http://www.ubuntu.com/download/desktop. We recommend downloading the LTS version.
2. R installation includes a step by step on the web. Alternatively, R studio can be used that contains graphical environment for R making designed to be as user-friendly environment. https://www.rstudio.com/.

3. Consider that total sperm profile obtained by Bioanalyzer present in spermatozoa does not show any profile as is detected in any cell type. It is known that in spermatozoa the 18S and 28S RNAs almost disappear in the sample but the amount of small RNAs is proportionally higher (Fig. 2).
4. For RNA-Seq sequencing different Massive Parallel Sequencing platforms are available (for example IIlumina sequencing by synthesis and Ion Torrent semiconductor sequencing platforms). NGS platforms have specific protocol for small RNA-Seq libraries construction and sequencing. Due to the peculiar characteristics of small RNA-seq libraries [12], we recommend performing the sequencing to get around ten millions of reads per library in single end sequencing assay (sequencing involves sequencing RNA from only one end) and read length of 50 cycles. Output format for bioinformatics protocol is FASTQ.
5. Go to miRBase web page, browse section and selected *Mus musculus*. A web-page with all pre- and mature miRNA forms is loaded. Go to the final of the web page, and select in dropped menus: *Steam-loop sequences*, *Unaligned fasta format*, *Select all button*, and finally *Fetch sequence* button. A txt web page will be loaded. Simply download the web page using the save button of your browser. We recommend rename the file as steam-loop_mmu_21.RNA.fasta, fasta extension is important for different phases of these protocols. Alternatively, it is possible to download the mature microRNA sequence by selecting "Mature sequence" in the drop menu.
6. Isomirage needs the Java Virtual Machine working in the system. In Ubuntu to check the Java Virtual Machine, type in the terminal:

```
java -version
```

If Java was installed, three lines will be shown similar to:

```
java version "1.8.0_77"
Java(TM) SE Runtime Environment (build
1.8.0_77-b03)
Java HotSpot(TM) 64-Bit Server VM (build
25.77-b03, mixed mode)
```

To install Java manually (http://askubuntu.com/questions/521145/how-to-install-oracle-java-on-ubuntu-14-04)

7. We recommend using the *cds* values obtained from initial trimmed libraries. For this purpose, it is possible to obtain the read values of all sequencing libraries. When obtaining the values of cds from the total library, type sizeFactors(cds) to obtain the values. These values can be used by typing in R:

cds$sizeFactors <- cds values.

Acknowledgments

We thank D. Fernández-Pérez for critical review of the manuscript. The research in JdM's laboratory was supported by a grant from MINECO (BFU2013-42164-R), Spain.

References

1. Schultz RM (1993) Regulation of zygotic gene activation in the mouse. BioEssays 15(8):531–538
2. Pauli A, Rinn JL, Schier AF (2011) Non-coding RNAs as regulators of embryogenesis. Nature Rev 12(2):136–149
3. Yuan S, Schuster A, Tang C, Yu T, Ortogero N, Bao J, Zheng H, Yan W (2016) Sperm-borne miRNAs and endo-siRNAs are important for fertilization and preimplantation embryonic development. Development 143(4):635–647
4. García-López J, Brieño-Enríquez MA, del Mazo J (2013) MicroRNA biogenesis and variability. Biomol Concepts 4(4):367–380
5. Tang F, Kaneda M, O'Carroll D, Hajkova P, Barton SC, Sun YA, Lee C, Tarakhovsky A, Lao K, Surani MA (2007) Maternal microRNAs are essential for mouse zygotic development. Genes Dev 21(6):644–648
6. Giraldez AJ (2010) microRNAs, the cell's nepenthe: clearing the past during the maternal-to-zygotic transition and cellular reprogramming. Curr Opin Genet Dev 20(4):369–375
7. Giraldez AJ, Mishima Y, Rihel J, Grocock RJ, Van Dongen S, Inoue K, Enright AJ, Schier AF (2006) Zebrafish MiR-430 promotes deadenylation and clearance of maternal mRNAs. Science 312(5770):75–79
8. García-López J, Hourcade Jde D, Alonso L, Cárdenas DB, del Mazo J (2014) Global characterization and target identification of piRNAs and endo-siRNAs in mouse gametes and zygotes. Biochim Biophys Acta 1839(6):463–475
9. García-López J, Alonso L, Cárdenas DB, Artaza-Alvarez H, Hourcade JD, Martínez S, Brieño-Enriquez MA, del Mazo J (2015) Diversity and functional convergence of small noncoding RNAs in male germ cell differentiation and fertilization. RNA 21(5):946–962
10. Gou LT, Dai P, Yang JH, Xue Y, Hu YP, Zhou Y, Kang JY, Wang X, Li H, Hua MM, Zhao S, Hu SD, Wu LG, Shi HJ, Li Y, Fu XD, Qu LH, Wang ED, Liu MF (2014) Pachytene piRNAs instruct massive mRNA elimination during late spermiogenesis. Cell Res 24(6):680–700
11. Muller H, Marzi MJ, Nicassio F (2014) IsomiRage: from functional classification to differential expression of miRNA isoforms. Front Bioeng Biotechnol 2:38
12. Conesa A, Madrigal P, Tarazona S, Gomez-Cabrero D, Cervera A, McPherson A, Szcześniak MW, Gaffney DJ, Elo LL, Zhang X, Mortazavi A (2016) A survey of best practices for RNA-seq data analysis. Genome Biol 17(1):13

Chapter 8

Purification of Zygotically Transcribed RNA through Metabolic Labeling of Early Zebrafish Embryos

Patricia Heyn and Karla M. Neugebauer

Abstract

Early embryonic development in all known metazoans is characterized by a transcriptionally silent phase, during which development is under control of maternally loaded protein and RNA. The zygotic genome becomes transcriptionally active after a series of rapid reductive cleavage divisions. In this chapter, we present a method to metabolically label, purify, and analyze newly transcribed RNAs in early zebrafish embryos. We previously used this method, which is adaptable to other embryos and systems, to determine the onset of zygotic transcription activation and identify the first zygotic transcripts.

Key words 4-sUTP, Biotinylation, Zebrafish, RNA, Transcription, Zygotic genome activation

1 Introduction

The early embryonic cell cycles of zebrafish (*Danio rerio*) are very short (15 min) and synchronous, consisting only of S and M phases and are accompanied by transcriptional silence of the zygotic genome [1]. Zygotic genome activation (ZGA) takes place in a gradual manner with the first transcripts being detected at the 64-cell stage and additional transcriptionally active genes at subsequent cell cycles [2, 3]. An exception to this rule is the mitochondrial genome, which is active from fertilization onward. Identification and characterization of the first zygotic transcripts in wild-type and mutant zebrafish embryos is an excellent strategy for determining the principles that govern early development. The precise knowledge of the earliest transcriptional active loci permits characterization of sequence composition, chromatin characteristics, transcription factor, and replication complex binding profiles at these loci before and after ZGA, crucial to understand how the genome acquires transcriptional competence [4–6].

Identification of the first nuclear zygotic transcripts is challenging as large amounts of maternal RNA are loaded into the egg and persist until after ZGA to support the first zygotic cell cycles.

Kiho Lee (ed.), *Zygotic Genome Activation: Methods and Protocols*, Methods in Molecular Biology, vol. 1605, DOI 10.1007/978-1-4939-6988-3_8, © Springer Science+Business Media LLC 2017

However, maternally loaded RNAs are also regulated by complex post-transcriptional mechanisms such as progressive cytoplasmic polyadenylation and maternal RNA clearance [7–11]. Therefore, simple subtractive comparisons of transcript levels determined by microarray analysis or RNA-Seq are not suitable to identify the first zygotic transcripts. Additionally, upon onset of transcription the amount of newly transcribed RNA is minuscule, and genes encoding maternally loaded RNAs can also be zygotically transcribed; subtractive strategies can therefore miss these transcripts. Hence, identification of zygotic transcripts produced during ZGA requires specific methods that distinguish newly transcribed from maternally transcribed genes. Recently, metabolic labeling has become an important tool to analyze different aspects of transcription and RNA processing [12–20]. Typically, nucleotide analogs, such as 4-thiouracil or 4-thiouridine precursors, are fed to tissue culture cells or yeast, taken up, converted to 4-sUTP, and incorporated through transcription.

Here, we provide a protocol that employs metabolic labeling of RNA with the nucleotide analog 4-thio-UTP (4-sUTP), for the purpose of isolating newly transcribed RNA in early zebrafish embryos (Fig. 1). Injecting the triphosphate 4-sUTP into the 1-cell embryo allows for immediate incorporation of the analog into RNA without relying on further metabolic conversion. Biotinylation of incorporated 4-sU provides stringent positive selection of labeled RNA by binding to streptavidin, which is superior to immunoprecipitation of, for example, BrU-labeled RNA due to the high affinity between biotin and streptavidin [21]. This method should be easily adaptable to embryos from other organisms, which can be microinjected with 4-sUTP at fertilization.

2 Materials

2.1 4-sUTP Microinjection

1. E3 embryo water: 5 mM NaCl, 0.17 mM KCl, 0.33 mM $CaCl_2$, 0.33 mM $MgSO_4$, 10^{-5}% Methyl blue in ddH_2O.
2. Glass capillaries (e.g., 1.0 mm O.D. × 0.58 mm I.D capillaries, Harvard Apparatus).
3. Needle puller (e.g., P-97 Flaming/Brown Micropipette Puller, Sutter instrument).
4. Microinjection system (e.g., from World Precision Instruments).
5. Stereomicroscope.
6. 4-sUTP (TriLink Biotechnologies) 50 mM solution in 10 mM Tris–HCl pH 7.4 (aliquot and keep at −20 °C).
7. 6-cm petri dishes.

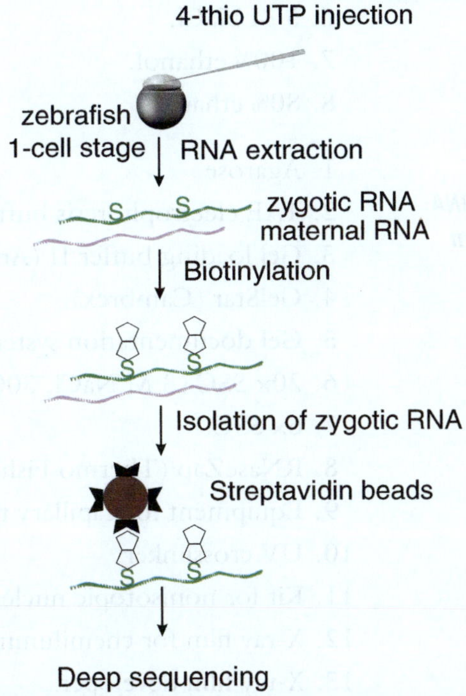

Fig. 1 Workflow. Embryos are microinjected with 4-sUTP at the 1-cell stage and grown to the desired developmental stage. Total RNA is extracted and biotinylated in a thiol-specific manner. Labeled, newly transcribed RNA with incorporated 4-sU is isolated via the biotin on magnetic streptavidin beads. Isolated RNA is used for downstream applications

2.2 Collection of Microinjected Embryos and RNA Extraction

1. 1.5 ml tubes.
2. 4 °C centrifuge.
3. H_2O (PCR-grade).
4. Trizol.
5. 22-guage needle.
6. 80% ethanol.
7. DEPC-H_2O/RNAse-free H_2O (Ambion).
8. DNase (preferably TURBO DNase from Thermo Fisher Scientific).
9. Nanodrop instrument.

2.3 Biotinylation of 4-sUTP Labeled RNA

1. DEPC-H_2O (Ambion).
2. 1 M Tris–HCl pH 7.4.
3. 0.5 M EDTA.
4. EZ-Link HPDP-Biotin (Thermo Fisher Scientific).
5. DMF.

6. 5 M NaCl.
7. 100% ethanol.
8. 80% ethanol.

2.4 Detection of Biotinylated RNA by North-Western

1. Agarose.
2. TAE electrophoresis buffer.
3. Gel loading buffer II (Ambion, Life Technologies).
4. GelStar (Cambrex).
5. Gel documentation system.
6. 20× SSC (3 M NaCl, 300 mM Trisodium Citrate pH 7).
7. 6× SSC.
8. RNaseZap (Thermo Fisher Scientific).
9. Equipment for capillary transfer.
10. UV crosslinker.
11. Kit for nonisotopic nucleic acid detection (for biotin).
12. X-ray film for chemiluminescent detection.
13. X-ray film developer.

2.5 Isolation of 4-sUTP Labeled and Biotinylated RNA

1. MyOne Streptavidin C1 beads (Thermo Fisher Scientific).
2. Magnet with sample rack suitable for 1.5 ml tubes.
3. 2× BWT buffer (10 mM Tris–HCl (pH 7.5), 1 mM EDTA, 2 M NaCl, 0.05% Tween20).
4. Thermomixer.
5. 4 °C Centrifuge.
6. 5% β-mercaptoethanol.
7. 5 M NaCl.
8. Glycogen (20 μg/μl).
9. 100% ethanol.
10. 80% ethanol.
11. DEPC-H_2O (Ambion).
12. Siliconized 1.5 ml tubes.

2.6 Quantitative RT-PCR and RNA-Seq

2.6.1 Quantitative RT-PCR

1. SuperScript® III Reverse Transcriptase (Thermo Fisher Scientific).
2. Real-Time PCR instrument.
3. ABsolute qPCR SYBR Green Mix (Thermo Fisher Scientific).
4. PCR primers to quantify RNA of interest.

2.6.2 RNA-Seq

1. Ovation® One-Direct System (*see* **Note 1**).
2. MinElute PCR Purification Kit (QIAGEN).
3. DEPC-H_2O (Ambion).

3 Methods

3.1 4-sUTP Microinjection into 1-Cell Stage Zebrafish Embryos

1. Set up several breeding tanks on the day before you want embryos.
2. Pull several glass needles suitable for microinjection. Load a glass needle for microinjection with 4-sUTP solution (50 mM). Load an additional spare glass needle in case the first needle breaks or clocks (*see* **Note 2**).
3. Load the needle in the microinjector and adjust the injection volume to 1 nl in a droplet of mineral oil with the help of a micrometer stage. Leave the tip of the needle inside the mineral oil until you proceed with the injections, to avoid drying and clocking of the needle while you collect the embryos.
4. Allow the breeding pair to mate and collect fertilized eggs from timed matings. The fertilized eggs should still be at 1-cell stage for injection (*see* **Note 3**).
5. Inject about 200 embryos into the middle of the yolk with 1 nl of 4-sUTP.
6. Distribute the 200 embryos in several smaller dishes with embryo E3 water and incubate at 28 °C until they reached the developmental stage of interest. Several dishes allow collection of multiple stages per day. In addition, prepare several dishes with uninjected embryos as control samples.

3.2 Collection of Microinjected Embryos and RNA Extraction

1. Remove one dish at a time from the incubator and carefully stage individual embryos under a stereomicroscope. Collect about 25 zebrafish embryos at a time per developmental stage into a 1.5 ml tube to ensure all embryos are correctly staged. Depending on the developmental stage studied, pooling of several collections might be necessary to obtain enough 4-sUTP labeled RNA for subsequent analysis steps. For blastula stages 80–100 zebrafish embryos will be sufficient. Collect both 4-sUTP injected and uninjected control embryos (*see* **Note 4**).
2. Remove as much E3 media as possible and quickly wash embryos three times with 1 ml of H_2O (PCR-grade). Add 500 μl Trizol per 100 embryos (blastula stage). Homogenize the embryos in Trizol by sucking multiple times through a 22-G needle. Make sure everything is homogenized. Embryos homogenized in Trizol can be frozen at −80 °C and stored for several weeks.
3. If samples were frozen, let them thaw and then incubate for 5 min at room temperature.
4. Follow Trizol manufacturer's instructions to extract RNA. Use 80% ethanol for wash steps to ensure that small RNAs are not lost. Remove as much ethanol as possible and air dry. Take care not to over dry the pellet.

5. Solve pellet in DEPC-H_2O--amount depends on method for DNase treatment.
6. Treat extracted RNA with a DNase of choice. Follow manufacturer's instructions. We prefer TURBO DNase digestion with subsequent Phenol-Chloroform extraction and ethanol precipitation (use 80% ethanol for wash steps). Solve DNase-treated RNA pellet in 26 μl DEPC-H_2O and use 1 μl for RNA concentration determination on a nanodrop instrument.

3.3 Biotinylation of 4-sUTP Labeled RNA

1. Add the following reagents in a microcentrifuge tube (*see* **Note 5**).

   DEPC-H_2O to a final volume of 250 μl.

   2.5 μl 1 M TRIS–HCl pH 7.4.

   0.5 μl 0.5 M EDTA.

   RNA (25 μl or volume according to μg desired max 1 μg/μl concentration).

   Mix well before adding the HPDP.

   Add 50 μl 1 mg/ml HPDP in DMF (freshly prepared).
2. Incubate the reaction mix for 3 h at room temperature in the dark.
3. Add 25 μl 5 M NaCl and 750 μl 100% ethanol and mix well. Leave the samples at −20 °C overnight to precipitate the biotinylated RNA.
4. Centrifuge at 20,000 × *g* for 30 min at 4 °C. Carefully remove the supernatant and keep the pellet.
5. Wash the pellet by adding 1 ml 80% ethanol and mix with a quick vortex. Make sure the pellet is dislodged from the tube wall to ensure the pellet is properly washed. Centrifuge again at 20,000 × *g* for 20 min at 4 °C.
6. Remove as much ethanol as possible and air dry. Take care to not over-dry.
7. Dissolve the pellet in DEPC-H_2O. For isolation of biotinylated RNA with streptavidin-beads (Subheading 3.5) use 50 μl of DEPC-H_2O. For the detection of bulk 4-sUTP labeled RNA by blotting onto a membrane (Subheading 3.4), dissolve the pellet in a small amount of DEPC-H_2O. The volume should be small enough to ensure the sample can easily be loaded on an agarose gel. Take a small aliquot to determine the RNA concentration on a nanodrop instrument (dilute in H_2O to obtain a volume that can be measured on a nanodrop).
8. Since the establishment of this protocol an alternative chemistry for biotinylation of 4-sU labeled RNA has been shown to be significantly more efficient, ensuring a higher percentage of biotinylated U's and therefore the better recovery of short transcripts [23].

3.4 Detection of Biotinylated RNA by North-Western Blots

1. Prepare a 1% agarose gel with 1xTAE electrophoresis buffer and 10 μl GelStar in 50 ml agarose. Equilibrate and run the gel in a cold room at 4 °C to avoid overheating.
2. Mix a desired amount of RNA with an equal amount of gel-loading buffer II (Ambion, Life Technologies). Keep the amount of RNA-loaded constant between samples to be able to compare the amount of labeled and therefore newly transcribed RNA between samples. Load the samples on the 1% agarose gel, include a RNA ladder (optionally include a biotinylated marker or positive control, *see* **Note 6**) and run the gel for 20–30 min at 14 V/cm at 4 °C.
3. Document the RNA gel with a gel documentation system. Transfer the RNA by overnight capillary transfer [22] onto a nylon membrane. Use 20× SSC as a transfer buffer. Clean the transfer tray with RNaseZap before you start. Following the transfer give the membrane a quick wash in 6× SSC buffer and crosslink the RNA to the membrane by UV exposure (auto crosslinking function, UV Stratalinker 2400, Stratagene). Detection of biotinylated RNA can be accomplished with commercially available kits for nonisotopic nucleic acid detection (Fig. 2). Alternatively, any Northern blot protocol, using biotinylated probes, can be adapted by omitting the probe hybridization step, as the RNA is already biotinylated.

3.5 Streptavidin-Bead Isolation of 4-sUTP Labeled and Biotinylated RNA

1. Prepare 50 μl MyOne Streptavidin C1 beads per reaction.
2. Wash beads three times with 200 μl 2×BWT buffer per 50 μl beads using a magnet (in 1.5 ml tubes) and finally resuspend beads in 50 μl 2×BWT.
3. Before adding the biotinylated RNA to the streptavidin-beads pellet any excess of biotin in a microcentrifuge (forms white pellet). Carefully take the supernatant (50 μl) and add to the beads. Mix gently by flicking the tube.

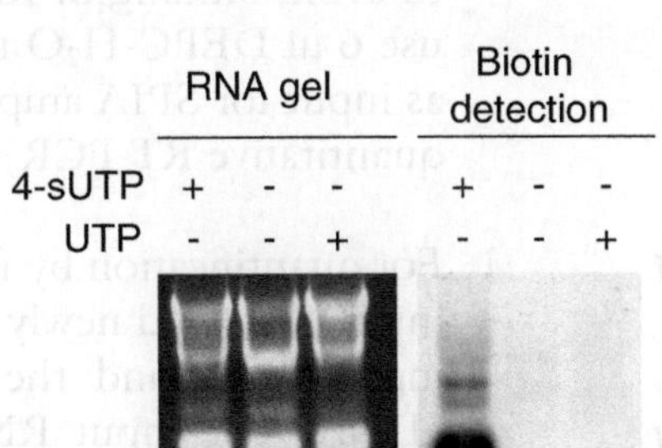

Fig. 2 Detection of biotinylated RNA. Northern blot from 4-sUTP-injected, UTP-injected, or uninjected wild-type control embryos (3 h post-fertilization [hpf]). The panel on the right shows "Northwestern" detection of the biotinylated 4-sU incorporated in the RNA

4. To allow binding of the biotin to streptavidin incubate RNA-bead-mix for 30 min at room temperature on a rotating wheel.
5. Place tubes on a magnet, ensure that all beads are collected at the tube wall before you remove the supernatant. Keep the supernatant and store on ice. This is supernatant 1 and contains any unbound RNA (maternally loaded RNA).
6. Wash beads by adding 100 μl 1×BWT and incubate in a thermomixer at 50 °C and 1400 rpm for 5 min.
7. Place tubes on magnet and remove supernatant.
8. Repeat washing in **steps 6** and 7 three more times.
9. Add 100 μl 1 × BWT and transfer RNA-bead-mix to a new tube and incubate one more time in the thermomixer, place on magnet, and remove supernatant.
10. To elute the RNA from the beads by reducing disulfide-bonds, add 50 μl 5% β-mercaptoethanol to beads (in a hood), mix gently by flicking the tube, then collect solution at the bottom of the tube; ensure that beads are not pelleted. Incubate for 5 min at room temperature. Now place the tube on a magnet and remove the supernatant. This is the eluate 1 containing the isolated RNA. Keep on ice.
11. To ensure all the RNA is eluted add another 50 μl 5% β-mercaptoethanol to the beads, mix as before, and incubate for 10 min at 50 °C (no shaking). Place the tubes on the magnet and remove the supernatant and combine with eluate 1.
12. Add 1/10 volume of 5 M NaCl and 0.5 μl Glycogen (20 μg/μl) and three times volume 100% ethanol to supernatant 1 and eluate 1. Precipitate at −20 °C overnight.
13. Spin at 20,000 × g for 45 min at 4 °C. Remove the supernatant and wash pellet twice with 1 ml of 80% ethanol.
14. Air dry the pellet (take care not to over-dry) and then dissolve pellet in desired volume of DEPC-H_2O (use siliconized tubes to avoid binding of RNA to tube wall). For RNA-seq analysis use 6 μl DEPC-H_2O to dissolve the pellet from eluate 1 (5 μl as input for SPIA amplification and 1 μl for quality control by quantitative RT-PCR).

3.6 Quantification of Isolated RNA by Quantitative RT-PCR or RNA-Seq

1. For quantification by RT-qPCR, continue by reverse transcribing the isolated newly transcribed RNA (eluate 1), 10% of the input RNA, and the unbound RNA (supernatant 1) into cDNA. The input RNA is used to assess the percentage of enrichment over input by quantitative RT-PCR (**step 3** of Subheading 3.6). The unbound RNA sample will reveal the levels of unbound transcripts after isolation compared to the input (either maternally loaded or not efficiently bound to the streptavidin-beads).

2. Prepare a RT-PCR mastermix according to manufacturer's instructions to analyze all cDNA samples in triplicates.
3. To quantify the amount of RNA isolated in comparison to the total input RNA, use the percent input method (%input = $100 * 2^{(Ct_{input} - Ct_{sample})}$ where Ct_{input} is corrected for dilutions used). Analyze the levels of some known zygotic and maternal transcripts to ensure that the isolation was successful (*see* **Note 7**).
4. For RNA-seq analysis of the isolated RNA, amplify and convert the isolated material into cDNA by SPIA-amplification (Ovation® One-Direct System). Amplification is necessary to ensure sufficient material for library preparation and avoid dropouts of low abundant molecules. The kit uses the Ribo-SPIA® technology that is a linear and isothermal amplification system. Follow the manufacturer's instructions to generate amplified cDNA.
5. Purify the resulting material with the MinElute PCR Purification Kit. Use two columns per sample and elute each in 20 μl DEPC-treated water. Combine the eluates afterward.
6. For quality control of the amplification, determine the relative levels of known zygotic and maternal transcripts for each developmental time point before and after amplification by quantitative PCR (*see* **Note 7**).
7. For RNA-seq library construction use a suitable commercial library construction kit for double-stranded cDNA for your deep-sequencing platform of choice.

4 Notes

1. The Ovation® One-Direct System does not allow generating a strand-specific deep sequencing library. In addition, a newer version of this system is now available Ovation® RNA-Seq System V2.
2. Tolerated 4-sUTP injection doses will depend on the species studied. Perform a toxicity curve to determine the optimal concentration of 4-sUTP, which is not adversely affecting development. We noticed also that 4-sUTP from different manufacturers seem to have different toxicity levels.
3. In vitro fertilization allows for even more synchronous samples [3].
4. Embryos might be snap-frozen to ease the collection process.
5. For the isolation of zygotic RNA from zebrafish early blastula stages 10 μg total RNA for the biotinylation step are sufficient. Ensure that an amount equivalent to 10% of the input RNA is kept for enrichment analysis by quantitative RT-PCR (Subheading 3.6).

6. Biotinylated markers can be generated either by 3′end biotinylation or in vitro transcription using Biotin-UTP of different sized fragments.

7. To test by quantitative RT-PCR whether labeling and isolation of the zygotic RNA was successful, known zygotic transcripts such as *klf4* and *vox*, as well as known maternal transcripts, such as *retsatl*, can be analyzed for their enrichment over input. In addition, the relative levels of these transcripts over the developmental time course studied can be assessed before and after SPIA amplification, to ensure that the amplification did not alter relative levels.

References

1. Kane DA, Kimmel CB (1993) The zebrafish midblastula transition. Development 119:447–456
2. Heyn P, Kircher M, Dahl A et al (2014) The earliest transcribed zygotic genes are short, newly evolved, and different across species. Cell Rep 6:285–292. doi:10.1016/j.celrep.2013.12.030
3. Harvey SA, Sealy I, Kettleborough R et al (2013) Identification of the zebrafish maternal and paternal transcriptomes. Development 140:2703–2710. doi:10.1242/dev.095091
4. Lee MT, Bonneau AR, Takacs CM et al (2013) Nanog, Pou5fl and SoxB1 activate zygotic gene expression during the maternal-to-zygotic transition. Nature 503:360–364. doi:10.1038/nature12632
5. Leichsenring M, Maes J, Mössner R et al (2013) Pou5fl transcription factor controls zygotic gene activation in vertebrates. Science 341:1005–1009. doi:10.1126/science.1242527
6. Vastenhouw NL, Zhang Y, Woods IG et al (2010) Chromatin signature of embryonic pluripotency is established during genome activation. Nature 464:922–926. doi:10.1038/nature08866
7. Giraldez AJ, Mishima Y, Rihel J et al (2006) Zebrafish MiR-430 promotes deadenylation and clearance of maternal mRNAs. Science 312:75–79. doi:10.1126/science.1122689
8. Tadros W, Lipshitz HD (2009) The maternal-to-zygotic transition: a play in two acts. Development 136:3033–3042. doi:10.1242/dev.033183
9. Aanes H, Winata CL, Lin CH et al (2011) Zebrafish mRNA sequencing deciphers novelties in transcriptome dynamics during maternal to zygotic transition. Genome Res 21:1328–1338. doi:10.1101/gr.116012.110
10. Lee MT, Bonneau AR, Giraldez AJ (2014) Zygotic genome activation during the maternal-to-zygotic transition. Annu Rev Cell Dev Biol 30:581–613. doi:10.1146/annurev-cellbio-100913-013027
11. Zhang M, Skirkanich J, Lampson M, Klein P (2016) Cell cycle remodeling and zygotic gene activation at the midblastula transition. In: Pelegri FJ, Danilchek M, Sutherland A (eds) Vertebrate development: maternal to zygotic control. Springer Nature, New York, pp 441–487
12. Melvin WT, Milne HB, Slater AA et al (1978) Incorporation of 6-thioguanosine and 4-thiouridine into RNA. Application to isolation of newly synthesised RNA by affinity chromatography. Eur J Biochem 92:373–379
13. Cleary MD (2008) Cell type-specific analysis of mRNA synthesis and decay in vivo with uracil phosphoribosyltransferase and 4-thiouracil. Methods Enzymol 448:379–406. doi:10.1016/S0076-6879(08)02619-0
14. Zeiner GM, Cleary MD, Fouts AE et al (2008) RNA analysis by biosynthetic tagging using 4-thiouracil and uracil phosphoribosyltransferase. Methods Mol Biol 419:135–146. doi:10.1007/978-1-59745-033-1_9

15. Miller C, Schwalb B, Maier K et al (2011) Dynamic transcriptome analysis measures rates of mRNA synthesis and decay in yeast. Mol Syst Biol 7:458. doi:10.1038/msb.2010.112
16. Rabani M, Levin JZ, Fan L et al (2011) Metabolic labeling of RNA uncovers principles of RNA production and degradation dynamics in mammalian cells. Nat Biotechnol 29:436–442. doi:10.1038/nbt.1861
17. Windhager L, Bonfert T, Burger K et al (2012) Ultrashort and progressive 4sU-tagging reveals key characteristics of RNA processing at nucleotide resolution. Genome Res 22:2031–2042. doi:10.1101/gr.131847.111
18. Rädle B, Rutkowski AJ, Ruzsics Z et al (2013) Metabolic labeling of newly transcribed RNA for high resolution gene expression profiling of RNA synthesis, processing and decay in cell culture. J Vis Exp. doi:10.3791/50195
19. Rabani M, Raychowdhury R, Jovanovic M et al (2014) High-resolution sequencing and modeling identifies distinct dynamic RNA regulatory strategies. Cell 159:1698–1710. doi:10.1016/j.cell.2014.11.015
20. Barrass JD, Reid JEA, Huang Y et al (2015) Transcriptome-wide RNA processing kinetics revealed using extremely short 4tU labeling. Genome Biol 16:282. doi:10.1186/s13059-015-0848-1
21. Green NM (1990) Avidin and streptavidin. Methods Enzymol 184:51–67
22. Rio DC (2015) Northern blots: capillary transfer of RNA from agarose gels and filter hybridization using standard stringency conditions. Cold Spring Harb Protoc 2015:306–313. doi:10.1101/pdb.prot081018
23. Duffy EE, Rutenberg-Schoenberg M, Stark CD et al (2015) Tracking distinct RNA populations using efficient and reversible covalent chemistry. Mol Cell 59:858–866. doi:10.1016/j.molcel.2015.07.023

Chapter 9

RNA FISH to Study Zygotic Genome Activation in Early Mouse Embryos

Noémie Ranisavljevic, Ikuhiro Okamoto, Edith Heard, and Katia Ancelin

Abstract

Characterizing the maternal-to-zygotic transition (MZT) is a central question in embryogenesis, and is critical for our understanding of early post-fertilization events in mammals. High-throughput RNA sequencing (RNA Seq) of mouse oocytes and early embryos has recently revealed that elaborate transcription patterns of genes and repeats are established post-fertilization. This occurs in the context of the gradually depleted maternal pool of RNA provided by the oocyte, which can confound the accurate analysis of the zygotic genome activation when the mRNA population is sequenced. In this context, and given the limited amounts of material available from embryos, particularly when studying mutants, as well as the cost of sequencing, an alternative, complementary single cell approach is RNA FISH. This approach can assay the expression of specific genes or genetic elements during preimplantation development, in particular during the MZT. Here, we describe how RNA FISH can be applied to visualize nascent transcription at specific genomic loci in embryos at different stages of preimplantation development and also discuss possible analytical methods of RNA FISH data.

Key words RNA FISH, Nascent transcripts, Mouse preimplantation development, Maternal-to-zygotic transition

1 Introduction

Fertilization of an oocyte by a sperm embarks the two highly differentiated parental genomes into a major epigenetic remodeling process and launches the formation of a totipotent zygote. In animals, oocytes are supplied with proteins and RNAs, designated as the maternal pool that ensures the very first stages of development. Zygotic products gradually replace this preloaded maternal material through transcriptional activation, known as zygotic genome activation (ZGA).

The maternal-to-zygotic transition (MZT) is crucial for developmental progression after fertilization [1] and its characterization is a central question in early embryogenesis. In mammals, this massive shift in the transcriptional program is achieved at different times after

Kiho Lee (ed.), *Zygotic Genome Activation: Methods and Protocols*, Methods in Molecular Biology, vol. 1605,
DOI 10.1007/978-1-4939-6988-3_9,

fertilization, from as early as the 2-cell stage (mouse), or later at the 8-cell stage (cow, human). This has been analyzed by microarray or RNA-seq analyses [2–6]. Notably in mouse embryos, a fine choreography of gene expression patterns is established post-fertilization, between the inherited maternal pool of RNA, the first wave of ZGA (also called minor ZGA), at the 1-cell stage, and then the second wave of ZGA (also called major ZGA), at the 2-cell stage [4].

One challenge for our understanding of the mechanisms underlying this complex series of important events has been to try to define the maternal factors that are pivotal during the MZT. In particular, the factors that unravel the specialized chromatin states of the differentiated gametes to a totipotent zygote, and those that underlie zygotic genome activation still remain largely unknown. Indeed, due to the sparse availability of material for biochemical approaches, the involvement at the molecular level of chromatin factors on gene expression dynamic in vivo during early development is under-investigated.

In recent years, genetic approaches have been developed that can specifically assess the role of maternal factors during female germline formation as well as during preimplantation development [7]. These tools are based on the induction of gene deletion (conditional knock-out) during oocyte growth, which can thus eliminate the RNA and protein products of the gene in question in the oocyte and in the early preimplantation embryo after fertilization. Thanks to such approaches, using conditional knock-out mouse models, several studies have at last started to define the key roles that certain chromatin factors play during early development and in particular in the critical steps of MZT and the first cleavage stages [8–10]. We recently used a maternal deletion of KDM1A (LSD1), a master regulator of histone methylation, to reveal its critical role in establishing the correct chromatin landscape upon fertilization and in initiating new patterns of genome expression in early mouse development [10].

In such a study, the use of a combination of RNA sequencing and RNA FISH was very powerful for assessing the impact of loss of this factor on ZGA in particular. RNA seq was used to analyze steady-state levels of mRNA pools (maternal, 1-cell stage; 2-cell stage), while RNA FISH could assess the de novo (active) transcription of genes and specific repeat elements through the visualization of nascent transcripts, at the single cell level. RNA FISH is also useful at the maternal-to-zygotic transition to discriminate between the absence of degradation of maternal RNAs and aberrant active transcription at later stage. When fluorescently labeled probes (DNA or RNA) are hybridized to fixed and permeabilized embryos that have not been denatured, then single-stranded RNA rather than double-stranded DNA can be detected. This technique can be applied to study the expression of specific genes as well as the repetitive fraction of the genome, such as LINEs or pericentromeric satellite sequences [10–14].

In this chapter, we outline RNA FISH methods applied to preimplantation mouse embryos, which might also be a starting point for optimization for oocytes stages. We present in particular (a) how embryos are fixed, permeabilized, and placed on coverslips; (b) the RNA FISH procedure and imaging using these coverslips. It should be noted that embryos analyzed by RNA FISH can subsequently be analyzed by DNA FISH, following denaturation [11]. The RNA FISH procedure can also be combined with immunofluorescence (for a protocol on fixed cells, *see* [15, 16]).

2 Materials

2.1 Preimplantation Embryo Isolation and Manipulation

1. Petri dish 10 mm × 35, sterile.
2. Tweezers for dissection (fine stainless steel Dumont #5).
3. Scissors (fine, stainless steel, dissection).
4. Dissection binocular with 100× magnification.
5. Pasteur pipette: pulled the thin extremity as shown in Fig. 1; will be referred to as transfer pipette.
6. Aspirator tube assemblies (Fig. 1).
7. M2 medium for embryo (Sigma).
8. Tyrod Acid.

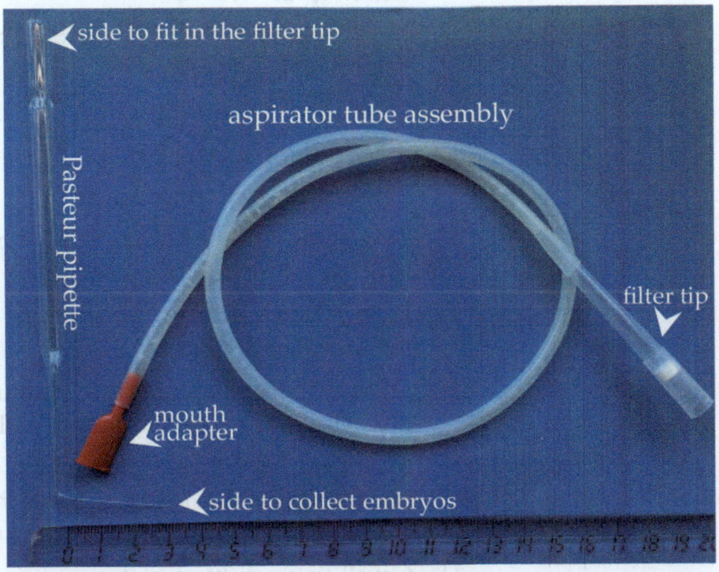

Fig. 1 Equipment setup for the mouth device used for handling by slightly suctions the mouse 2-cell stage embryos. An aspirator tube assembly as shown in the center is connected by the filter tip to a modified Pasteur pipette. The Pasteur pipette (on the *left*) is drawn to a diameter of ~100 μm with an angle to the rest of the pipette. Numbers on the ruler are indicated in centimeter

9. Coverslips (18 mm diameter, 1.5 mm thickness; borosilicate glass; Marienfield).
10. 6-well tissue culture plates.
11. Liquiport Liquid pump.

2.2 RNA FISH

1. Coating solution for coverslip: 3× SSC, 0.2 mg/ml BSA, 0.2 mg/ml Ficoll-400, 0.2 mg/ml Polyvinylpyrrolidone 40 (PVP40).
2. Tri-ethanolamine solution: 0.0025% Glacial acetic acid, 0.013% TE.
3. Methanol/glacial acetic acid (v/v) 3:1.
4. Fixation solution: 3% paraformaldehyde in PBS (freshly prepared, pH adjusted).
5. Permeabilization solution: 0.5% Triton X-100 in 1× PBS supplemented with 2-mM vanadyl ribonucleoside complex (VRC).
6. 70, 80, 95% ethanol in double-processed tissue-culture water; 100% ethanol.
7. Nick translation kit (Abbott) containing nick translation enzyme, dNTP solutions, and 10× nick translation buffer.
8. Green-dUTP or Orange-dUTP (Enzo) or Cyanine 5-dUTP (Roche).
9. Formamide (FA; Sigma). Upon opening, aliquot immediately and keep at −20 °C.
10. Mouse Cot-1 DNA (Life Technologies, ThermoFisher Scientific).
11. DNA, molecular biology grade from fish sperm (Roche).
12. 3 M sodium acetate pH 5.2.
13. 20× SSC buffer concentrate.
14. 2× Hybridization buffer: 20%, dextran sulfate, 2 mg/ml BSA (NEB), 20 mM VRC; 4× SSC.
15. Post-hybridization wash buffer: freshly prepared 50% formamide/2× SSC pH 7.2–7.4.
16. DNA counterstaining solution: 4′,6-diamidino-2-phenylindodihydrochloride (DAPI) in 2× SSC at 1 μg/ml.
17. Mounting solution: 90% v/v glycerol, 0.1% (w/v) *p*-phenylenediamine (Sigma), pH 9 in PBS; or else non-hardening Vectashield (Vector Laboratories) with DAPI.
18. 6-well tissue culture plates.
19. Fine wiper tissues (e.g., Kimwipes).
20. Dark chamber light tight plastic, capable of holding microscope slides (*see* Fig. 2).

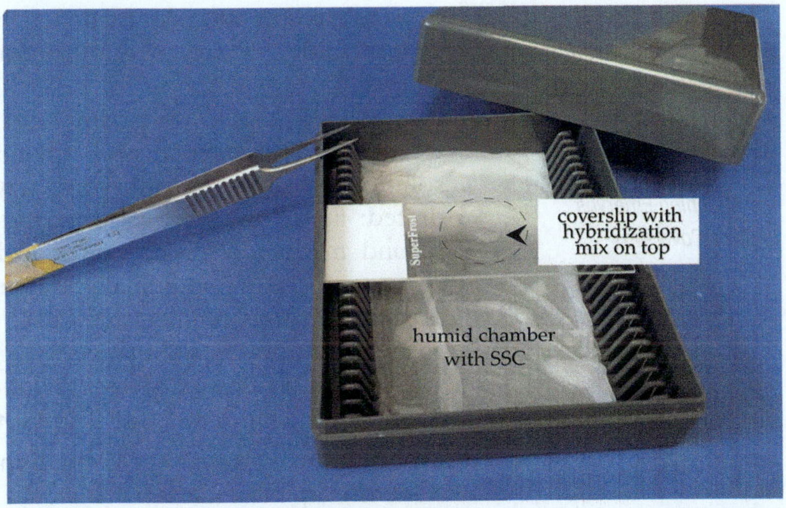

Fig. 2 Equipment setup for the hybridization incubation. A slide holder is used to create a humid chamber where the slides are positioned to hold the coverslips with the attached embryos facing up. Coverslips can be recovered from the slide using fine forceps

21. Microscope slides (always use gloves to handle them out of the package and through the procedure).
22. Nail varnish for sealing.
23. Eppendorf Centrifuge 5417R.
24. Eppendorf Concentrator plus.
25. Eppendorf Thermomixer comfort.
26. Liquiport Liquid pump.
27. Shake'N'Bake Hybridization Oven.

2.3 Microscopy and Image Analysis

1. Inverted confocal microscope (Zeiss LSM700).
2. Plan apo DICII (numerical aperture 1.4) 63× oil immersion objective.
3. ImageJ software.

3 Methods

The following methods described here were adapted from [12, 15]. Our study focused on mouse 2-cell stage embryos [10], but they can be applied to any preimplantation stages from zygote to 3.5 days *post-coitum* (early blastocyst) embryos, provided some adjustments, such as the time of embryo collection, or the permeabilization conditions (see details below). We recommend that RNA FISH is started on the day of embryo recovery, as soon as possible after fixation of embryos on coverslips, to minimize the risk of RNA degradation (*see* **Notes 1–3**). Conditions should be RNAse free throughout.

3.1 Preparation and Fixation of Mouse 2-Cell Stage Embryos on Coverslips

Affixing embryos to coverslip facilitates their subsequent processing through the different steps of our procedure.

3.1.1 Coverslip Coating and Attaching Embryos to Glass Coverslips

1. The coverslips used for preimplantation embryo analyses are treated by three successive incubations in the following solutions to allow the embryo to stay fixed and partially dehydrated: first treat coverlips in coating solution 3 h at 65 °C, second methanol/glacial acetic acid solution for 20 min at room temperature, and finally in tri-ethanolamine solution for 10 min at room temperature. Each coverslip is rinsed twice with distilled water, then twice with 100% ethanol. Coverslips are dried on a Kimwipe at room temperature and stock at room temperature in a closed tube (*see* **Note 4**).
2. Embryos are collected in M2 medium as described [17], using a mouth aspirator device connected to a transfer pipette (*see* Fig. 1). After the removal of the *Zona pellucida* with tyrode acid, embryos are rinsed in the M2 medium.
3. Using a transfer pipette, embryos are quickly moved into a 0.5 μl drop of PBS 1× placed at the center of a coated coverslip (around five embryos per coverslip; see the previous section). Using the transfer pipette, liquid is carefully removed around the embryos. Blastomeres will spread out, but should not be destroyed. Good attachment will occur only if PBS does not contain proteins such as BSA or serum.
4. The coverslips should be dried for 30 min at room temperature, after placing them into a 6-well plate (one per well).

3.1.2 Fixation and Permeabilization

1. In the following steps, we routinely work with the coverslips in the 6-well plate, using 2 ml per well for each solution. Replacement of solutions is done always removing liquid (without touching the coverslip) by rapid aspiration (using Liquiport Liquid pump) and rapidly, but gently replacing the solution (with a 10 ml pipette for example) to prevent the coverslips from drying.
2. Embryos on coverslips are fixed with 3% paraformaldehyde in 1× PBS for 10 min at room temperature.
3. Coverslips are washed twice with 1× PBS.
4. Embryos are permeabilized for 1 min on ice with ice-cold 0.5% Triton X-100, 2 mM vanadyl-ribonucleoside complex in PBS. The length of treatment of the coverslips with the permeabilization solution is critical and is dependent on the developmental stage, e.g., 2 min for zygotes, and 2-cell stage; 8 min for 8-cell stage, 15 min for blastocyst stage (*see* **Note 5**).
5. In order to test the specificity of a probe for nascent transcripts (and not DNA in case of partial recognition with the DNA probes), we recommend incubation of one coverslip with

RNAse A as a control. Incubation can be done in PBS 1× with RNAse A (40 μg/ml final) at 37 °C for 15 min; after the permeabilization and prior to dehydration.

6. The permeabilization solution is replaced by ice-cold 70% ethanol, and the coverslips can then be stored at −20 °C, until the hybridization mix preparation is ready. Ideally, they should be processed within the same day, but can be kept overnight.

3.2 Generating and Labeling Probes for RNA FISH by Nick Translation

The design of the probes needs to cover introns and exons for specific genes to detect the nascent transcripts (as a punctuate signal at the genomic locus). The use of cDNA probes is not recommended as this will largely detect mRNA and not nascent transcripts. Several strategies are currently available to generate FISH probes that enable efficient visualization of nascent transcripts that are based on genomic probes that are several kilobase pairs long, or plasmid probes or else labeled oligos [16, 18–20]. The procedure we describe here relies on generating fluorescently labeled DNA probes by nick translation on BAC or plasmid DNA, using fluorescent dUTPs (Spectrum Green SG, Spectrum Red SR) to label genomic probes (plasmids, fosmids, or BACs). 1–2 μg of DNA is labeled in a 50 μl reaction.

1. 1 μg of BAC or plasmid DNA is mixed with water up to a volume of 17.5 μl. Add 2.5 μl of 0.2 mM SR dUTP or SG-dUTP, 10 μl of 0.1 mM each dNTP mix (dGTP, dATP, dCTP), 5 μl of 0.1 mM dTTP, 5 μl of 10× nick translation buffer, and 10 μl of nick translation enzyme.
2. The enzymatic reaction is incubated for 16 h at 15 °C in the dark (*see* **Notes 6** and 7).
3. The reaction is inactivated by freezing at −20 °C. Probes can be stored for up to 6 months at −20 °C.

3.3 RNA FISH Procedure on 2-Cell Stage Embryos

3.3.1 Probe Mix Preparation

1. 200 ng (per coverslip) of each probe to be used is precipitated by adding 10 μg salmon sperm DNA (as carrier), 1/10 volume 3 M sodium acetate pH 5.2, and 3 volumes of 100% ethanol; and spun at 4 °C, 16,100 × *g* for 30 min. In order to ensure probe specificity, Cot-1 DNA competition is required for most probes based on fosmid or BACs as they can contain repeat sequences that will cross hybridize and increase background unless competed away prior to hybridization. It is often not the case for probes based on plasmid (if they do not contain repetitive elements) and it is also often not the case if the aim is to assess specifically the expression level of repetitive fraction of the genome (e.g., LINEs, SINEs, Satellite sequences). Cot-1 DNA should be added at this stage in the precipitation mix (*see* [16, 21], **Note 8**).
2. The pellet is washed in 70% ethanol and spun down at 16,100 × *g* and 4 °C for 5 min.

3. The pellet is dried for 1 min in a concentrator/speed vacuum.
4. The pellet is resuspended in an appropriate volume of formamide (half the volume required for hybridization if final % is 50% FA). For example, for 5 μl hybridization mix the pellet needs to be dissolved in a 2.5 μl of FA.
5. The tube is placed in a heat block preferably with shaking for at least 15 min at 37 °C to dissolve the pellet in FA.
6. The probe is denatured for 10 min at 75 °C by placing the tube in a thermomixer.
7. After denaturation, the tube is rapidly transferred to ice, or if competition is to be performed with Cot-1 DNA, put directly at 37 °C for at least 1 h in a thermomixer.
8. An equal volume of 2× hybridization buffer (ideally freshly prepared or else a fresh aliquot from −20 °C) is added to the probe mix, according to the amount of FA used in the previous step (*see* **Note 9**).

3.3.2 Hybridization

1. Coverslips with embryos (in a 6-well plate) are dehydrated by sequential incubation in 80, 95, and 100% ethanol (this last step is repeated twice) for 5 min each at room temperature.
2. Coverslips are then air dried by being placed on fine tissues (Kimwipes). Remember the embryo side should always be facing up.
3. During the air-drying step for coverslips, the hybridization chamber is prepared with tissue soaked in FA/2× SSC (*see* Fig. 2). A drop of water is spotted on a clean slide, in order to receive the coverslip (with the *embryo side still facing up*) (*see* **Note 10**).
4. Hybridization: 5 μl probe hybridization mix containing the probe (*see* **step 8** in subheading 3.3.1) is spotted, just on the side of the dried embryos, carefully monitoring this step under the binocular to place it as close as possible of the embryos, avoiding bubbles.
5. The humid chamber containing the slides and coverslips is covered and placed at 37 °C overnight (*see* **Note 11**).

3.3.3 Post-hybridization Washes

1. The following day, 1 ml of pre-warmed, freshly prepared 50%FA/2× SSC is added onto the coverslip on the slide to loosen it, and to allow removing it carefully from the slide. It is then placed (with embryos facing upward), into a 6-well plate containing pre-warmed 50%FA/2× SSC (2 ml/well) at 42 °C for 7 min.
2. Washes in 2× SSC are repeated twice at 42 °C for 5 min (*see* **Note 12**).
3. Nuclei are counterstained by washing in 2× SSC with DAPI (1 μg/ml) for 5 min at 42 °C.

4. Coverslips are then rinsed three times with 2× SSC at room temperature.
5. 10 μl mounting medium is spotted onto a glass slide and the coverslip is placed on the top of the drop with embryos facing down, avoiding bubbles. Excess mounting medium can be carefully removed with a fine tissue, without moving the coverslip. The coverslip is sealed on the slide with a minimal amount of nail polish.
6. Slides should ideally be imaged immediately, as signal can sometimes fade quickly but can also be stored at −20 °C until imaging, or at −80 °C for longer storage.

3.4 Image Acquisition and Analysis

3.4.1 Microscopy

In the examples we show in Fig. 3, we imaged the embryos following RNA FISH with an inverted confocal microscope Zeiss (Germany) LSM700 with a Plan apo DICII (numerical aperture 1.4) 63× oil objective, and Z sections were taken every 0.4 μm. It is critical to have used the correct coverslip thickness and the correct oil adapted to the lens to obtain correct images.

3.4.2 Analysis

Depending on the target for the hybridization (i.e., a specific gene or else specific repetitive elements), the pattern of hybridization will be different, as displayed in Fig. 3b. For a unique gene (Fig. 3a), one can expect two pinpoints for each allele of an autosomal gene in a diploid cell (if the two alleles are both expressed). For X chromosome one or two alleles can be detected depending on sex or also the expression status of the X chromosome [12]. When testing a new probe it is important to compare it to a previously tested probe, and if possible to test it on cultured cells [16, 21] in which the gene is known to be expressed to ensure the good detection of the nascent transcript. Finally, a critical point in RNA FISH is that, in case of suboptimal RNA FISH conditions or probes, fewer cells with a signal may be detected, leading to an underestimation of the degree to which a gene is expressed. In the case of mono versus biallelic expression, this can be critical as one can conclude monoallelic expression when in fact it is biallelic.

After fertilization, embryos from a same female might not be at the same stage of the cell cycle. Depending on whether they have passed S phase or not, and the degree to which the sister chromatids are separated, the signal corresponding to the nascent transcript can be seen as a singlet or a doublet (*see* Fig. 3a lower signal) corresponding to transcription of well-separated sister chromatids.

In Fig. 3b an example of hybridization following hybridization to repetitive elements such as LINE-1 (a retrotransposon family) is shown. In this case, the pattern is more complex as it is associated with many different pinpoints of different intensities reflecting the many sites of ongoing transcription. A treatment with the RNAse A,

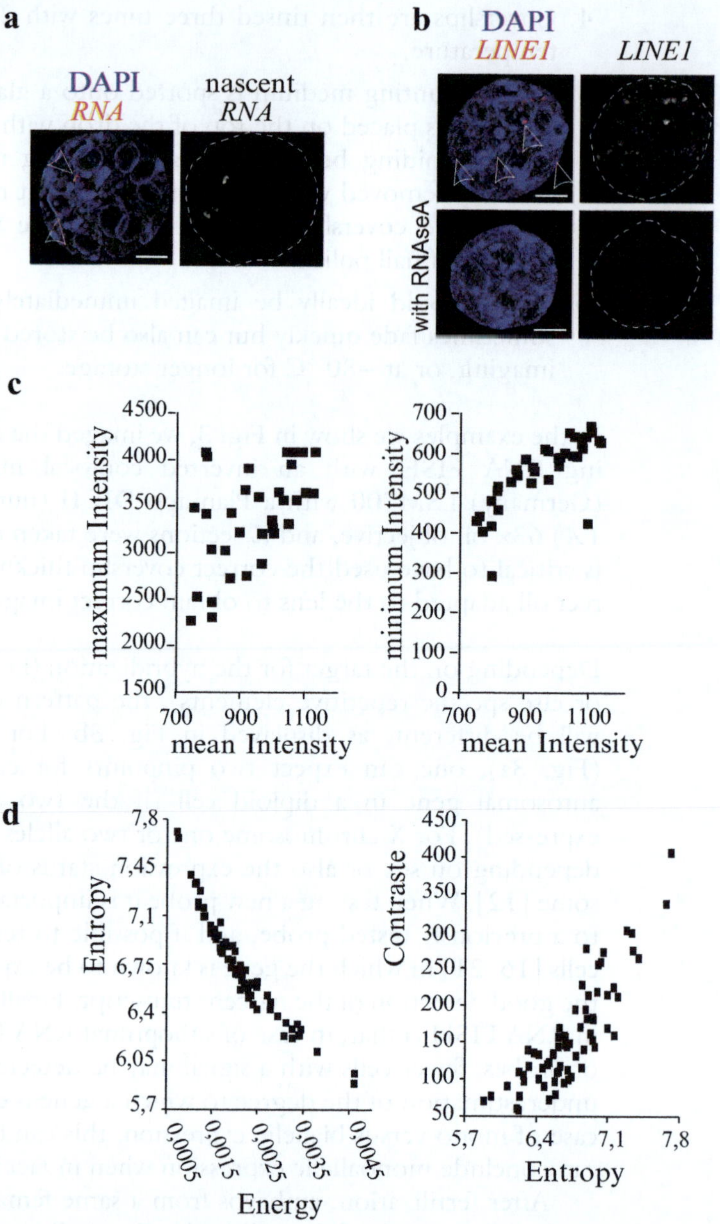

Fig. 3 Examples of RNA FISH and analysis performed on mouse 2-cell stage embryos (reproduced by permission of eLife Sciences Publications). (**a**) A nucleus showing the DNA probe hybridized (in *red*) to a single loci on the X chromosome (from a female embryo). The nascent transcripts arising from these two spots can be seen. (**b**) A nucleus showing plasmid probes hybridizing to repetitive elements *LINE1* transcripts (*top*) and a nucleus after RNAse A treatment using the same probe (*bottom*). The DNA is counterstained with DAPI. Scale is 10 μm. (**c**) Scatter plots showing the distribution according to the fluorescence intensity measurement of nuclei as the one shown in (**b**). Criteria such as maximum or minimum intensity values can be used to discriminate the hybridization results from one nucleus to another (or from one embryo to another). (**d**) Scatter plots showing analysis of hybridization results according to the pattern/texture of the obtained image such as in (**b**). Examples in (**c**) and (**d**) indicate means to analyze complex patterns of RNA FISH results such as in (**b**), rather than in (**a**)

prior to the hybridization, attests for the probe specificity and the RNA nature of the signal. When such complex patterns are to be analyzed, one alternative is to study fluorescent intensities within each nucleus of every embryo as displayed in Fig. 3c or else the fluorescence arrangement as shown in Fig. 3d.

1. Images are opened in Image J software.
2. For fluorescence intensity measurements, the 3D object counter plugin can be used [22].
3. For fluorescence arrangement analysis, we used a custom-made ImageJ macro [10] based on descriptors defined as [23] that quantitatively study the texture and structure of images.

4 Notes

1. Wear gloves throughout the whole procedure, from drying the embryos onto coverslips up to the hybridization step of the RNA FISH procedure. This is critical to minimize risk of RNases contamination.
2. Use clean sterile forceps for handling coverslips, and process them either in Petri dishes or in 6-well plates.
3. All solutions are prepared with double-processed (i.e., distilled and sterile) tissue culture water in RNase-free conditions (although no specific treatment to eliminate RNAses needs to be used).
4. Prepare the coating solution for coverslips receiving embryos the night before the preparation of the coverslips or autoclave it and store at 4 °C for 6 months. Always transfer coverslips one by one from one solution to another. Use 50 ml Falcon tubes for each solution. Coverslips can be stored at room temperature up to 4 months.
5. Permeabilization solution must be prepared fresh every time and kept on ice before collecting the preimplantation embryos.
6. Following nick translation reaction, the size range of the labeled DNA is always checked by electrophoresis on an agarose gel (1.8%). The optimal range size of a FISH probe is between 50 and 300 base pairs, short enough to enter the nucleus and long enough to be specific. Larger size fragments will not be able to enter the nucleus efficiently. Note that the efficiency of incorporation is specific for each nucleotide conjugated to a fluorophore and needs to be tested.
7. Test run should be performed on culture cells that are known to express the gene targeted for RNA FISH to test the quality of new probes. Ideally, one should always test new sets of probes along with probes for which the hybridization pattern is known.

8. Cotl DNA: competition is required for most probes as they can contain repeat sequences that will cross hybridize and increase background, unless competed away prior to hybridization. We routinely use between 2 and 5 μg of Cotl DNA per 5 μl of nick translation mix/per probe type.
9. We recommend preparation of small aliquots of hybridization buffer (about 50 μl and kept at −20 °C), and using each of them two to three times only. A final volume of 5 μl hybridization mix/coverslip is ideal for overnight incubation.
10. The chamber, once closed, should protect the hybridization from light (*see* material in Fig. 2 or else use foil to cover the box).
11. Hybridization temperature: 37 °C is a routinely used temperature in our conditions. But higher temperature (up to 45 °C) can also give good results, depending on the probe.
12. If stringency of detection needs to be increased, additional washes with lower concentration in SSC can be performed (down to 0.2× SSC).

Acknowledgments

Research in the Heard lab is supported by "La Ligue Contre Le Cancer"; by an ERC Advanced Investigator award (ERC-2014-AdG no. 671027); Labex DEEP ANR-11-LBX-0044.

References

1. Flach G, Hjohnson M, Braude PR, Taylor RAS, Bolton VN (1982) The transition from maternal to embryonic control in the 2-cell mouse embryo. EMBO J 1:681–686
2. Hamatani T, Carter MG, Sharov AA, Ko MSH (2004) Dynamics of global gene expression changes during mouse preimplantation development. Dev Cell 6:117–131. doi:10.1016/S1534-5807(03)00373-3
3. Wang QT, Piotrowska K, Ciemerych MA, Milenkovic L, Scott MP, Davis RW, Zernicka-Goetz M (2004) A genome-wide study of gene activity reveals developmental signaling pathways in the preimplantation mouse embryo. Dev Cell 1:133–144
4. Park S-J, Komata M, Inoue F, Yamada K, Nakai K, Ohsugi M, Shirahige K (2013) Inferring the choreography of parental genomes during fertilization from ultralarge-scale whole-transcriptome analysis. Genes Dev 27:2736–2748. doi:10.1101/gad.227926.113
5. Dufort I, Robert C, Sirard MA (2015) Studying bovine early embryo transcriptome by microarray. Methods Mol Biol 1222:197–208. doi:10.1007/978-1-4939-1594-1_15
6. Petropoulos S, Edsgärd D, Reinius B, Deng Q, Panula SP, Codeluppi S, Plaza Reyes A, Linnarsson S, Sandberg R, Lanner F (2016) Single-cell RNA-seq reveals lineage and X chromosome dynamics in human preimplantation embryos. Cell 165:1012–1026. doi:10.1016/j.cell.2016.03.023
7. De Vries WN, Binns LT, Fancher KS, Dean J, Moore R, Kemler R, Knowles BB (2000) Expression of Cre recombinase in mouse oocytes: a means to study maternal effect genes. Genesis 26:110–112
8. Bultman SJ, Gebuhr TC, Pan H, Svoboda P, Schultz RM, Magnuson T (2006) Maternal BRG1 regulates zygotic genome activation in the mouse. Genes Dev 20:1744–1754. doi:10.1101/gad.1435106

9. Andreu-Vieyra CV, Chen R, Agno JE, Glaser S, Anastassiadis K, Stewart AF, Matzuk MM (2010) MLL2 is required in oocytes for bulk histone 3 lysine 4 trimethylation and transcriptional silencing. PLoS Biol 8:e1000453. doi:10.1371/journal.pbio.1000453
10. Ancelin K, Syx L, Borensztein M, Ranisavljevic N, Vassilev I, Briseño-Roa L, Liu T, Metzger E, Servant N, Barillot E, Chen C-J, Schüle R, Heard E (2016) Maternal LSD1/KDM1A is an essential regulator of chromatin and transcription landscapes during zygotic genome activation. Elife 5. pii: e08851. doi: 10.7554/eLife.08851.
11. Okamoto I, Arnaud D, Le Baccon P, Otte AP, Disteche CM, Avner P, Heard E (2005) Evidence for de novo imprinted X-chromosome inactivation independent of meiotic inactivation in mice. Nature 438:369–373. doi:10.1038/nature04155
12. Patrat C, Okamoto I, Diabangouaya P, Vialon V, Le Baccon P, Chow J, Heard E (2009) Dynamic changes in paternal X-chromosome activity during imprinted X-chromosome inactivation in mice. Proc Natl Acad Sci U S A 106:5198–5203. doi:10.1073/pnas.0810683106
13. Probst AV, Okamoto I, Casanova M, El Marjou F, Le Baccon P, Almouzni G (2010) A strand-specific burst in transcription of pericentric satellites is required for chromocenter formation and early mouse development. Dev Cell 19:625–638. doi:10.1007/s00412-007-0106-8
14. Fadloun A, Le Gras S, Jost B, Ziegler-Birling C, Takahashi H, Gorab E, Carminci P, Torres-Padilla M-E (2013) Chromatin signatures and retrotransposon profiling in mouse embryos reveal regulation of LINE-1 by RNA. Nat Struct and Mol Biol 20:332–338. doi:10.1038/nsmb.2495
15. Okamoto I, Otte AP, Allis CD, Reinberg D, Heard E (2004) Epigenetic dynamics of imprinted X inactivation during early mouse development. Science 303:644–649. doi:10.1126/science.1092727
16. Chaumeil J, Okamoto I, Heard E (2004) X-chromosome inactivation in mouse embryonic stem cells: analysis of histone modifications and transcriptional activity using immunofluorescence and FISH. Methods Enzymol 376:405–419. doi:10.1016/S0076-6879(03)76027-3
17. Hogan BL, Beddington R, Costantini F, Facy E (1994) Manipulating the mouse embryo. Cold Spring Harbor Laboratory Press, New York, NY
18. Giorgetti L, Piolot T, Heard E (2015) High-resolution 3D DNA FISH using plasmid probes and computational correction of optical aberrations to study chromatin structure at the sub-megabase scale. Methods Mol Biol 1262:37–53. doi:10.1007/978-1-4939-2253-6_3
19. Lai LT, Meng Z, Shao F, Zhang LF (2016) Simultaneous RNA-DNA FISH. Methods Mol Biol 1402:135–145. doi:10.1007/978-1-4939-3378-5_11
20. Orjalo AV Jr, Johansson HE (2016) Stellaris® RNA fluorescence in situ hybridization for the simultaneous detection of immature and mature long noncoding RNAs in adherent cells. Methods Mol Biol 1402:119–134. doi:10.1007/978-1-4939-3378-5_10
21. Pollex T, Piolot T, Heard E (2013) Live-cell imaging combined with immunofluorescence, RNA, or DNA FISH to study the nuclear dynamics and expression of the X-inactivation center. Methods Mol Biol 1042:13–31. doi:10.1007/978-1-62703-526-2_2
22. Bolte S, Cordelières FP (2006) A guided tour into subcellular colocalization analysis in light-microscopy. J Microsc 224(Pt 3):213–232
23. Haralick RM (1979) Statistical and structural approaches to texture. Proc IEEE 67:786–804. doi:10.1109/PROC.1979.11328

Chapter 10

Detection of RNA Polymerase II in Mouse Embryos During Zygotic Genome Activation Using Immunocytochemistry

Irina O. Bogolyubova and Dmitry S. Bogolyubov

Abstract

Mammalian pre-implantation embryos represent a highly dynamic experimental model for comparative studies of nuclear structure and functions in the context of gradual reactivation of transcription. Here, we present details of the methods that allow localizing RNA polymerase II in mouse pre-implantation embryos with specific antibodies, using fluorescent/confocal and electron microscopy. We stress the special aspects of immunolabeling protocols in respect to the embryonic material. We made a special emphasis on the essential steps preceding the immunocytochemical experiments. In particular, we consider the procedures of female hormonal stimulation and embryo collection. The described approaches are also applicable to study other nuclear proteins.

Key words Mouse embryos, Nucleus, Immunocytochemistry, Confocal microscopy, Immunogold electron microscopy, RNA polymerase II

1 Introduction

The nuclei of mammalian early embryos represent a peculiar experimental model providing a good opportunity to study nuclear structure and functions [1]. They are characterized by a specific nuclear ultrastructure [2] that obviously reflects the unique functional state of the nucleus during embryonic early development. At the initial period, the nucleus of mammalian embryos is transcriptionally silent, whereas transcription is reactivated at the appointed stage later, and the nucleus undergoes a series of successive structural and transcriptional changes known as zygotic gene activation (ZGA) [3]. As a result, the mammalian early embryos allow studying dynamics of the nuclear apparatus at both transcriptionally active and inert stages in natural conditions. In particular, significant alterations of the pattern of RNA polymerase II (RNAPII)-dependent transcription occur during the 1- and 2-cell stages of mouse development.

Kiho Lee (ed.), *Zygotic Genome Activation: Methods and Protocols*, Methods in Molecular Biology, vol. 1605,
DOI 10.1007/978-1-4939-6988-3_10, © Springer Science+Business Media LLC 2017

In mammalian cells, the largest subunit of RNAP II contains 52 copies of the consensus heptapeptide repeat YSPTSP, constituting the C-terminal domain (CTD) that can be phosphorylated. The unphosphorylated and hyperphosphorylated forms of RNAP II have distinct roles in the transcription cycle: RNAP II containing an unphosphorylated CTD assembles into a preinitiation complex on the promoter, whereas transcript elongation is catalyzed by the hyperphosphorylated form of RNAP II [4]. It is suggested that phosphorylation of the CTD might control ZGA in mammalian embryos [5].

Here, we reproduce the methods of fluorescent and immunoelectron microscopy that routinely are used in our laboratory to localize RNA polymerase II (RNAP II) or other nuclear proteins in mouse early embryos. We will consider the key steps of sample processing: from collecting of desired-age embryos up to the receiving of confocal or electron-microscopical images. Although the methods have been adapted for the nuclei of mouse early embryos, similar approaches are generally applicable for different cell types. We also pay a special attention to several hidden pitfalls that may arise during the preparation of the specimens for immunostaining/labeling.

2 Materials

2.1 Mouse Embryo Collection

1. Pregnant mare serum gonadotrophin: Folligon (Intervet) or another trademark. Dissolve the liophilized hormone in the required volume of Ringer's solution or ddH_2O. As a rule, we use 20 ml of a solvent for a pack of 1000 IU. The gonadotrophins in the form of veterinary medicinal products are designed for use in domestic and farm animals that are larger than the mouse. To reduce the expenditure of the hormones, we prepare the aliquots 50 IU/ml that are stored at −18 °C before use. The volume of aliquots is defined by the number of animals in each experiment. Avoid repeated freeze/thawing.
2. Human chorionic gonadotrophin: Chorulon (Intervet) or another trademark. Dissolve and store in the same manner. As a rule, we use 30 ml of a solvent for a pack of 1500 IU.
3. Disposable 1 ml syringes to inject the hormones.
4. Handling medium: nutrient mixture with HEPES (e.g., Ham's medium HEPES-modification).
5. Nutrient mixture supplemented with $NaHCO_3$: e.g., Ham's F-10 or Ham's F-12 if it is necessary to cultivate embryos in a CO_2 incubator.
6. 0.1% hyaluronidase to remove cumulus masses.
7. Acidified Tirode's solution or handling medium (pH 2.5) to remove *zona pellucida*. The pH of handling medium is adjusted by 1 N HCl.

8. Petri dishes Ø 40 and 90 mm.
9. 100 μl glass capillaries (micro-sampling pipettes): the capillaries are pulled over the flame of a burner.
10. Surgical and microsurgical tools (scissors, tweezers, microscalpels, etc.) to eviscerate the reproductive tract and to dissect the oviducts.

2.2 Immunofluorescent Microscopy

1. Phosphate-buffered saline (PBS): tablets dissolving in the required volume of ddH_2O are available.
2. 20% formaldehyde freshly prepared from paraformaldehyde: this stock is used to prepare the fixatives for both immunofluorescent and immunoelectron microscopy. When preparing the solution, we use 4 mM Na_2CO_3, since paraformaldehyde is a slow-soluble compound and alkaline medium is required. Others add a trace of NaOH for this purpose. Paraformaldehyde should be dissolved under a ventilation hood in a closed vessel at 60 °C. Allow the mixture to cool to room temperature. Minor suspended matter can be filtered using a 0.45 μm membrane filter.
3. Fixative 1: 4% formaldehyde in PBS.
4. Fixative 2: 2% formaldehyde in PBS.
5. 0.5% Triton X-100 in PBS.
6. 10% fetal bovine serum containing 0.02% NaN_3 in PBS.
7. Primary antibodies directed against different forms of RNAP II (*see* **Note 1**).
8. Secondary antibodies: conjugated with appropriate fluorochrome (*see* **Note 2**).
9. Moist chambers: may be prepared from Petri dishes Ø 40 mm.
10. Glass wells.
11. Microscope slides and coverslips.
12. Mounting medium for fluorescence: we conventionally use Vectashield. Instead, 50% glycerol containing 1 mg/ml of *p*-phenylenediamine [6] is also suitable as an antifade solution. A DNA-specific dye may be applied additionally (*see* **Note 2**).

2.3 Fixation and Embedding for Immunoelectron Microscopy

1. Glutaraldehyde: It should be of EM grade without evidence of polymerization. The stock (25% or 50%) should be stored in vacuum-sealed ampoules at 4 °C.
2. Phosphate-buffered saline (PBS).
3. Fixative 1: 4% formaldehyde, 0.5% glutaraldehyde in PBS, pH 7.4.
4. Fixative 2: 2% formaldehyde in PBS (without glutaraldehyde).
5. 0.05 M NH_4Cl in PBS.

6. Ethanol series (35%, 50%, 75%, 85%).
7. Embedding medium: e.g., LR white acrylic resins of medium grade. Others use Lowycril or similar hydrophilic resins. Follow the manufacturer's instructions because additional supplements such as benzoyl peroxide for LR white may be required to activate.
8. Gelatin capsules of desired size.
9. Tools for manipulation with fixed embryos: glass wells, capillaries, etc.
10. Thermostat for polymerization of embedded samples.

2.4 Post-Embedding Immunogold Labeling

1. Ultrathin sections, mounted on grids (*see* **Note 3**): the sections are prepared using an ultramicrotome and a diamond or glass knife by standard methods.
2. Blocking buffer: 0.5% gelatin, 0.02% Tween-20 in PBS, pH 7.4.
3. Washing buffer: 0.1% gelatin, 0.02% Tween-20 in PBS, pH 7.4.

   Both buffers should be filtered through a syringe-driven membrane filter and stored at 4 °C in syringes equipped by new filters. To prepare buffers, we use stock solution of gelatin (5%) prepared from 40% to 50% fish skin gelatin.
4. Primary antibodies (*see* **Note 1**).
5. Secondary antibodies: conjugated with colloidal gold particles of appropriate size (*see* **Note 4**).
6. Uranyl acetate: we use saturated solution in 70% ethanol.
7. Grid mats: silicone rubber mats with numbered compartments to rinse/stain the grids with ultrathin sections.
8. Moist chambers (*see* **Note 5**).
9. Special forceps to hold grids.
10. Storage boxes for grids.

3 Methods

3.1 Hormonal Stimulation of Mouse Females

Hormonal stimulation allows synchronizing the estrous cycle in females to obtain the embryos of appointed age. Besides, hormonal stimulation allows increasing yield of the embryos from one female. This procedure involves injection of pregnant mare serum gonadotropin with subsequent injection of chorionic gonadotropin. Administration of serum gonadotropin, possessing the follicle-stimulating activity, leads to expansion in the number of developing follicles, while injection of chorionic gonadotropin, with its luteinizing effect, results in ovulation of the follicles against the backdrop of higher concentration of serum gonadotropin. Veterinary drugs are used for stimulation.

1. Administrate mouse females intraperitoneally or subcutaneously with Folligon (10 IU per female) (*see* **Note 6**).
2. After 44–48 h, administrate the females intraperitoneally or subcutaneously with Chorulon (10 IU per female).
3. Place females in cages with fertile males for insemination just after injection of Chorulon (*see* **Note 7**).
4. Based on the presence of a vaginal plug, check whether copulation has occurred and separate the mated females for further use (*see* **Note 8**).

3.2 Embryo Collection

The age of embryos is counted upward from the time of Chorulon injection. Therefore, the experiment must be planned so that the step of embryo collection was a convenient. During ZGA, mouse embryos are located in oviducts. At the later stages of cleavage, the embryos (blastocysts and sometimes also morulae) are in the uterine cavity. This should be taken into account at the initial stages of isolation of the fragments of the female reproductive tract.

1. Sacrifice the stimulated and mated females by cervical dislocation.
2. Laparotomize and isolate either oviducts alone or the oviducts with uterine horns. In the former instance, retrieve oviducts together with the neighboring ovary and with a fragment of the uterine horn to avoid damage and loss of the embryos. Place the isolated fragments in a Petri dish with handling medium.
3. Remove excessive tissues and prepare oviducts (and also uterine horns, if necessary) under a binocular microscope for explantation of embryos.
4. Flush the oviducts by a syringe (*see* **Note 9**).
5. Collect the embryos using a glass capillary or a special pipette for gamete/embryo manipulation. Dispose embryos with failed morphology. Zygotes can be distinguished from unfertilized oocytes by means of the second polar body. If necessary, the remnants of cumulus masses can be removed with warm solution of hyaluronidase (approx. 37 °C) for 3–4 min.
6. Transfer the collected embryos to appropriate medium (*see* Subheading 2.1), if it is necessary to cultivate them further in vitro (e.g., for transcription inhibition). The medium previously must be placed in a CO_2 incubator for stabilization of pH. Embryo cultivation is performed at 5% CO_2 environment (37 °C).

3.3 Indirect Immunofluorescent Microscopy

1. Fix embryos in the fixative 1 for 40 min at room temperature.
2. Place the embryos in the fixative 2 overnight at 4 °C. The embryos may be stored in this fixative at 4 °C for several days, if necessary. We try to do so that the period of storage of the fixed samples did not exceed 1 week.

3. Rinse in PBS.
4. Permeabilize the embryos with 0.5% Triton X-100 in PBS for 10 min at room temperature.
5. Rinse in PBS.
6. Incubate the embryos in 10% fetal bovine serum in PBS for 10 min at room temperature to prevent unspecific antibody binding.
7. Incubate the embryos in primary antibodies overnight at 4 °C. Put a coverslip into a moist chamber made from a Petri dish (∅ 40 mm) and place a droplet (10 μl) of antibody solution on the coverslip.
8. Rinse in PBS.
9. Incubate the embryos in secondary antibodies, as in **step 7**, in a moist chamber for 90 min at room temperature.
10. Rinse in PBS.
11. Mount embryos in the mounting medium (*see* **Note 10**).

3.4 Preparation of Embryo for Immunoelectron Microscopy

Preparation of the specimens for immunoelectron microscopy involves the same steps as the routine electron microscopy (fixation, dehydration, impregnation, embedding). However, the formaldehyde fixatives used for immunoelectron microscopy poorly preserve the fine morphology of the samples, whereas osmium tetroxide may strictly diminish the antigenicity of proteins. High concentrations of glutaraldehyde in fixative mixtures also make worse the quality of immunolabeling. We recommend preparing two kinds of the samples in parallel: for routine electron microscopy (double-fixed with glutaraldehyde and osmium tetroxide) and for immunoelectron microscopy as described below.

1. Remove *zona pellucida* with an acidified Tirode's solution (or with handling medium at pH 2.5). It is necessary to control the embryos under a binocular microscope when placed in acidic solution. After *z. pellucida* is removed, the embryos must be transferred immediately in standard handling medium (*see* **Note 11**).
2. Rinse embryos, transferring them through several drops of fresh medium.
3. Fix the embryos in the fixative 1 for 1.5 h at room temperature.
4. Additionally fix in the fixative 2 overnight at 4 °C. The embryos may be stored in the fixative 2 for several days, but additional washing is required after more prolonged fixation.
5. Wash 2× 5 min in PBS, then in freshly prepared PBS containing 0.05 M NH_4Cl, and finally rinse briefly in ddH_2O.

6. Dehydrate in an ethanol series: 35% (2× 5 min), 50% (5 min), 75% (5 min), and 85% (3× 5 min).
7. Impregnate in mixtures of 85% ethanol and LR white 2:1, 1:1, 1:2 for 1 h in each and then in pure resin (2× 1 h).
8. Place an each embryo in a gelatin capsule entirely filled with fresh pure resin, cover with a cap firmly, and then polymerize in a thermostat at 57 °C for at least 24 h.; too prolonged heating is not recommended. Before cutting, polymerized blocks must be left at room temperature for at least several hours.
9. Prepare ultrathin (50–70 nm) sections. After cutting and before mounting on grids, the sections must be smoothed out in xylene vapors using a wooden stick, moistened with xylene. Do not touch the sections!

3.5 Post-Embedding Immunogold Labeling

Immunogold labeling at the ultrastructural level requires more steps as compared with routine electron microscopy. For immunoelectron microscopy, processing of the ultrathin sections mounted on special grids is performed on a silicone rubber grid mat, e.g., available from Electron Microscopy Sciences. The mat is placed in a moist chamber, and a series of droplets separated from each other is applied by a syringe for each grid. Never allow even a slightly drying of the sections during any step of their processing. The excess of a fluid may be removed by a brief touching of grid edge to a piece of filter paper before placing the grid in antibody solution to avoid a superfluous antibody diluting. It is better slightly to dilute antibody, than to allow the grid drying.

1. Place a grid into a droplet of blocking buffer for 10 min to prevent nonspecific antibody binding.
2. Transfer the grid in a droplet (10–15 μl, also *see* **Note 5**) of primary antibody (*see* **Note 12**). Leave in a moist chamber overnight at 4 °C.
3. Wash the grid in washing buffer (6× 1 min).
4. Transfer the grid in secondary antibody solution and incubate in a moist chamber for 1.5 h at room temperature.
5. Wash the grid in washing buffer (2× 1 min), rinse in ddH_2O, and allow drying on a high-quality paper filter in dust-free conditions.
6. Contrast ultrathin sections with uranyl acetate. The time for contrasting is selected empirically depending on the material. For mouse embryos, it is about 10 min.
7. Rinse thoroughly in ddH_2O and allow drying.

4 Notes

1. We routinely use a set of antibodies for detection of different forms/epitopes of RNAP II. To detect the unphosphorylated RNAP II, we use mAb 8WG16 [7] available from Abcam, for example. The given antibody cross-reacts with a wide range of species, from plants to mice and human. To detect the hyperphosphorylated RNAP II, we use affinity-purified rabbit polyclonal serum [8] that reacts predominantly with the phosphorylated CTD. For some instances, mAb H5 available from Santa Cruz may be used. This antibody recognizes the hyperphosphorylated CTD of RNAP II [9]. However, mAb H5 also cross-reacts with phospho-epitopes of one or more SR proteins [10], and this should be considered when interpreting the results of staining. To localize RNAP II, mAb ARNA-3 (Chemicon) that reacts with a 175 kDa subunit of RNAP II, detecting RNAP II regardless of its phosphorylation state, is also used.

   In Figs. 1, 2, and 3, immunocytochemical staining/labeling results are presented for the two-cell mouse embryos with different functional status of their nuclei. The images were obtained after two antibodies against the different forms of RNAP II have been applied for laser-scanning confocal (Figs. 1 and 2) and immunoelectron microscopy (Fig. 3).

2. The choice of fluorescent conjugate is a matter of personal preference. For double staining, a combination of green and red fluorescence is preferred, since the sites of colocalization where the colors are mixed will appear yellow. Secondary antibodies must be raised against the host species. For double staining, secondary antibodies must be produced in a third species, other than first two. A broad range of secondary antibodies developed in goat, rabbit, or dunkey and conjugated with different fluorescent dyes are now available. If it is necessary to use a far red dye such as Alexa 633, the signal may be false-colored blue in merged images.

   To reveal DNA in the same preparations, we routinely use a mounting medium supplemented with 0.5 μg/ml of 4′,6-diamidino-2-phenylindole (DAPI). When an appropriate laser is not available for the UV-exitable dye DAPI, chromatin may be stained with a far red dye such as To-Pro-3 (1 μg/ml).

3. The grids must be made of a chemically inert material. As a rule, we use nickel grids. In our practice, there was an unpleasant experience when conspicuous precipitates appeared on standard copper grids during the procedure.

4. Secondary antibodies for electron labeling microscopy must be conjugated with colloidal gold particles of preferred size (e.g., 5, 10, or 15 nm). For double/triple labeling, particle

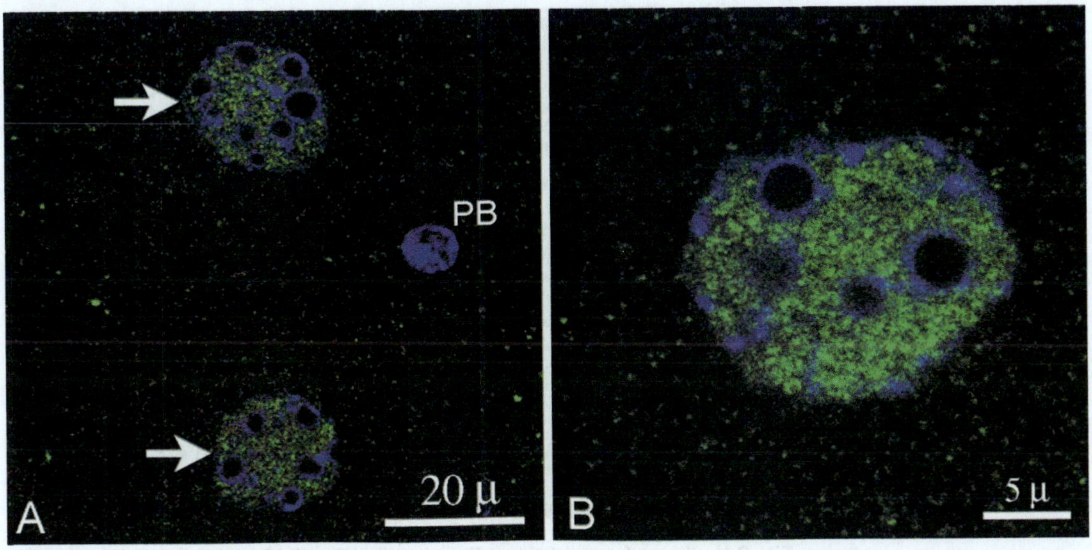

Fig. 1 Mouse early embryos after fluorescent staining for the unphosphorylated RNAP II and DNA. Confocal merged images. (**a**) Late two-cell embryo (46 h post injection of human chorionic gonadotrophin). Two nuclei of blastomeres (white arrows) and the nucleus of polar body (PB) are seen. (**b**) Nucleus of a four-cell embryo (55 h post injection of human chorionic gonadotrophin) at higher magnification. RNAP II was revealed with mAb 8WG16, dilution 1:200. Secondary antibody was goat anti-mouse conjugated with Alexa Fluor 488 (green fluorescence), dilution 1:200. DNA was stained with 1 µg/ml To-Pro-3 (false-colored blue)

size should be clearly distinguished. We prefer 10- and 15 nm conjugates for double labeling. In our opinion, such a difference in particle size (5 nm) is sufficient to distinguish different particles in micrographs. Others prefer a combination of 5 and 15 nm, but 5 nm particles seem too small when a micrograph is reproduced in a paper. The particles larger than 18 nm are not recommended for electron microscopy, since they will mask a few ultrastructural "units" of the cell nucleus such as different types of fibrils and granules, whose size is on average 15–30 nm.

5. Covered plate/box with a standard grid mat supplemented with a piece of wet filter paper may be used as a moist chamber. The expense of antibodies may be reduced to 5–6.5 µl if the moist chambers will be prepared from a small Petri dish (Ø 40 mm) with a piece of filter paper sticked to the reverse side of a cover. The paper must be moistened with water. The chambers (dishes) must be sealed with parafilm.
6. The best effect of stimulation is reached when using young mice at the age of 5–8 weeks. When using elder animals it is possible to increase a dose of the injected hormone a bit (up to 12.5 IU). At the same time, it is necessary to remember that a redundant dose of hormone, on the contrary, can slow down follicle development.

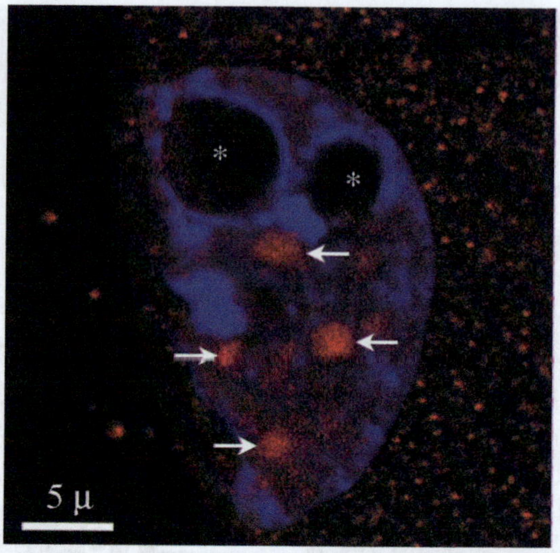

Fig. 2 Nucleus of an embryo at the so-called 2-cell block in vitro [11] stained for the hyperphosphorylated RNAP II and DNA. Confocal merged image. RNAP II was revealed with affinity-purified rabbit polyclonal serum [8] against the hyperphosphorylated CTD, dilution 1:600. Secondary antibody was goat anti-rabbit conjugated with Alexa Fluor 568 (red fluorescence), dilution 1:200. Accumulation is observed in the enlarged domains corresponding to nuclear speckles (arrows). Enlargement of the speckles is characteristic for cells with suppressed transcriptional activity [12]. Large dark "holes" (asterisks) are the nucleolus precursor bodies, devoid of RNAP II and surrounded by peripheral heterochromatin stained with 1 μg/ml To-Pro-3 (false-colored blue)

7. As a rule, we use the following proportion: one female on two to three males. It is not recommended to do more than two matings with the same males during 1 week because this can lead to decrease in male sexual activity and fertility.

   Copulation may not happen if the males have been kept separately from females, owing to insufficiently fixed pattern of male sexual behavior. To prevent such failures, we recommend using the males that have been kept together with females for some time. It is recommended to monitor the success of mating, replacing males if their sexual activity decreases.

8. The precise time cannot be prescribed when the vaginal plugs have to be checked, because this depends on the experiment design. However, it is necessary to remember that a vaginal plug disappears gradually during 24 h after mating.

   Our practice has shown that vaginal plugs can be poorly expressed or even absent in rare instances. However, the large number of viable embryos of the required age can be found in the female genital tract. It seems this can be explained by either insufficient functional activity of the accessory male glands or by too high rate of lysis of the vaginal plugs that happens for unknown reasons.

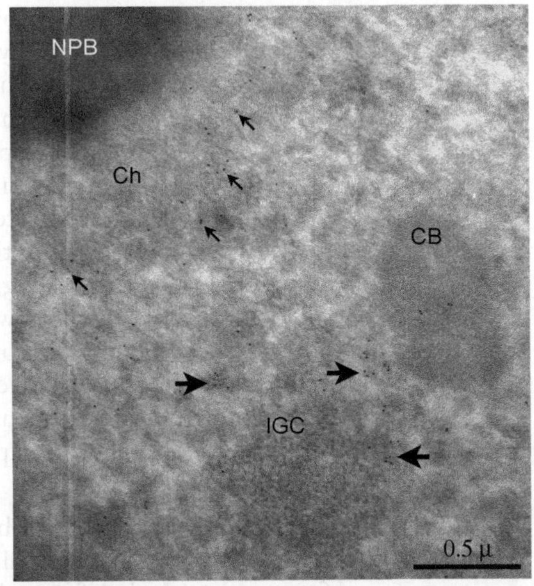

Fig. 3 A fragment of the nucleus of the two-cell embryo in the end of ZGA after electron immunogold labeling with mAb 8WG16 to reveal the unphosphorylated RNAP II, dilution 1:200. Secondary antibody was goat anti-mouse conjugated with 10-nm colloidal gold particles, dilution 1:20. The following nuclear structures are observed: a fragment of a nucleolus precursor body (NPB) peripherally surrounded by chromatin (Ch); an interchromatin granule cluster (IGC); and a Cajal body (CB). Small arrows indicate labels in the perichromatin nuclear compartment [13] located near the NPB. Large arrows point to labeled fibrillar areas in the vicinity of the IGC. These areas obviously resemble the so-called perispeckles [14]—intermediate territories between the sites of transcription and splicing factor reservoirs (IGC/speckles). In transcriptionally active cells, both places are known to be enriched in accumulations of the perichromatin fibrils which represent in situ forms of nascent transcripts [15] and contain, on top of everything else, active RNA polymerase II

9. To flush oviduct, it is convenient to use a syringe of a small volume (e.g., 1 ml) equipped with a dull needle. The needle is introduced into the oviduct ampulla under a binocular microscope. This manipulation demands certain skills. Alternatively, the oviducts can be dissected into small fragments and this will lead to release of the embryos. However, in the latter instance, searching/collecting of embryos among tissue debris is more difficult.
10. We place a drop of mounting medium on a standard microscope slide, transfer the embryos into a drop, and then cover them with a coverslip. To avoid sliding, especially when an inverted microscope is used, the coverslip edges are sealed with nail polish. To make searching the embryos easier during microscopy, their location on the microscope slide may be marked with a permanent marker under a binocular microscope.

There is a high probability to lose samples at the step of covering the embryos by a coverslip. Most frequently, it occurs when a large volume of medium is used. To avoid loss of the embryos, we use the volumes of the medium as small as possible (approx. 7 μl per a 18 × 18 coverslip). Also, it is necessary to minimize transfer of buffer between drops during embryo transfer. Since a mounting medium and buffer possess different density, the embryos initially emerge at the droplet surface and only gradually gravitate to the surface of the microscope slide. So, we recommend waiting for 3–5 min after embryo transfer before covering them with a coverslip. This also reduces a chance to loss the embryo and prevents their movement to the marginal zones that are inconvenient for microscopy. By our experience, it is optimal to place 2–3 embryos on a slide.

11. The adhesiveness of embryos increases significantly after removing of *z. pellucida*. Siliconizing of the glass wells is therefore recommended for embryo manipulation in media other than those containing BSA.

    We manipulate with embryos transferring them from one solution to another with the use of glass capillaries. Others do not transfer the embryos renewing the solutions themselves. In our opinion, this method increases the risk of embryo loss, since the replacement of fluids (especially with different concentrations of alcohol) causes strong turbulent fluid flows.

12. The optimal dilution of each antibody should be determined empirically. A manufacturer's datasheet usually provides a wide range of antibody concentrations. As a rule, higher concentrations of antibodies are required for microscopy than for immunoblotting. For fluorescent and immunoelectron microscopy, these concentrations may be either the same or higher for the immunogold labeling. The optimal is the dilution when an antibody still stains without loss of the intensity. Maximal possible dilution should be fitted to avoid unnecessary expense of high-priced reagents.

    For double labeling, the use of an antibody mixture is possible (e.g., [16]) to perform reaction in a single step. Mix two primary antibodies produced in different hosts, in a ratio of 1:1 to the desired final concentration of both. Always prepare fresh antibody solutions only in necessary quantities for each experiment, since diluted antibodies are unstable and cannot be stored for a long time.

Acknowledgment

This work was supported by the granting program "Molecular and Cell Biology" of the Russian Academy of Sciences and the Russian Foundation for Basic Research (grant No. 15-04-01857).

References

1. Bogolyubova IO, Bogolyubov DS (2014) Nuclear distribution of RNA polymerase II and mRNA processing machinery in early mammalian embryos. Biomed Res Int 2014:681596
2. Fakan S, Odartchenko N (1980) Ultrastructural organization of the cell nucleus in early mouse embryos. Biol Cell 37:211–218
3. Li L, Lu X, Dean J (2014) The maternal to zygotic transition in mammals. Mol Asp Med 34:919–938
4. Dahmus ME (1996) Reversible phosphorylation of the C-terminal domain of RNA polymerase II. J Biol Chem 271:19009–19012
5. Bellier S, Chastant S, Adenot P, Vincent M, Renard JP, Bensaude O (1997) Nuclear translocation and carboxyl-terminal domain phosphorylation of RNA polymerase II delineate the two phases of zygotic gene activation in mammalian embryos. EMBO J 16:6250–6262
6. Gall JG, Murphy C, Wu Z (1991) Lampbrush chromosomes. Methods Cell Biol 36:149–166
7. Thompson NE, Steinberg TH, Aronson DB, Burgess RR (1989) Inhibition of in vivo and in vitro transcription by monoclonal antibodies prepared against wheat germ RNA polymerase II that react with the heptapeptide repeat of eukaryotic RNA polymerase II. J Biol Chem 264:11511–11520
8. Kim W-Y, Dahmus ME (1986) Immunocytochemical analysis of mammalian RNA polymerase II subspecies. Stability and relative in vivo concentration. J Biol Chem 261:14219–14225
9. Bregman DB, Du L, van der Zee S, Warren SL (1995) Transcription-dependent redistribution of the large subunit of RNA polymerase II to discrete nuclear domains. J Cell Biol 129:287–298
10. Doyle O, Corden JL, Murphy C, Gall JG (2002) The distribution of RNA polymerase II largest subunit (RPB1) in the *Xenopus* germinal vesicle. J Struct Biol 140:154–166
11. Bogolyubova IO, Bogoliubova NA, Bogolyubov DS, Parfenov VN (2006) Nuclear structure in early mouse embryos: a comparative ultrastructural and immunocytochemical study with special emphasis on the "2-cell block in vitro". Tissue Cell 38:389–398
12. Misteli T, Cáceres JF, Spector DL (1997) The dynamics of a pre-mRNA splicing factor in living cells. Nature 387:523–527
13. Fakan S, van Driel R (2007) The perichromatin region: a functional compartment in the nucleus that determines large-scale chromatin folding. Semin Cell Dev Biol 18:676–681
14. Daguenet E, Baguet A, Degot S, Schmidt U, Alpy F, Wendling C, Spiegelhalter C, Kessler P, Rio MC, Le Hir H, Bertrand E, Tomasetto C (2012) Perispeckles are major assembly sites for the exon junction core complex. Mol Biol Cell 23:1765–1782
15. Fakan S (1994) Perichromatin fibrils are in situ forms of nascent transcripts. Trends Cell Biol 4:86–90
16. Cmarko D, Verschure PJ, Martin TE, Dahmus ME, Krause S, Fu X-D, van Driel R, Fakan S (1999) Ultrastructural analysis of transcription and splicing in the cell nucleus after bromo-UTP microinjection. Mol Biol Cell 10:211–223

Chapter 11

Immunological Staining of Global Changes in DNA Methylation in the Early Mammalian Embryo

Yan Li and Christopher O'Neill

Abstract

The structural complexity of chromatin can make antibody access to some nuclear antigens energetically unfavorable. This can limit the usefulness and reliability of immunostaining unless validated methods of epitope retrieval are applied. We found that denaturation of chromatin by sequential use of acidification and tryptic digestion of fixed cells is required to reliably detect DNA methylation in the embryo. Using this method to unmask the epitope revealed an unexpected pattern of reprogramming of global patterns of DNA methylation in the preimplantation embryo. This paper provides a detailed description of the procedures required for immunological detection of 5-methylcytosine in the early embryo.

Key words DNA methylation, Zygote, Immunostaining, Embryo, Epigenetics, Development, Immunostaining of DNA methylation

1 Introduction

Methylation of DNA at cytosine-phosphate-guanine (CpG) dinucleotides (5meC) is generally considered to be a stable epigenetic modification that is commonly associated with repression of gene expression, heterochromatin formation, and the maintenance of genomic stability [1]. Immunolocalization of 5meC has been used for many years for the investigation of major epigenetic transitions during development and in cellular pathologies, such as cancer [2]. The earliest approaches to immunological detection of 5meC used UV irradiation of DNA, heat treatment, or acid treatment to elicit epitope retrieval [3, 4]. In recent decades, brief acidification has been the most common method used, primarily due to its convenience. We have recently shown that a large but variable pool of 5meC remains undetectable after common acidification protocols [5]. The addition of tryptic digestion revealed a further large pool of 5meC that remained masked after acidification alone. This trypsin-sensitive pool varied greatly depending on the ontological state, cell-cycle status, and age of cells in culture

Kiho Lee (ed.), *Zygotic Genome Activation: Methods and Protocols*, Methods in Molecular Biology, vol. 1605, DOI 10.1007/978-1-4939-6988-3_11,

[5–8]. This procedure for 5meC retrieval also identified dynamic changes in global masking of methylation patterns after DNA damage in the living cells [9].

While some level of trypsin-sensitive masking seems a general feature of this epitope, the level of masking detected in the zygote is exceptionally high [5], and greater than we have observed in other cellular settings [5, 7]. The extent of this masking increased through the first cell-cycle so that just prior to syngamy little immunodetectable 5meC is observed after conventional acid-based epitope retrieval [5]. Figure 1 shows the marked changes in relative levels of epitope retrieval when the antibody is applied without any retrieval steps, with acid only and with acid followed by trypsin in the mouse zygote. The level of masking can vary between the paternally and maternally

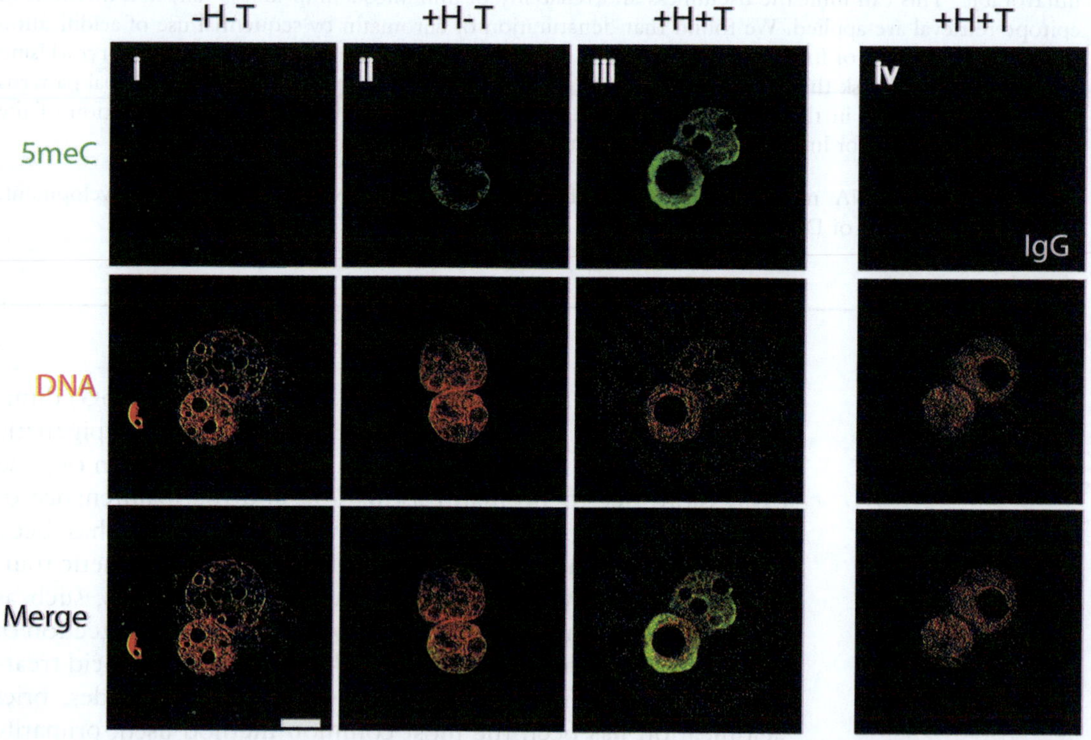

Fig. 1 5meC (green) staining of PN5 stage mouse zygotes collected directly from the oviduct after (i) no antigen retrieval process; (ii) acid treatment only; and (iii) acid treatment followed by brief tryptic digestion. (iv) nonimmune IgG-negative control of zygotes treated in the same manner as (iii). Chromatin is counterstained with propidium iodide (red). Denaturation of chromatin caused by trypsin digestion results in a marked reduction in the intensity of PI staining. The 5meC and PI images are merged to show the extent of colocalization of these signals. The images from each treatment were captured and processed under the same conditions. Bars = 10 μm. H HCl, T Trypsin. Thus, +H−T means epitope retrieval with acid but not tryptic digestion. +H+T means both treatments were applied

acquired genomes, and the generation of embryos by in vitro fertilization causes less masking of 5meC to occur within the female pronucleus compared with the male (creating the illusion of asymmetric loss of methylation in the male relative to female pronucleus) [5]. Such findings from immunolocalization using acid-only denaturation have led to a widely accepted view of the global demethylation of the genome soon after fertilization (reviewed [10]). The addition of a tryptic digestion step allowed the recovery of a large additional pool of 5meC which revealed that 5meC levels are relatively similar across the first cell-cycle and not markedly different between the paternally and maternally inherited genomes prior to syngamy [5]. By contrast, a loss of 5meC commenced in the inner cells of the embryo soon after compaction at the 8-cell stage, leading to a marked hypomethylated state within the emerging pluripotent inner cell mass and a relatively hypermethylated state within the differentiated trophectoderm of the blastocyst (Fig. 2A) [6].

This method can also be successfully applied to metaphase chromosome preparations (Fig. 2B). We have found that other CpG modifications, such as 5′-hydroxymethylcytosine, are subject to much less trypsin-sensitive masking than 5meC. Yet, the methods described here can be successfully applied to the colocalization of both these antigens within the nucleus, and in the early embryo this shows that they have characteristic but different patterns of localization within the genome (Fig. 2C) [6, 8]. This staining method can also be applied while costaining for a range of nuclear protein antigens [5–7]. This approach, however, requires that one first validates that the detection of the protein is not adversely affected by either the acidification or tryptic digestion of cells, and this is not always the case [6]. This approach for epitope retrieval has also recently been validated for quantitative use with fluorescence-activated flow cytometry [11].

Here, we provide a detailed description of the methodology developed for the faithful retrieval and immunolocalization of global levels of 5meC within mouse zygotes and preimplantation embryos.

2 Materials

2.1 Reagents

1. Dulbeccos Phosphate-Buffered Saline (DPBS).
2. Bovine Serum Albumin (BSA).
3. Polysorbate 20 (Tween 20).
4. Triton X-100.
5. Paraformaldehyde (PFA).

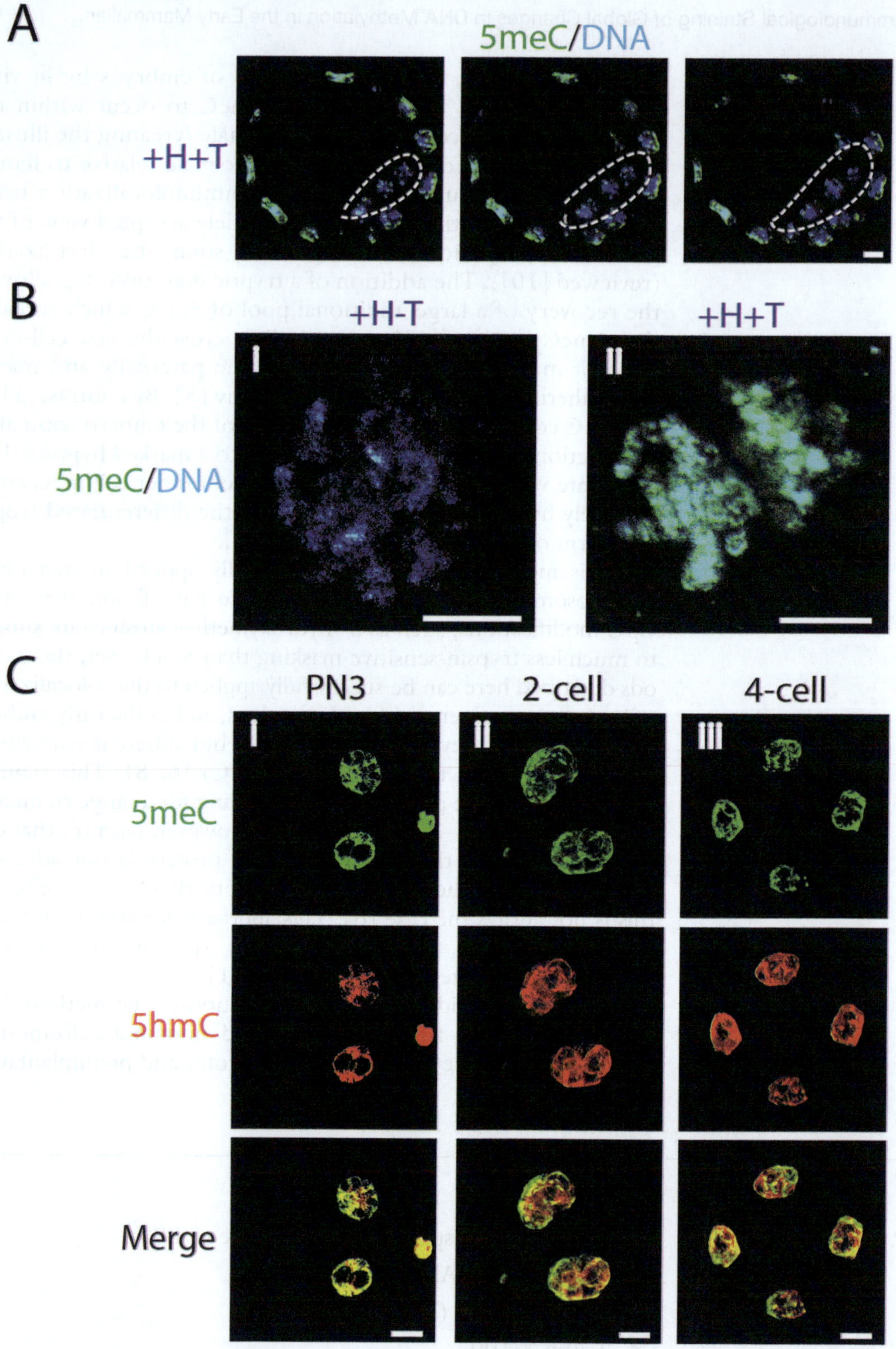

Fig. 2 (**A**) Merged images of 5meC (green) and DNA (blue) show the inner cell mass and the trophectoderm of the mouse blastocyst had marked differences in the levels of global 5meC. The inner cell mass (inside dotted line) is hypomethylated and the trophectoderm is hypermethylated. Adapted from Fig. 1 within Li et al. [6].

6. Hydrochloric acid (HCl).
7. 0.25% (w/v) Trypsin-EDTA (Invitrogen).
8. Mineral oil (embryo culture grade).
9. Propidium iodide (PI).
10. Primary antibody: mouse anti 5-methylcytidine antibody (AbD Serotec, UK).
11. Fluorophore-labeled secondary antibody: sheep anti-mouse IgG-FITC (Sigma).
12. Mouse IgG (Sigma)
13. Antifade mounting solution: (VECTASHIELD).

2.2 Solutions and Media

1. Phosphate Buffer—Tween (PBT): 0.05% (v/v) Tween 20 in DPBS.
2. PBT/BSA: PBT with 2% (w/v) BSA.
3. Embryo handling media: 102 mM NaCl, 4.6 mM KCl, 0.4 mM KH_2PO_4, 0.2 mM $MgSO_4$, 4 mM $NaHCO_3$, 0.24% (v/v) Lactic acid (60%), 0.4 mM Na Pyruvate, 2.8 mM Glucose, 0.04% (v/v) Phenol Red, 0.06 mg/ml Penicillin, 2.04 mM $CaCl_2{\cdot}2H_2O$, 21 mM HEPES, 0.3% (w/v) BSA, pH 7.4.
4. 4% (w/v) PFA in DPBS pH 7.4 (*see* **Note 1**).
5. Permeabilization solution: 0.5% (v/v) Tween 20, 0.5% (v/v) Triton X-100 in DPBS.
6. Trypsin Stop solution: 10% (v/v) sheep serum (always use serum from the same species as that in which the secondary antibody was raised) in embryo handling media.
7. Blocking solution: 30% (v/v) sheep serum (same species as secondary antibody) in PBT.

2.3 Equipment

1. Confocal or UV Epifluorescence microscope.
2. Dissecting microscope.
3. Warming plate or incubator (37 °C).
4. Hotplate Stirrer.

Fig. 2 (continued) **(B)** Merged images of 5meC (green) and DNA (blue) in 1-cell metaphase chromosomes collected directly from the reproductive tract after antigenic unmasking by (i) acid alone or (ii) acid plus trypsin. Little 5meC is detected after acid treatment alone but after acid plus trypsin treatment caused extensive staining of all chromosomes. Adapted from Fig. 6 within Li and O'Neill [5]. **(C)** Costaining of 5meC (green) and 5′-hydroxymethylcytosine (red) in a (i) PN3 stage zygote, (ii) a 2-cell stage embryo; and (iii) 4-cell embryo. Merged images of each stain are also shown. Adapted from Fig. 1 within Li et al. [6] All Bars = 10

3 Method

This procedure was developed specifically for staining 5meC in mouse preimplantation embryos (*see* **Note 2**). Its use in other settings requires optimization and validation in each case.

1. Collect zygotes and thoroughly wash through three changes of 1 ml DPBS to remove all extraneous protein and cellular material.
2. Transfer embryos in a minimum volume to 1 ml of 4% (w/v) PFA at 22 °C for 30 min with mixing, followed by three washes each in 1 ml DPBS at 22 °C. PFA must be prepared fresh each day of use.
3. Transfer to Permeabilization solution for 40 min at 22 °C with gentle mixing.
4. Place in 4 M HCl + 0.1% (v/v) Triton X-100 at 22 °C for 10 min, followed by three washes, each in 1 ml PBT at 22 °C.
5. Transfer in minimum volume to 50 μl drops (for ten embryos) of 0.25% Trypsin-EDTA at 37 °C for 40 s (*see* **Note 3**).
6. Embryos are rapidly transferred in minimum volume to 3 ml of Trypsin stop solution to terminate trypsinization. Embryos are quickly mixed with this solution to dilute trypsin and saturate the remaining enzyme with excess serum protein. The embryos are then transferred to another dish of Trypsin stop solution to further dilute the enzyme and it is left for 2 min at 37 °C. The Trypsin Stop solution is then removed by washing three times in 1 ml PBT/BSA at 22 °C.
7. An essential step in immunofluorescence is to minimize non-specific antibody-binding sites within the target cells. This is achieved by incubation in Blocking solution at 4 °C for 18 h, followed by washing three times in 1 ml PBT/BSA at 22 °C.
8. Primary antibody binding: Embryos are transferred to 50 μl drops (for 20 embryos) of anti 5-methylcytidine antibody at 4 °C 18 h. The antibody is diluted to a working concentration of 10 μg/ml in 5% (v/v) sheep serum (same species as secondary antibody) in PBT (*see* **Note 4**). Nonimmune negative controls are prepared in the same manner by preparing an equal concentration of mouse IgG. The embryos are then washed three times in 1 ml PBT/BSA at 22 °C (*see* **Note 5**).
9. Incubate in 300 μl (for 20 embryos) of 1:300 dilution FITC-labeled secondary antibody in PBT/BSA at 22 °C for 1 h. They were then washed three times in 1 ml PBT/BSA at 22 °C.
10. Incubate embryos in 0.5 ml of 10 μg PI/ml PBT at 22 °C for 10 min (*see* **Note 6**).

11. Wash twice in PBT and once in DPBS.
12. Transfer embryos to antifade mounting solution. In order to prevent the collapse of embryos due to osmotic difference between PBS and antifade solution (this is especially important for the blastocyst stage embryos), the embryos are passed through graded concentrations of antifade solution as follows: 25% (v/v) Antifade in DPBS, 50%, 75%, and 100%. The embryos are left in each step for 1 min with gentle mixing. Cover the dish with a light-tight box during each of these incubations.
13. For 18 mm^2 coverslips, place 12 μl of mounting media onto a microscope slide and transfer embryos in minimal volume to the middle of this drop (*see* **Note** 7).
14. Place a small amount of petroleum jelly on each corner of the coverslip and then gently place the cover slip over the drop of mounting media ensuring that no bubbles are formed.
15. While observing the embryos under a dissecting microscope, gently press down each corner of the coverslip until it just makes contact with the embryos. Immediately remove any excess media from the edge of the coverslip with an absorbent lint-free tissue. Seal the edges of the coverslip with nail varnish and leave to dry.
16. The slides are held at 4 °C in the dark until imaging that is always performed on the same day as preparation.

4 Notes

1. 4% PFA preparation: Warm 10 ml of analytical grade water to 50 °C on a hotplate-stirrer. Add 0.4 g of PFA and stir for 2 mins and then add 15 μl of 1 M NaOH, which assists in the solubilization of PFA. After a solution is achieved add 0.096 g of DPBS. Allow the solution to cool to 22 °C and then titrate to pH 7.4 with 1 M HCl.
2. Embryo collection was as previously described [12, 13]. All embryo transfers are performed with fine drawn glass pipettes, ensuring minimal transfer of extra media (< 1 μl). All washing and incubation steps are performed in glass watch face dishes. The dishes are covered with PARAFILM and set in a sealed humidified box during each incubation period.
3. Set up a dish with 50 μl drops of trypsin, cover with mineral oil, and pre-warm to 37 °C for at least 2 h before use. Transfer groups of ten embryos into each drop of Trypsin. The length of trypsin treatment needs to be calibrated for the source of trypsin used and the type of cell being treated. The digestion time required varies with differing cell types, presumably due to differences in

chromatin structure and complexity. Excessive trypsinization can cause chromatin to be so significantly denatured so as to become difficult to manage during staining procedures, while insufficient treatment will leave some antigen masked. Careful validation and calibration of this step must be applied for each new batch of trypsin used and each new tissue type under study. This validation involves varying the digestion time so that the antigen retrieval is maximized while maintaining sufficient chromatin integrity for processing. Excessive trypsinization can also lead to poor counterstaining with typical nuclear stains such as PI or DAPI. Careful calibration of the level of trypsinization is required if these counterstains are used.

4. We have not validated how long the 5meC epitope is stable after fixation nor how long the staining remains stable. We therefore always immuno-stain embryos immediately after fixation and they are imaged and recorded within 3 h of staining. We are not aware of any validation of the stability of the 5meC antigen and this should not be assumed.
5. Reliable immunolocalization requires that the staining procedures proceed under saturating conditions. Thus, for any given tissue and antibody preparation, antibody dilution and time-response curves must be constructed to find the appropriate conditions where binding and hence staining approximates saturation [8]. This should be repeated for any new batches of antibody, even when from the same supplier. Performing staining at non-saturation conditions will provide artifactual results and erroneous conclusions. The conditions for staining will vary for the tissue, species, and fixation procedures used and must be newly validated for each experimental condition under investigation.
6. The DNA counterstain used can be changed to be compatible with different fluorophores on the secondary antibody.
7. All microscope slides are acid washed (0.3 M HCl overnight and washed five times with MilliQ water). Then these are stored in 95% ethanol and cleaned with lint-free tissue immediately prior to use. The cover slides are cleaned with 95% ethanol before use.

Acknowledgments

This work is supported by Project grants from the Australian National Health and Medical Research Foundation. We thank Dr. Romanthi Madawala for comments on the manuscript.

References

1. Klose RJ, Bird AP (2006) Genomic DNA methylation: the mark and its mediators. Trends Biochem Sci 31(2):89–97
2. Yang B, Sun H, Lin W, Hou W, Li H, Zhang L et al (2013) Evaluation of global DNA hypomethylation in human prostate cancer and prostatic intraepithelial neoplasm tissues by immunohistochemistry. Urol Oncol 31(5):628–634
3. Miller OJ, Schnedl W, Allen J, Erlanger BF (1974) 5-methylcytosine localised in mammalian constitutive heterochromatin. Nature 251(5476):636–637
4. Miniou P, Jeanpierre M, Blanquet V, Sibella V, Bonneau D, Herbelin C et al (1994) Abnormal methylation pattern in constitutive and facultative (X inactive chromosome) heterochromatin of ICF patients. Hum Mol Genet 3(12):2093–2102
5. Li Y, O'Neill C (2012) Persistence of cytosine methylation of DNA following fertilisation in the mouse. PLoS One 7(1):e30687
6. Li Y, Seah MKY, O'Neill C (2016) Mapping global changes in nuclear cytosine base modifications in the early mouse embryo. Reproduction 151(2):83–95
7. Celik S, Li Y, O'Neill C (2014) The exit of mouse embryonic fibroblasts from the cell-cycle changes the nature of solvent exposure of the 5′-methylcytosine epitope within chromatin. PLoS One 9(4):e92523
8. Li Y, O'Neill C (2013) 5-Methylcytosine and 5-hydroxymethylcytosine each provide epigenetic information to the mouse zygote. PLoS One 8(5):e63689
9. Çelik S, Li Y, O'Neill C (2015) The effect of DNA damage on the pattern of immune-detectable DNA methylation in mouse embryonic fibroblasts. Exp Cell Res 339(1):20–34
10. Auclair G, Weber M (2012) Mechanisms of DNA methylation and demethylation in mammals. Biochimie 94(11):2202
11. Çelik-Uzuner S, Li Y, Peters L, O'Neill C (2017) Measurement of global DNA methylation levels by flow cytometry in mouse fibroblasts. In Vitro Cell Dev Biol Anim 53(1):1–6
12. Behringer R, Gertsenstein M, Nagy KV, Nagy A (2014) Manipulating the mouse embryo: a laboratory manual, 4th edn. Cold Spring Harbor Laboratory Press, Cold Spring Harbor, NY
13. Jin X, O'Neill C (2014) Systematic analysis of the factors that adversely affect the rate of cell accumulation in mouse embryos during their culture in vitro. Reprod Biol Endocrinol 12:35

Chapter 12

Single Cell Restriction Enzyme-Based Analysis of Methylation at Genomic Imprinted Regions in Preimplantation Mouse Embryos

Ka Yi Ling, Lih Feng Cheow, Stephen R. Quake, William F. Burkholder, and Daniel M. Messerschmidt

Abstract

The methylation of cytosines in DNA is a fundamental epigenetic regulatory mechanism. During preimplantation development, mammalian embryos undergo extensive epigenetic reprogramming, including the global erasure of germ cell-specific DNA methylation marks, to allow for the establishment of the pluripotent state of the epiblast. However, DNA methylation marks at specific regions, such as imprinted gene regions, escape this reprogramming process, as their inheritance from germline to soma is paramount for proper development. To study the dynamics of DNA methylation marks in single blastomeres of mouse preimplantation embryos, we devised a new approach—single cell restriction enzyme analysis of methylation (SCRAM). SCRAM allows for reliable, fast, and high-throughput analysis of DNA methylation states of multiple regions of interest from single cells. In the method described below, SCRAM is specifically used to address loss of DNA methylation at genomic imprints or other highly methylated regions of interest.

Key words Epigenetics, DNA methylation, Single cell, Oocyte, Blastomere, SCRAM, Imprinted genes, MSRE

1 Introduction

1.1 DNA Methylation

DNA methylation is a key epigenetic regulatory mechanism involved in multiple developmental processes, including transcriptional control, X-chromosome inactivation, and genomic imprinting (for review *see* [1]). In mammals, DNA methylation most commonly refers to the addition of a methyl group to the fifth carbon of cytosine nucleotides to form 5-methylcytosine (5mC). Yet, more recently, other forms of DNA methylation in mammals, e.g., N^6-methyladenosine [2], have been described. Cytosine methylation mostly occurs in mammals in the context of CpG dinucleotides and is introduced by the enzymatic activity of DNA methyltransferases (DNMTs) [3–9]. The removal of DNA

Kiho Lee (ed.), *Zygotic Genome Activation: Methods and Protocols*, Methods in Molecular Biology, vol. 1605,
DOI 10.1007/978-1-4939-6988-3_12, © Springer Science+Business Media LLC 2017

methylation occurs either passively (through repression of the maintenance mechanisms) [10–12] or actively by ten-eleven translocation (TET) enzymes, which converts 5mC to 5-hydroxymethylcytosine (5hmC) and further to 5-formylcytosine (5fC) and 5-carboxycytosine (5caC) [13–15] (for review *see* [1]).

1.2 DNA Methylation Dynamics in Preimplantation Embryos

The genomes of mammalian embryos undergo dramatic DNA methylation changes during preimplantation development (Fig. 1) (for review *see* [16]). At fertilization (Fig. 1a), both the maternally—and paternally—contributed genomes show remarkably high 5mC levels [17–19]. These high 5mC levels are efficiently removed by the time the pluripotent epiblast of the blastocyst is established (Fig. 1b) [18, 20, 21]. This reprogramming of parental epigenomes occurs asymmetrically, with the paternal genome undergoing immediate, active demethylation at fertilization [20, 21], and the maternal genome undergoing passive demethylation over several rounds of DNA replication (Fig. 1a) [8, 22, 23]. Although reprogramming of parental epigenomes is a genome-wide process, it is nevertheless exquisitely selective in retaining DNA methylation marks, e.g., at sites of specific endogenous retrotransposons and genomic imprints (Fig. 1a and b) [18, 24–27].

Genomic imprinting drives parent-of-origin-specific gene expression and is regulated by differential DNA methylation of parental alleles (for review *see* [28]). This differential methylation originates from oocyte and sperm, which acquired specific maternal or paternal marks respectively during germ cell differentiation that must be maintained throughout life (for review *see* [1]). Failure to maintain proper DNA methylation marks at these parental alleles results in mis-regulation of imprinted genes and can lead to abnormal development or even embryonic lethality (for reviews *see* [29]; [30]). Due to the genome wide demethylation in the preimplantation embryo, genomic-imprinted regions are at high risk of aberrant loss of methylation and special mechanisms are required to prevent this loss [16, 26, 27].

1.3 DNA Methylation Analysis

Since DNA methylation and demethylation dynamics are intimately linked to transcriptional control and affect developmental and/or pathological outcomes (for reviews *see* [1, 31]), the analysis of DNA methylation is of great interest. Over the years, a range of analysis methods has been developed to assay the methylation state of a genome, a genomic region, or even individual cytosines. The gold standard for DNA methylation analysis today is bisulfite conversion, a chemical treatment that converts unmethylated cytosine bases to uracil, while leaving methylated cytosines unchanged [32]. The changes in DNA sequence ($C \rightarrow U$ conversion) can then be analyzed by multiple means [32–37] and offer single-base pair resolution and genome-wide scalability. Other methods of 5mC analysis depend on recognition or binding to the 5mC epitope itself [38–40]. For example, enrichment of methylated DNA by 5mC-specific antibodies (MeDIP) [41] allows a qualitative estimate of locus-specific

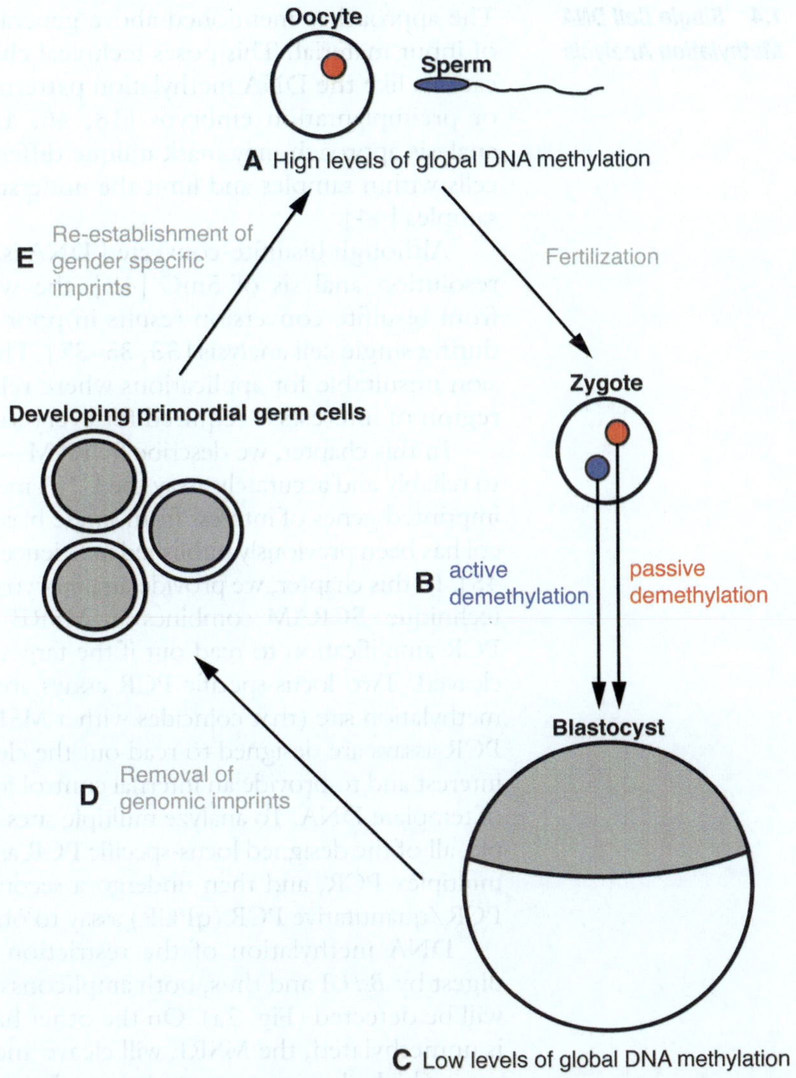

Fig. 1 DNA methylation patterns change throughout development. (**a**) Parental genomes are highly methylated. (**b**) The paternal (blue) and maternal (red) genome undergoes demethylation upon fertilization, except at genomic imprinted regions. (**c**) By the blastocyst stage, the global genome is largely demethylated. However, methylation at imprinted regions is conserved. (**d**) As primordial germ cells continue to differentiate and migrate to the gonads, demethylation of genomic imprints begins. (**e**) Upon acquiring gender identity, germ cells in the gonads reestablish sex-specific imprints

methylation levels on a genome-wide scale, but lacks single-base resolution. A third alternative uses methylation-sensitive restriction enzymes (MSREs) that cut DNA in a sequence-specific manner depending on the presence or absence of DNA methylation [42–45]. This approach allows the methylation state of single CpGs contained within restriction sites to be determined and, depending on the MSRE used, might be limited in its scalability.

1.4 Single Cell DNA Methylation Analysis

The approaches mentioned above generally require large amounts of input material. This poses technical challenges for certain applications like the DNA methylation pattern analysis of rare cell types or preimplantation embryos [18, 46, 47]. Furthermore, a bulk analysis approach may mask unique differences between individual cells within samples and limit the understanding of heterogeneous samples [44].

Although bisulfite-converted DNA is the benchmark for high-resolution analysis of 5mC [48], the widespread DNA damage from bisulfite conversion results in poor whole genome coverage during single cell analysis [33, 35–37]. This makes bisulfite conversion unsuitable for applications where reliable coverage of specific region of interests is required for every sample.

In this chapter, we describe SCRAM—a technique that allows us to reliably and accurately assay the DNA methylation state of multiple imprinted genes of interest from single blastomeres [44]. This protocol has been previously published in Science and Nature Protocol [44, 45]. In this chapter, we provide an embryo-focused application of the technique. SCRAM combines an MSRE digest with locus-specific PCR amplification to read out if the target DNA restriction site was cleaved. Two locus-specific PCR assays are designed for each DNA methylation site (that coincides with a MSRE restriction site). These PCR assays are designed to read out the cleavage status of the site of interest and to provide an internal control for the presence or absence of template DNA. To analyze multiple sites of interest in a single sample, all of the designed locus-specific PCR assays are pre-amplified in a multiplex PCR, and then undergo a second round of locus-specific PCR/quantitative PCR (qPCR) assay to obtain the final readout.

DNA methylation of the restriction site inhibits the MSRE digest by *BstUI* and thus, both amplicons of the locus-specific assay will be detected (Fig. 2a). On the other hand, if the restriction site is unmethylated, the MSRE will cleave and only the control amplicon, which does not cover the restriction site, will be detectable (Fig. 2b). Previous studies have shown that DNA methylation is relatively homogeneous within a genomic region [49]. Thus, the methylation state of a single CpG can be used as a proxy for the methylation state of the surrounding region in the SCRAM assay.

Enzymatic digest by MSRE is gentle and results in minimal DNA damage [43]. The steps from cell lysis to multiplex preamplification of regions of interest are carried out in a single tube, minimizing loss of DNA material. Furthermore, multiplexed PCR reactions are more cost-friendly and time efficient compared to whole genome sequencing [35–37]. Despite the many benefits of SCRAM of specific target loci, SCRAM is not suitable for whole-genome discovery analysis.

Although SCRAM described here is designed to identify the methylation status of target 5mC at imprinted regions, other modifications such as N^6-methyladenosine (with endonuclease

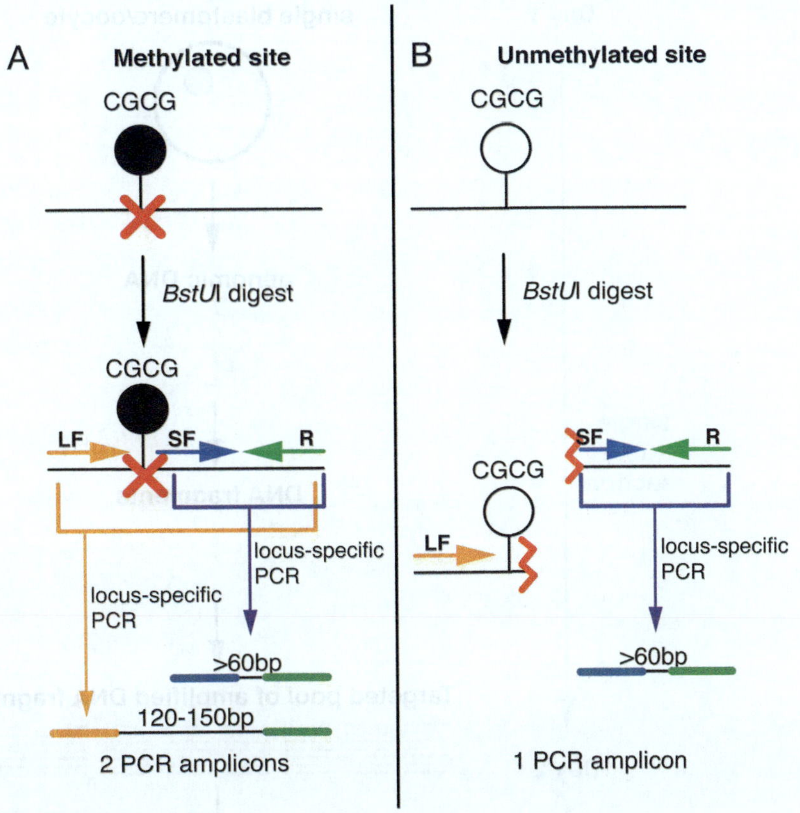

Fig. 2 Summary of Single Cell Restriction Analysis of Methylation (SCRAM). Two PCR reactions are designed for each BstUI restriction site. BstUI recognizes the restriction site CGCG and only cuts non-methylated sites. (**a**) When the restriction site is methylated, BstUI fails to digest and both amplicons will be amplified. (**b**) If the BstUI site is unmethylated, BstUI digests at the restriction site and only the short product is detected. Abbreviations are as follows: bp base pairs, LF long forward primer, SF short forward primer, R reverse primer

*Dpn*I or *Dpn*II) and 5mC derivative 5hmC (with *PvuRts*1I endonucleases) can also be analyzed by combining this method with other methylation-sensitive/dependent restriction enzymes. In addition, high-throughput analysis of multiple samples and target sites can be easily facilitated with proper readout systems (e.g., 96 × 96 Fluidigm Dynamic Arrays) [44, 45]. Finally, SCRAM is compatible with concurrent analysis of gene expression, genotyping/SNP analysis alongside DNA methylation detection [50].

In summary, SCRAM is a high-throughput method with a binary readout to analyze DNA methylation status of various targets in single cells.

After the initial assay design and preparation of control λ-DNA (Subheadings 3.1 and 3.2), SCRAM can be completed in 2-days (Fig. 3):

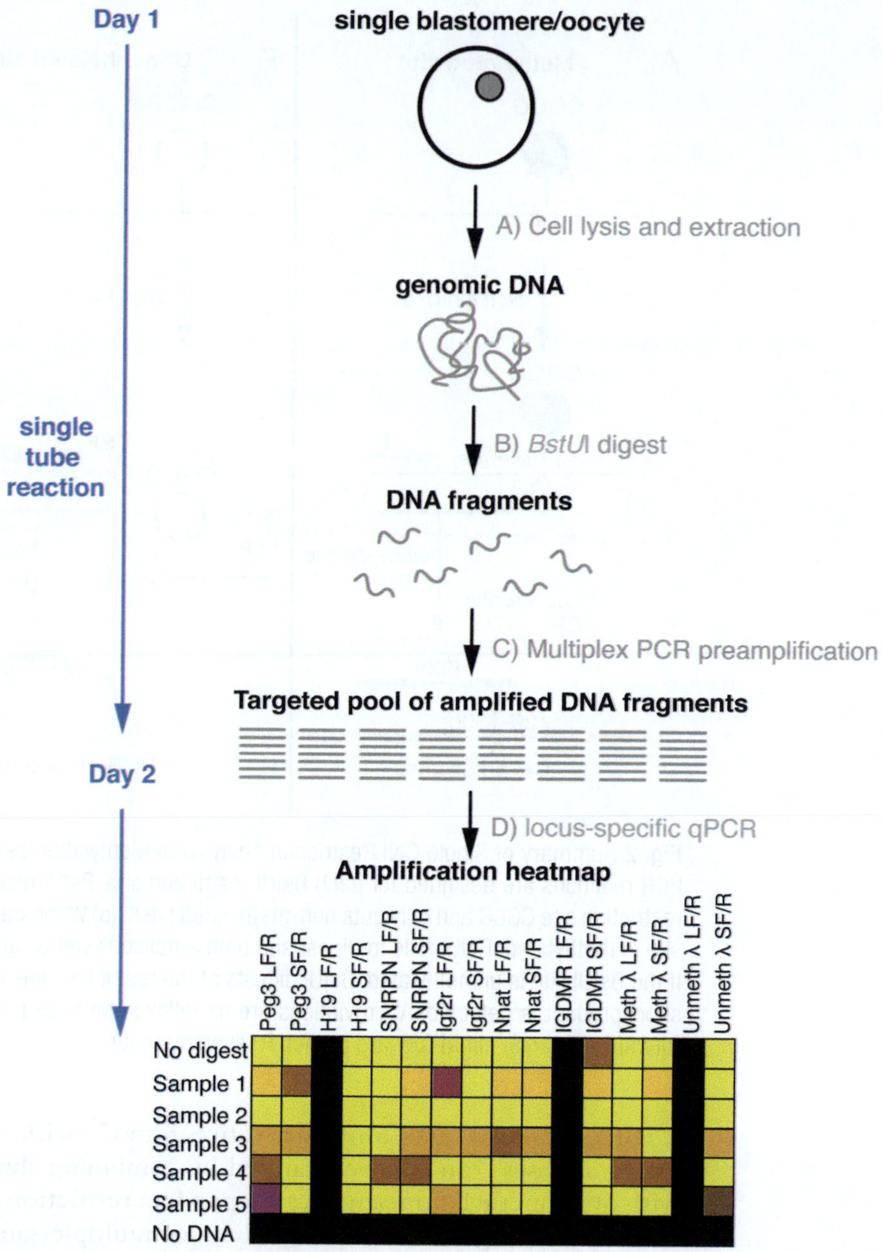

Fig. 3 Workflow of SCRAM assay consists of four major steps: (**a**) single cell isolation and extraction; (**b**) enzymatic digest by BstUI; (**c**) PCR preamplification of all targets of interest; (**d**) locus-specific analysis by qPCR on the fluidigm platform. Abbreviation are as follows: LF long forward primer, Meth methylated, SF short forward primer, R reverse primer, Unmeth unmethylated

Day 1

1. Isolation of single cells and DNA extraction (2 hour [h] 15 minutes [min]; Fig. 3a).
2. MSRE digest and inactivation (2 h 20 min; Fig. 3b).

3. Multiplex preamplification PCR (2.5 h; Fig. 3c).
Day 2
4. Locus-specific (qPCR) assay (3 h 45 min; Fig. 3d).
5. Data analysis.

2 Materials

2.1 Preparation of Methylated and Unmethylated Spike-in Controls

1. GoTaq Flexi DNA polymerase (Promega).
2. Magnesium chloride, 25 mM (Promega).
3. dNTP mix, 10 mM (Promega).
4. Primers, 100 μM (Integrated DNA Technologies in LabReady format; *see* **Note 1**).
5. CpG DNA methyltransferase M.SssI, 4 U/μL (New England BioLabs).
6. S-Adenosylmethionine (SAM), 32 mM (New England BioLabs).
7. Unmethylated λ-DNA (Promega). For preparation of controls (Subheading 3.2).
8. QIAQuick PCR purification kit (Qiagen).
9. NanoDrop 8000 spectrophotometer (Thermo Scientific).

2.2 Cell Lysis and DNA Extraction

1. CellsDirect resuspension and lysis buffer (Invitrogen).
2. Qiagen protease, 7.5 Anson units (AU) (Qiagen; *see* **Note 2**).

2.3 MSRE Digest

1. *BstU*I, 10 U/μL (New England BioLabs).
2. Diluent A (New England BioLabs).
3. NEBuffer 4 (New England BioLabs).
4. Proteinase K (recombinant; PCR grade), 20 mg/mL (Thermo Scientific).

2.4 Multiplex Preamplification PCR

1. GoTaq Flexi DNA polymerase (Promega).
2. Magnesium chloride, 25 mM (Progmega).
3. dNTP mix, 10 mM (Promega).
4. Nuclease-free water (Ambion).
5. Exonuclease I, 20 U/μL (Enzymatics).

2.5 Locus-Specific Microfluidic qPCR Analysis

1. Integrated fluidic circuit (IFC) controller MX (Fluidigm).
2. Biomark system (Fluidigm).
3. Fluidigm real-time PCR analysis software (Fluidigm).
4. SsoFast EvaGreen Supermix with Low ROX (Bio-Rad).
5. Assay-loading reagent, 2× (Fluidigm).

6. Syringes of Control loading fluid (Fluidigm; 2 per chip).
7. Fluidigm 48.48 Dynamic Array chip (Fluidigm).
8. Mx300P 96-well plates (Agilent Technologies).

2.6 General Supplies

1. PCR strip tubes (Axygen).
2. Barrier-type pipette tips.
3. Nuclease-free water (Ambion).
4. TAE buffer, 50×, pH 8.0.
5. Agarose.
6. DNA-binding dye sample-loading reagent, 20× (Fluidigm).
7. TE buffer, 1× (Promega).
8. Primers, 100 μM (Integrated DNA Technologies in LabReady format; *see* Table 1, *see* **Note 1**).
9. DNA Engine Tetrad 2 Peltier thermal cycler (Bio-Rad).

2.7 List of Verified and Designed Primers for BstUI MSRE Sites

See Table 1.

Table 1
List of verified and designed primers for *BstU*I MSRE sites

Gene	Primer	Sequence (5′ to 3′)	Chromosome location
Peg3	Short forward	GCTCCCAAGGGTAACTGACA	Chr7:6730081–6730206
	Long forward	CGTCTGCAGAGTTCAGATGG	
	Reverse	GTGGTGATTCCCCCTTCC	
H19	Short forward	GATTGCGCCAAACCTAAAGA	Chr7:60005064–60005205
	Long forward	GACCATGCCCTATTCTTGGA	
	Reverse	ACAGCATTGCCATTTGTGAA	
Snrpn	Short forward	ACTAGCGCAGAGAGGAGAGG	Chr7:60005064–60005205
	Long forward	CATTGCGGCAAAAATGTG	
	Reverse	CTCCTCAGAACCAAGCGTCT	
Igf2r	Short forward	TTTGAGCTTGCCTCTCTTGC	Chr17:12742534–12742678
	Long forward	ATAGCCAGGATAGCGCCAAA	
	Reverse	GTTCTGTGATCAGGGCCAAC	

(continued)

Table 1 (continued)

Gene	Primer	Sequence (5′ to 3′)	Chromosome location
Nnat	Short forward	GTAGGCTTGGGTGTGTCGAA	Chr2:157559738–157559882
	Long forward	GTAACTTCGGCTTTCCATCC	
	Reverse	AGGGAAGTCGTGGCTCTACA	
IGDMR	Short forward	AGCCGCTATGCTATGCTGTT	Chr12:109528377–109528536
	Long forward	CCGTGTACTAATGCCGCTTC	
	Reverse	CAAAATAATGCAGCCCTTCC	
Methylated λ	Short forward	TAGGCATCACCGAAAATTCA	
	Long forward	CTGCGAAAACTTGACCTTTCT	
	Reverse	TGTGGTGATATAGGACAGACAAAAA	
Unmethylated λ	Short forward	ATGCCCACACAAGTGGTTTA	
	Long forward	GATTACGGCACCAAATCGAC	
	Reverse	TCCTGTAATGGTTGCTGTTCC	

3 Methods

3.1 Design of Methylation Site Assays

Here, we provide the details for a SCRAM assay to address DNA methylation states at six imprinted gene regions, using the above-mentioned primers (Subheading 2.7) designed based on *BstUI* digest. For the establishment of new, custom assays, proceed with the following assay design (*see* **Note 3**):

1. Identify regions of interest and determine exact chromosome location of methylated regions by reviewing literature and/or annotated genome browser (e.g., University of California, Santa Cruz genome browser).
2. Identify an MSRE site that overlaps with a CpG that lies within the region of interest. Ideally, based on previously published bulk DNA methylation data, it should be determined whether there is a high correlation between the methylation status of that site and other CpG sites within the locus of interest.
3. Design, with Primer3 software (default parameters in version 0.4.0; http://primer3.ut.ee), a set of two forward primers and one reverse primer (Fig. 2) that flank the restriction site of interest. PCR amplicon size should be less than 150 base pairs (bp) for the long primer set and more than 60 bp for the short primer set, i.e., total product size range field: 60–150 bp (*see* **Notes 4** and **5**). Ideally, the size range should be between 80 and 120 bp.

4. Design the primers for the long amplicon, ideally the forward primer should be closely adjacent to the restriction site (the restriction site should be highlighted in the source sequence field with "[ ]"; Fig. 2a), while the reverse primer is at the other extreme of the amplicon.
5. Copy the reverse primer into the right primer field, and exclude the amplicon extending up to the restriction site by indicating the region with "< >" (Fig. 2b). Remove the "[ ]" that flanks the restriction site from the source sequence field in the previous step. Indicate product size range in field as 60–120 bp.
6. Primer specificity can be verified with free programs available online.
7. Verify designed primers via traditional single PCR and multiplexed PCR/qPCR on bulk genomic DNA for correct amplicon sizes of PCR products on agarose gels (*see* **Notes 5** and **6**).
8. Also confirm that primers are working in SYBR green qPCR assays.

3.2 Spike-in DNA (~3 days)

1. Two regions suitable for methylated and unmethylated controls containing a *BstUI* site, can be amplified from λ-DNA with primers described in table above (Subheading 2.7; *see* **Note 7**).
2. Prepare PCR master mix:

Reagent	Volume (μL)	Final concentration
GoTaq buffer (5×)	10	1×
dNTPs (10 mM)	1	0.2 mM
$MgCl_2$ (25 mM)	5	2.5 mM
GoTaq DNA polymerase (5 U/μL)	0.25	1.25 U
Forward and reverse primers (5 μM)	5	0.5 μM
Unmethylated λ-DNA	1	200 ng
Nuclease-free water	27.75	

3. Run the PCR reaction on the thermocycler with these conditions:

Cycle number	Denature	Anneal	Extend	Final
1	95 °C, 10 min			
2–31	95 °C, 30 s	60 °C, 1 min	72 °C, 1 min	
32			72 °C, 10 min	
33				4 °C, hold

4. Clean up products with QIAGEN PCR purification kit and elute with 40 μL of elution buffer. Measure concentration of product with Nanodrop.
5. Prepare reaction as described below, and incubate at 37 °C overnight (~14 h). Keep a frozen aliquot of each amplicon. These will be the unmethylated controls.

Reagent	Volume (μL)	Final concentration
Purified product and nuclease-free water	40.5	1.2 μg
M.SssI	2	40 U
10× NEBuffer 4	5	
20× SAM (3.2 mM)	2.5	

6. Incubate the DNA methylation reaction 37 °C overnight.
7. On day 2, combine the overnight methylated reaction with 2 μL of M.SsssI and 2.5 μl of 20× SAM.
8. Incubate for 6 h at 37 °C and 20 min at 65 °C to inactivate enzymes.
9. Clean up methylated products with PCR purification kit and elute with 40 μL of elution buffer (*see* **Note 8**).
10. To verify methylation efficiency, digest all methylated and unmethylated control products with *BstUI*. Run digest products on 2% (wt/vol) agarose gel. If a digested band is present in methylated control, then **steps 5–10** need to be repeated.
11. Measure concentration of the unmethylated and methylated control products on Nanodrop.

3.3 Cell Lysis and DNA Extraction (4 h)

1. Separate cells of interest into single cells (if necessary) (*see* **Note 9**).
2. For each sample, prepare lysis buffer by adding 4.5 μL of frozen aliquot of CellsDirect Resuspension Solution to 0.5 μL of CellsDirect Lysis Enhancer just prior to use (*see* **Note 10**). Aliquot 5 μL of this lysis buffer mixture into individual PCR tubes. Transfer single blastomeres/oocytes directly into lysis buffer (*see* **Notes 11–13**).
3. Include a negative control with no cell added, and a no enzyme control.
4. Lyse immediately in using a thermocycler (*see* **Note 12**): ~28 °C, 5 min; 75 °C, 10 min; hold at 4 °C.
5. Mix 0.5 μL protease (3.75 AU/mL) to 0.5 μL nuclease-free water. Add 1 μL of diluted protease reagent (1.875 AU/mL; *see* **Notes 14** and **15**) to 5 μL of lysed cells. Spin down briefly.

6. Carry out protease reaction in thermocycler:

Time	Temperature
90 min	50 °C
30 min	70 °C
Hold	4 °C

3.4 MSRE Digest (4.5 h)

1. Prepare *BstUI* (CpG Methylation-sensitive) digestion reaction for samples (from Subheading 3.3) as follows:

Component	Volume (μL)
BstUI (10 U/μL)	0.5
NEB buffer 4 (10×)	1
Methylated control DNA (1 attogram/μL)	1
Unmethylated control DNA (1 attogram/μL)	1
Nuclease-free water	0.5
DNA/protease mix (Subheading 3.3)	6

2. Prepare a reaction for no digest control as above, but by replacing *BstUI* with 0.5 μL of Diluent A (*see* Subheading 2.3, **item 2**).
3. Briefly spin down reactions and carry out digest in thermocycler:

Time	Temperature
2 h	60 °C
Hold	4 °C

4. To inactivate the digest reaction, add 1 μL of proteinase K (10 mg/mL) and incubate in thermocycler:

Time	Temperature
2 h	50 °C
10 min	95 °C
Hold	4 °C

3.5 Multiplex Preamplification PCR

1. Prepare multiplex PCR reaction as described in table (*see* **Notes 14**, **16**, and **17**):

Component	Volume (μL)
GoTaq clear buffer (5×)	5
$MgCl_2$ (25 mM)	2.5
dNTPs	0.5
Preamplification primer mix (500 nM)	2.5
GoTaq DNA polymerase	0.125
Nuclease-free water	4.375
DNA/protease/*BstUI* mix (Subheading 3.4)	10

2. Briefly spin down and carry out reaction in thermocycler:

Cycle	Denature	Anneal	Final
1	95 °C 10 min		
2—27	95 °C 30s	60 °C 4 min	
28			4 °C hold

3. (*PAUSE point*) The product can be stored frozen for several months.

3.6 Locus-Specific Microfluidic qPCR Analysis (Fluidigm Analysis)

1. To eliminate remaining primers from the previous preamplification step, add 2 μL of exonuclease I (enzymatics; 4 U/uL) per 5 μL of PCR reaction (Subheading 3.5; *see* **Note 2**).
2. Carry out reaction on thermocycler: 30 min 37 °C; 15 min 80 °C.
3. To prepare for Fluidigm qPCR assay, dilute sample by adding 43 μL of nuclease-free water per 7 μL of exonuclease-treated sample.
4. Arrange and prepare sample mixes as described in 96-well PCR plate: A1–A12, B1–B12, C1–C12, and D1–D12. Due to the large number of multiplex reactions, loading of reactions needs to be carried out in a specific manner to help keep track of samples. Reaction volume for each sample (*see* **Notes 2, 17,** and **18**):

Component	Volume (μL)
SsoFast EvaGreen Supermix with low ROX (2×)	3.5
DNA-binding dye sample loading reagent (20×)	0.35
Diluted exonuclease I-treated sample (Subheading 3.6, **step 1**)	3.15

5. Use a sealing sticker or aluminum foil to cover the 96-well and keep on ice after spinning down. Also include control samples without MSRE digest or without DNA/cell added.
6. Prepare qPCR assay mix in a 96-well plate in loading order (similar to Subheading 3.6, **step 4**) using locus-specific primer mixes prepared (*see* **Note 19**).
7. Prepare qPCR assay mix reagent mix:

Component	Volume (μL)
Assay-loading buffer (2×)	5
Locus-specific primer mix (mix 1 or 2; 50 μM each)	1
Nuclease-free water	4

8. Keep on ice after spinning down reactions. Cover plate with sealing sticker or aluminum foil until loading.
9. Prime Fluidigm chip by loading accompanying control line fluid into Fluidigm 48.48 Dynamic Array and selecting the "prime chip" program on the IFC controller.
10. After priming step, load samples and assay mix onto Fluidigm chip. Add 5 μL assay mix from qPCR assay plate (Subheading 3.6, **step 7**) into corresponding "assay inlet" (left side of chip) and add 5 μL of sample mix (Subheading 3.6, **step 4**) into corresponding "sample inlet."
11. Return array into IFC controller. Run "load mix" program to sort sample and assay reactions into individual chambers.
12. Take array out of IFC controller and remove sticker below the chip (*see* **Note 20**).
13. Place array into Biomark reader and carry out the locus-specific qPCR as described:

Cycle	Denature	Anneal	Melting Curve
1	95 °C, 10 min		
2–31	95 °C, 15 s	60 °C, 1 min	
32			60–95 °C, ramp

3.7 Data Analysis and Visualization

1. Results will be available in the Biomark reader. Transfer files into flash drive to facilitate analysis of results at remote location.
2. Download Fluidigm real-time PCR analysis software (https://www.fluidigm.com/software) (*see* **Note 21**).
3. Create heat map with downloaded software (*see* **Notes 22–28**).
4. The following criterion will be helpful for data analysis: (1) eliminate assays where measured melting temperature differs

from expected by more than 1.5 °C; (2) eliminate samples where unmethylated-λ control assay has both products amplified; (3) eliminate samples where the short fragment is not amplified: this suggests that the assay was not efficient and not a reliable readout on the long assay; and (4) the cutoff value for qPCR C_t values is 30 cycles, beyond which, it is assumed that there is no amplification of product.

4 Notes

1. Primers can be stored frozen or in the fridge for short-term purposes (6 months).
2. Carry out locus-specific qPCR at a separate area with separate equipment to prevent contamination.
3. We have successfully used *HpaII* and *BstUI* for SCRAM.
4. If primers cannot be designed as described (Fig. 2; Subheading 3.1), the order of primers can be reversed such that the long forward is on the right of the restriction site while the short forward and reverse primer are on the left of the restriction site.
5. In the case where primers fail to amplify in traditional PCR, multiplex PCR, or qPCR assays, redesign primers for the same or alternative site according to Subheading 3.1. Optimizing annealing temperature for multiplex PCR is not ideal because of reduced specificity for the primers in the reaction.
6. Individual qPCR products can be visualized on agarose gels to verify specificity of assay.
7. Other sources of DNA can be used for methylation controls, but new primers that flank the *BstUI* site need to be designed.
8. Keep frozen aliquots of methylated and unmethylated controls. Stock solution can be kept in the fridge for a month. Make up fresh working solution to avoid adsorption of control DNA.
9. Single cell isolation in early embryos is straightforward in eggs, sperm, or zygotes. However, beyond these cell types, micromanipulation will be needed to separate the growing embryo. Micromanipulation is only suitable for larger cells, like preimplantation embryos, and for smaller cell numbers [44, 45]. We have previously managed to separate up to 8-cell blastomeres [44]. Beyond 8-cell stage, cells may need to be separated by quick and gentle enzyme treatment with trypsin [51]. For blastocysts, antibody complement treatment can be used to separate the extraembryonic trophoblast layer from the embryonic inner cell mass [52]. Flow cytometry (FACS) and microfluidics-based platforms are other common ways to separate a large group of cells [53, 54]. Laser capture can isolate single cells in specific locations from histological sections [55].

10. It is recommended not to freeze/thaw resuspension solution, so prepare frozen aliquots for storage. Add lysis enhancer only just prior use. Minimize movement in buffer/embryo mix upon addition of sample.
11. Visually verify that only a single cell is added into the lysis buffer under the microscope.
12. Overnight storage of cells or lysates increased dropouts and decreased sample quality in MSRE assay [45].
13. Thermal and mechanical stress should be avoided to prevent DNA double-stranded breaks, which leads to false negatives and affects digest and PCR efficiency [45].
14. Filter tips need to be changed between each sample to avoid cross-contamination.
15. Prepare protease stock (3.75 AU/mL) by adding 2 mL of nuclease-free water into the provided 7.5 AU vial. Dissolve well by pipetting and divide into 100 μL aliquots. Store frozen aliquots at −20 °C; the stock is good in the fridge for 6 months.
16. For the multiplex PCR reaction, prepare a preamplification primer mix by combining primers for all loci, including control sites, at 500 nM each. Top up total volume to 1 mL with nuclease-free water. The preamplification mix is good for 6 months in the fridge.
17. Include an undigested single cell DNA control in multiplex PCR and locus-specific qPCR assay to analyze specificity of PCR reaction.
18. Technical replicates are not required for locus-specific qPCR analysis because the results are binary and C_t values are not quantified like a traditional qPCR assay.
19. Prepare 2 primer stocks (50 μM) for each microfluidics-based qPCR assay. Primer stock 1 combines the long forward and reverse primers while primer stock 2 combines the short forward and reverse primers. Aliquots of primer mixes can be stored at 4 °C (6 months) or frozen.
20. Do not touch the area covered by the protective film. Remove any dust particles from chip surface.
21. Fluidigm real-time PCR analysis software is only compatible with windows interface. It is available for free on the fluidigm website.
22. If PCR reaction is unspecific from the multiplex step (i.e., broad, multiple peaks or incorrect melting temperature), design nested reverse primers to replace reverse primer and improve specificity of locus-specific amplification.
23. If heatmap from Fluidigm is mostly black or has low amplification values, it is likely that the assay failed to amplify properly. Primers need to be redesigned or multiplex cycle step needs to be optimized.

24. In the case where long assay is amplified for unmethylated λ-control, either increase amount of MSRE or digest incubation time. For example, after 2 h of initial digest, spiking of a second dose of MSRE and subsequently leaving for overnight digest before carrying out multiplex preamplification can be attempted.
25. If long assay is not amplified for methylated λ-control, there might be excessive digest of template. Reduce amount of restriction enzyme and shorten digest time.
26. If high C_t values are observed in locus-specific qPCR analysis, increase cycles in multiplex PCR or redesign assay primers.
27. If amplification is observed in no cell/DNA control, check specificity of product in melting curve and repeat locus-specific qPCR with new reagents.
28. If C_t value is low for short assay, verify that cell is placed in lysis buffer and minimize splashes in handling. Passive reference dye (ROX) can help indicate whether loading was properly done on fluidigm chip.

References

1. Messerschmidt DM, Knowles BB, Solter D (2014) DNA methylation dynamics during epigenetic reprogramming in the germline and preimplantation embryos. Genes Dev 28(8):812–828
2. Wu TP et al (2016) DNA methylation on N6-adenine in mammalian embryonic stem cells. Nature 532(7599):329–333
3. Li E, Bestor TH, Jaenisch R (1992) Targeted mutation of the DNA methyltransferase gene results in embryonic lethality. Cell 69(6):915–926
4. Okano M, Xie S, Li E (1998) Cloning and characterization of a family of novel mammalian DNA (cytosine-5) methyltransferases. Nat Genet 19(3):219–220
5. Okano M et al (1999) DNA methyltransferases Dnmt3a and Dnmt3b are essential for de novo methylation and mammalian development. Cell 99(3):247–257
6. Bourc'his D, Bestor TH (2004) Meiotic catastrophe and retrotransposon reactivation in male germ cells lacking Dnmt3L. Nature 431(7004):96–99
7. Bourc'his D et al (2001) Dnmt3L and the establishment of maternal genomic imprints. Science 294(5551):2536–2539
8. Hirasawa R et al (2008) Maternal and zygotic Dnmt1 are necessary and sufficient for the maintenance of DNA methylation imprints during preimplantation development. Genes Dev 22(12):1607–1616
9. Kaneda M et al (2004) Essential role for de novo DNA methyltransferase Dnmt3a in paternal and maternal imprinting. Nature 429(6994):900–903
10. Gjerset RA, Martin D (1982) Presence of a DNA demethylating activity in the nucleus of murine erythroleukemic cells. J Biol Chem 257(15):8581–8583
11. Kurimoto K et al (2008) Complex genome-wide transcription dynamics orchestrated by Blimp1 for the specification of the germ cell lineage in mice. Genes Dev 22(12):1617–1635
12. Kagiwada S et al (2013) Replication-coupled passive DNA demethylation for the erasure of genome imprints in mice. EMBO J 32(3):340–353
13. He Y-F et al (2011) Tet-mediated formation of 5-carboxylcytosine and its excision by TDG in mammalian DNA. Science 333(6047):1303–1307
14. Inoue A et al (2011) Generation and replication-dependent dilution of 5fC and 5caC during mouse preimplantation development. Cell Res 21(12):1670–1676
15. Ito S et al (2011) Tet proteins can convert 5-methylcytosine to 5-formylcytosine and 5-carboxylcytosine. Science 333(6047):1300–1303

16. Okamoto Y et al (2016) DNA methylation dynamics in mouse preimplantation embryos revealed by mass spectrometry. Sci Rep 6:19134
17. Popp C et al (2010) Genome-wide erasure of DNA methylation in mouse primordial germ cells is affected by AID deficiency. Nature 463(7284):1101–1105
18. Santos F et al (2002) Dynamic reprogramming of DNA methylation in the early mouse embryo. Dev Biol 241(1):172–182
19. Rougier N et al (1998) Chromosome methylation patterns during mammalian preimplantation development. Genes Dev 12(14):2108–2113
20. Mayer W et al (2000) Embryogenesis: demethylation of the zygotic paternal genome. Nature 403(6769):501–502
21. Oswald J et al (2000) Active demethylation of the paternal genome in the mouse zygote. Curr Biol 10(8):475–478
22. Howell CY et al (2001) Genomic imprinting disrupted by a maternal effect mutation in the Dnmt1 gene. Cell 104(6):829–838
23. Ratnam S et al (2002) Dynamics of Dnmt1 methyltransferase expression and intracellular localization during oogenesis and preimplantation development. Dev Biol 245(2):304–314
24. Peaston AE et al (2004) Retrotransposons regulate host genes in mouse oocytes and preimplantation embryos. Dev Cell 7(4):597–606
25. Macfarlan TS et al (2012) Embryonic stem cell potency fluctuates with endogenous retrovirus activity. Nature 487(7405):57–63
26. Messerschmidt DM et al (2012) Trim28 is required for epigenetic stability during mouse oocyte to embryo transition. Science 335(6075):1499–1502
27. Quenneville S et al (2012) The KRAB-ZFP/KAP1 system contributes to the early embryonic establishment of site-specific DNA methylation patterns maintained during development. Cell Rep 2(4):766–773
28. Ferguson-Smith AC (2011) Genomic imprinting: the emergence of an epigenetic paradigm. Nat Rev Genet 12(8):565–575
29. Tomizawa S-i, Sasaki H (2012) Genomic imprinting and its relevance to congenital disease, infertility, molar pregnancy and induced pluripotent stem cell. J Hum Genet 57(2):84–91
30. Lee JT, Bartolomei MS (2013) X-inactivation, imprinting, and long noncoding RNAs in health and disease. Cell 152(6):1308–1323
31. Bird A (2002) DNA methylation patterns and epigenetic memory. Genes Dev 16(1):6–21
32. Colella S et al (2003) Sensitive and quantitative universal pyrosequencing™ methylation analysis of CpG sites. BioTechniques 35(1):146–151
33. Guo H et al (2013) Single-cell methylome landscapes of mouse embryonic stem cells and early embryos analyzed using reduced representation bisulfite sequencing. Genome Res 23(12):2126–2135
34. Bibikova M et al (2006) Human embryonic stem cells have a unique epigenetic signature. Genome Res 16(9):1075–1083
35. El Hajj N et al (2011) Limiting dilution bisulfite (pyro) sequencing reveals parent-specific methylation patterns in single early mouse embryos and bovine oocytes. Epigenetics 6(10):1176–1188
36. Smallwood SA et al (2014) Single-cell genome-wide bisulfite sequencing for assessing epigenetic heterogeneity. Nat Methods 11(8):817–820
37. Farlik M et al (2015) Single-cell DNA methylome sequencing and bioinformatic inference of epigenomic cell-state dynamics. Cell Rep 10(8):1386–1397
38. Eads CA et al (2000) MethyLight: a high-throughput assay to measure DNA methylation. Nucleic Acids Res 28(8):E32
39. Ehrich M et al (2005) Quantitative high-throughput analysis of DNA methylation patterns by base-specific cleavage and mass spectrometry. Proc Natl Acad Sci U S A 102(44):15785–15790
40. Down TA et al (2008) A Bayesian deconvolution strategy for immunoprecipitation-based DNA methylome analysis. Nat Biotechnol 26(7):779–785
41. Weber M et al (2005) Chromosome-wide and promoter-specific analyses identify sites of differential DNA methylation in normal and transformed human cells. Nat Genet 37(8):853–862
42. Melnikov AA et al (2005) MSRE-PCR for analysis of gene-specific DNA methylation. Nucleic Acids Res 33(10):e93
43. Kantlehner M et al (2011) A high-throughput DNA methylation analysis of a single cell. Nucleic Acids Res 39(7):e44
44. Lorthongpanich C et al (2013) Single-cell DNA-methylation analysis reveals epigenetic chimerism in preimplantation embryos. Science 341(6150):1110–1112
45. Cheow LF et al (2015) Multiplexed locus-specific analysis of DNA methylation in single cells. Nat Protoc 10(4):619–631
46. Smith ZD et al (2012) A unique regulatory phase of DNA methylation in the early mammalian embryo. Nature 484(7394):339–344

47. Smallwood SA et al (2011) Dynamic CpG island methylation landscape in oocytes and preimplantation embryos. Nat Genet 43(8):811–814
48. Olek A, Oswald J, Walter J (1996) A modified and improved method for bisulphite based cytosine methylation analysis. Nucleic Acids Res 24(24):5064–5066
49. Barrera V, Peinado MA (2012) Evaluation of single CpG sites as proxies of CpG island methylation states at the genome scale. Nucleic Acids Res 40(22):11490–11498
50. Cheow LF et al (2016) Single-cell multimodal profiling reveals cellular epigenetic heterogeneity. Nat Methods 13(10):833–836
51. Evans MJ, Kaufman MH (1981) Establishment in culture of pluripotential cells from mouse embryos. Nature 292(5819):154–156
52. Handyside A, Barton S (1977) Evaluation of the technique of immunosurgery for the isolation of inner cell masses from mouse blastocysts. J Embryol Exp Morphol 37(1):217–226
53. Shapiro HM (2005) Practical flow cytometry. John Wiley & Sons, Hoboken, NJ
54. Suarez-Quian C et al (1999) Laser capture microdissection of single cells from complex tissues. BioTechniques 26(2):328–335
55. Suarez-Quian C et al (1999) Laser capture microdissection of single cells from complex tissues. BioTechniques 26(2):328–335

Chapter 13

Use of Chemicals to Inhibit DNA Replication, Transcription, and Protein Synthesis to Study Zygotic Genome Activation

Kyungjun Uh and Kiho Lee

Abstract

Maternal-to-zygotic transition is an event that developmental control of early embryos is switched from oocyte-derived factors to the zygotic genome. Ability to inhibit DNA replication, transcription, and translation is an important tool in studying events, such as zygotic genome activation, during embyogenesis. Here, we describe approaches to block DNA replication, transcription, and translation using chemical inhibitors. Then we also demonstrate how the transcript level of a maternally inherited gene, ten-eleven translocation methylcytosine dioxygenase 3, responses to the chemical treatments.

Key words Pig embryos, Oocytes, Zygotic genome activation, α-Amanitin, Cycloheximide, *TET3*

1 Introduction

Embryo development begins at fertilization by two different haploid germ cells, sperm and oocyte, forming diploid zygote. mRNAs and proteins accumulated in the oocytes drive early embryo development because early embryos are transcriptionally inactive until zygotic genome activation (ZGA) [1]. During ZGA, maternal mRNAs and proteins that are necessary for oocyte maturation and early embryogenesis are degraded, and embryo development is programmed by newly synthesized products from activated zygotic genome [2]. This transition occurs at diverse cleavage stages by species, 2-cell stage in the mouse [3] 4- to 8-cell stage in the human [4] and pig [5] and 8- to 16-cell stage in the cow [6] and sheep [7].

The amount of maternally derived transcripts greatly outnumbers the zygotic RNA molecules even after ZGA [8]. This limits distinguishing zygotic expression of genes from the maternal RNAs. To follow dynamic events during ZGA, pharmaceutical approaches have been used. For instance, in early 1970, α-amanitin, an inhibitor of RNA polymerase II, was used to block mRNA synthesis during preimplantation stage [9, 10]. Most of mouse embryos treated with α-amanitin at 1–2-cell stage were not able to

Kiho Lee (ed.), *Zygotic Genome Activation: Methods and Protocols*, Methods in Molecular Biology, vol. 1605,
DOI 10.1007/978-1-4939-6988-3_13, © Springer Science+Business Media LLC 2017

develop beyond 2-cells. Similarly, when α-amanitin was introduced into 4-cell pig embryos, their development was arrested at between 4- and 8-cells [11]. Use of these inhibitors has been critical in elucidating mechanism of ZGA.

Ten-eleven translocation methylcytosine dioxygenase 3 (*TET3*), a member of ten-eleven translocation family, is an enzyme responsible for demethylating paternal genome after fertilization by initiating hydroxylation of 5-methylcytosine (5mC) [12]. The expression level of Tet3 is high in zygotes, then reduced in 2-cell stage in the mouse. Based on the mouse study, Tet3 expression was suggested to be zygotic-specific phenomenon. However, we found that *TET3* expression was highly maintained until 2-cell stage and the level was decreased at 4-cell stage in pig embryos [13]. The study indicates that TET3 expression is not zygotic specific but the gene is active until ZGA; major ZGA is at the 2-cell stage in the mouse and 4-cell stage in the pig.

Here, a method to inhibit transcription and translation using chemical inhibitors onset of ZGA is described. Expression of *TET3*, a maternally derived transcript, was measured from embryos that were incubated in the presence of inhibitors compared to a control group. This chapter also describes in vitro production of pig embryos through in vitro fertilization (IVF), in vitro culture (IVC), and quantitative real-time RT-PCR.

2 Materials

2.1 Materials Required for in Vitro Maturation

2.1.1 Lab Equipment

1. CO_2 incubator.
2. Stereo microscope with warm plate.
3. Microneedles (18-gauge).
4. Syringes (10 ml).
5. Centrifuge tubes (15 and 50 ml).
6. Petri dishes (30 × 10 mm and 100 × 25 mm).
7. 4-Well culture dishes.
8. Glass capillary tubes.
9. Captrol III® micropipet (Drummond Scientific Company).

2.1.2 Reagents

1. Saline: 0.9% NaCl with 100 U/ml penicillin-streptomycin.
2. Hepes-buffered Tyrode's Lactate (TL-Hepes) medium: 2.0 mM $CaCl_2{\cdot}2H_2O$, 114.0 mM NaCl, 3.2 mM KCl, 2.0 mM $NaHCO_3$, 0.4 mM NaH_2PO_4, 10.0 mM Na Lactate (60% syrup), 0.5 mM $MgCl_2{\cdot}6H_2O$, 10.0 mM Hepes, 12.0 mM sorbitol, 0.2 mM sodium pyruvate, 0.075 g/L penicillin, 0.05 g/L streptomycin, 1 ml phenol red (0.5%), 0.1 g/L polyvinyl alcohol (PVA); pH 7.4.

3. IVM medium: Medium 199 supplemented with 3.05 mM glucose, 0.91 mM sodium pyruvate, 0.57 mM cysteine, 10 ng/ml epidermal growth factor (EGF), 0.5 μg/ml luteinizing hormone (LH), 0.5 μg/ml follicle stimulating hormone (FSH), 10 ng/ml gentamicin, and 0.1% polyvinyl alcohol (PVA); pH 7.4 (*see* **Note 1**).
4. IVM wash medium: IVM medium without LH and FSH; pH 7.4.
5. Mineral oil (suitable for embryo culture).

2.2 Materials Required for in Vitro Fertilization

2.2.1 Lab Equipment

1. CO_2 incubator.
2. O_2/CO_2 incubator.
3. Stereo microscope with warm plate.
4. Benchtop centrifuge.
5. Microcentrifuge tubes (1.5 ml).
6. Vortex mixer.
7. Petri dishes (30 × 10 mm).
8. Glass capillary tubes.
9. Captrol III® micropipet (Drummond Scientific Company).

2.2.2 Reagents

1. IVF medium (mTBM): modified Tris-buffered medium with 113.1 mM NaCl, 3 mM KCl, 7.5 mM $CaCl_2$, 11 mM glucose, 20 mM Tris, 2 mM caffeine, 5 mM sodium pyruvate, and 2 mg/ml BSA; pH 7.4.
2. Manipulation medium: medium 199 supplemented with 0.6 mM $NaHCO_3$, 2.9 mM Hepes, 30 mM NaCl, 10 ng/ml gentamicin, and 3 mg/ml BSA; pH 7.4.
3. Denuding medium: 0.3 M mannitol, 0.001% BSA, 0.03% hyaluronidase, 5% TL-Hepes medium in distilled water; pH 7.4.
4. AndroPRO® Plus semen extension medium (Mofa®).
5. Sperm wash medium: DPBS with 0.1% BSA and 10 ng/ml gentamicin.
6. Mineral oil (suitable for embryo culture).

2.3 Materials Required for in Vitro Culture

2.3.1 Lap Equipment

1. O_2/CO_2 incubator.
2. Stereo microscope with warm plate.
3. Petri dishes (30 × 10 mm).
4. Glass capillary tubes.
5. Captrol III® micropipet (Drummond Scientific Company).

2.3.2 Reagents

1. PZM3 culture medium [14].
2. Mineral oil (suitable for embryo culture).

2.4 Materials Required for Inhibition of RNA Transcription and Protein Translation at the Stage of ZGA

2.4.1 Lab Equipment

1. O_2/CO_2 incubator.
2. Stereo microscope with warm plate.
3. Petri dishes (30 × 10 mm).
4. Glass capillary tubes.
5. Captrol III® micropipet (Drummond Scientific Company).

2.4.2 Reagents

1. Culture medium to inhibit DNA replication: PZM3 supplemented with 3 μg/ml aphidicolin.
2. Culture medium to inhibit transcription: PZM3 supplemented with 20 μg/ml α-amanitin.
3. Culture medium to inhibit protein synthesis: PZM3 supplemented with 10 μg/ml cycloheximide.

2.5 Materials Required for Examining TET3 Expression at the Stage of ZGA

2.5.1 Lab Equipment

1. Heat block (adjustable to 65–80 °C).
2. Thermocycler.
3. Real-time PCR instrument.
4. 96-Well plates for real-time PCR instrument.
5. Adhesive films for 96-well plates.

2.5.2 Reagents

1. Dynabeads® mRNA DIRECT™ Micro Kit (Thermo Fisher Scientific).
2. Superscript™ III First-Strand cDNA Synthesis System (Invitrogen™).
3. IQ SYBR Green Supermix (Bio-Rad Laboratories).
4. Nuclease-free water.

3 Methods

3.1 In Vitro Maturation (IVM)

Since the first full-term development in the mouse from oocytes matured and fertilized in vitro [15], there have been great efforts to improve IVM in other species. In 1989, Mattiloli and his colleagues generated live piglets from oocytes maturated and fertilized in vitro [16].

1. 4-Well dishes containing 500 μl IVM medium covered with 450 μl mineral oil per well and two 30 mm IVM wash dishes with 3 ml IVM wash medium per dish are prepared in the morning and incubated for 4–6 h at 38.5 °C in an atmosphere containing 5% CO_2 and 100% humidity.
2. Porcine ovaries of pre-pubertal gilts or sows are obtained from slaughterhouse and transported to the laboratory.
3. The ovaries are washed with pre-warmed saline at 38.5 °C. Immature oocytes are aspirated from follicles (>2 mm

in diameter) using an 18-guage microneedle attached to a 10 ml syringe, and harvested into 50 ml centrifuge tube.

4. Oocytes aspirates are incubated for 20 min at 38.5 °C until oocytes and tissue pellets are formed on the bottom of the tube.
5. Supernatant is removed from the 50 ml conical tube and 40 ml TL-Hepes medium (pre-warmed at 38.5 °C) is added to wash and maintain physiological pH.
6. After incubating for 20 min, the washing step is repeated twice using the TL-Hepes medium.
7. Washed aspirates are placed in the 100 mm petri dish.
8. Cumulus-oocyte complexes (COCs) with evenly granulated cytoplasm and intact surrounding cumulus cells are collected using a finely drawn glass pipet under a microscope.
9. The collected COCs are washed with pre-incubated IVM wash medium in the 30 mm dish.
10. Approximately 50 COCs are placed in a well (4-well dish) containing 500 μl IVM medium and incubated for 42–44 h at 38.5 °C, 5% CO_2, and 100% humidity.

3.2 In Vitro Fertilization

Sperm penetration triggers oocyte activation and subsequently elevated calcium level induces cortical granule exocytosis [17, 18]. Zona pellucida is hardened by the released cortical granule to prevent polyspermy. In pigs, modified Tris-buffered medium (mTBM) is known to reduce polyspermy but not affecting the penetration rate [19–21].

3.2.1 Preparation of Oocytes

1. IVF medium (20 ml) is incubated at 38.5 °C in an atmosphere containing 5% CO_2 and 100% humidity 2 days before IVF.
2. IVF drops (each 50 μl) covered with mineral oil and two 3 ml washing media are prepared in 30 mm petri dishes using the incubated IVF medium on the day before the IVF. These plates and the rest of media are incubated at the same condition until the IVF day.
3. After the maturation for 42–44 h, COCs are transferred to 1.5 ml tube containing 1 ml denuding medium pre-warmed at 38.5 °C, and the expanded cumulus cells surrounding oocytes (Fig. 1a) are removed by vortexing for 3 min (*see* **Note 2**).
4. Three manipulation dishes are prepared during the vortexing: A 30 mm dish with 2.5 ml and another two dishes with 3.5 ml manipulation medium pre-warmed at 38.5 °C.
5. Denuded oocytes are moved to the first manipulation dish (2.5 ml) and collection of matured oocytes with a visible polar body (Fig. 1b) and washing are sequentially conducted in the other two dishes (*see* **Note 3**).

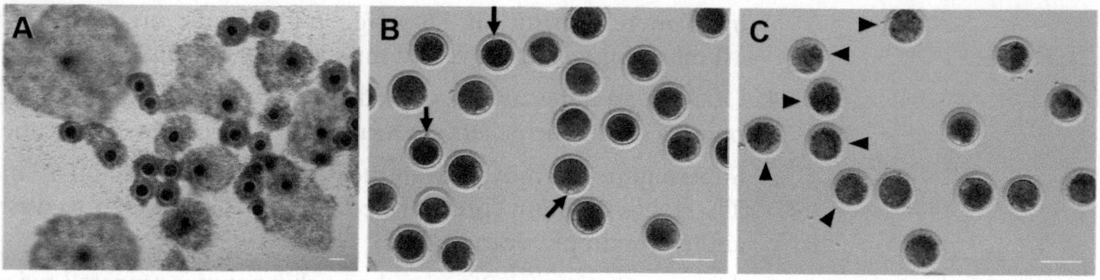

Fig. 1 Morphologies of pig oocytes after 40 h of IVM. (**A**) Oocytes surrounded with expanded cumulus cells after culture with chemically defined IVM medium. (**B**) Collected mature oocytes showing an obvious first polar body in the perivitelline space (black arrows). (**C**) Oocytes that failed to extrude the first polar body and dead oocytes (black arrow heads) lacking a plasma membrane. Scale bar = 100 μm

6. After washing of the collected oocytes with the incubated IVF medium (two 30 mm dishes), 25–30 collected oocytes are placed in 50 μl droplets of IVF medium (*see* **Note 4**).

3.2.2 Sperm Preparation

Fresh semen collected from boar is mixed with extension medium and stored at 17 °C up to 7 days.

1. For IVF, 1 ml semen is washed in 9 ml sperm wash medium by centrifugation at 750 × *g* for 3 min.
2. Sperm pellet is resuspended and washed twice more with 10 ml wash medium at the same centrifugation condition. Washed sperm pellet is resuspended and diluted to 2.5×10^5 using the incubated IVF medium (*see* **Note 5**).
3. Then, 50 μl sperm suspension is added to the prepared IVF droplets containing oocytes (final sperm concentration is 1.25×10^5) and incubated for 5 h at 38.5 °C in an atmosphere of 5% CO_2 and 100% humidity (Fig. 2).

3.3 In Vitro Culture

Developmental competency of in vitro produced (IVP) pig embryos is low compared to those of in vivo embryos. The first IVC medium to consistently support blastocyst development was NCSU23 [22], developed in North Carolina State University. Then porcine zygote medium (PZM) has emerged [14] by mimicking the composition of oviductal fluid.

1. PZM3 droplets (20 μl each) placed on 30 mm dish are covered with mineral oil and another two 30 mm dishes for washing are filled with 3 ml PZM3, followed by incubation at 38.5 °C, 5% O_2, and 5% CO_2 in humidified air during the 5 h of IVF.
2. Oocytes fertilized for 5 h are transferred to the incubated wash dish containing PZM3 using a glass pipet.
3. The remains of sperms unfertilized and attached to the oocytes are removed by repetitive passage of medium through glass pipet in the wash dish.

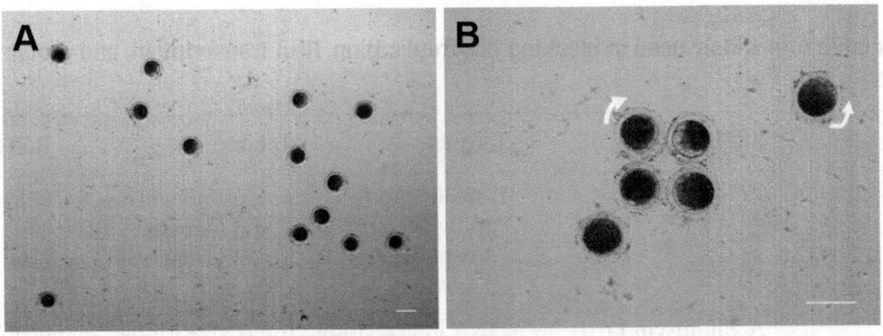

Fig. 2 Sperm and oocytes during the IVF procedure. (**A**) Addition of sperm to the IVF drop containing oocytes at a final concentration of 1.25 × 10^5. (**B**) A number of sperm bind to and rotate (white arrows) the oocytes immediately after the addition of sperm to the IVF drop. Scale bar = 100 μm

4. The oocytes are moved to the second wash dish.
5. 20–25 oocytes are placed in the pre-incubated PZM3 droplets and incubated at 38.5 °C, 5% O_2, and 5% CO_2 in humidified air.

3.4 Inhibition of DNA Replication, Transcription, and Translation at the Time of ZGA

The importance of ZGA was initially investigated in mouse embryos using α-amanitin, a specific inhibitor of RNA polymerase II [9, 10]. When this inhibitor was treated at 1- to 2-cell stage, mouse embryos could not develop beyond 2-cell stage suggesting ZGA occurs at 2-cell stage in the mouse. The ZGA is observed at 4–8-cell stage in the pig [11] and human [4]. Four-cell stage embryos treated with α-amanitin cannot complete the third cell cycle in the pig. Addition of cycloheximide, an inhibitor of protein synthesis, to culture medium blocks ZGA because not only zygotic proteins are not produced but also translation of maternally inherited transcripts required for ZGA is blocked [23, 24]. In this experiment, aphidicolin, α-amanitin, and cycloheximide are used to inhibit DNA replication, RNA transcription, and protein synthesis, respectively. The summary of chemical inhibitors widely used is described in Table 1.

1. Three different PZM3 media containing 3 μg/ml aphidicolin, 20 μg/ml α-amanitin, and 10 μg/ml cycloheximide, respectively, are prepared (*see* **Note 6**).
2. A 30 mm wash dish and 20 μl droplets covered with mineral oil in 30 mm dish are prepared with each PZM3 medium containing each inhibitor.
3. The wash and culture dishes are incubated at 38.5 °C, 5% O_2, and 5% CO_2 in humidified air for 4–6 h.
4. Embryos at 2-cell stage are collected after 24–36 h of IVF (*see* **Note 7**) and transferred to each pre-incubated wash dish containing the inhibitors. Assigned glass pipet should be used for handling each medium with different inhibitors to prevent contamination.

Table 1
Chemical inhibitors widely used in blocking DNA replication, RNA transcription, and protein translation

	Inhibitors	Targets	Actions	References
DNA replication	Aphidicolin	B-family DNA polymerases (α-polymerase)	Bind to polymerase α and inducing conformational changes	[25]
RNA transcription	Actinomycin D	DNA intercalation	Interfere with the elongation by the RNA polymerase	[26, 27]
	α-Amanitin	RNA polymerase II	Interfere with translocation of RNA polymerase II	[28, 29]
	DRB	CDK9 (a subunit of P-TEFb)	Interfere with the elongation by the RNA polymerase	[30]
	Flavopiridol	CDK9 (a subunit of P-TEFb)	Interfere with the elongation by the RNA polymerase	[31]
	Triptolide	XPB (a subunit of TFIIH)	Block RNA polymerase II-mediated transcription initiation	[32]
Protein translation	Cycloheximide	E site of the large ribosomal subunit	Block translation elongation	[33]
	Lactimidomycin	E site of the large ribosomal subunit	Block translation elongation	[33]

5. Embryos are placed in the droplets of PZM3 medium with inhibitors and incubated at 38.5 °C, 5% O_2, and 5% CO_2 in humidified air.
6. After 20 h, some embryos are collected for *TET3* expression assay and others are cultured for 20 h more to see their development (Fig. 3).

3.5 TET3 Expression at the Stage of ZGA

In mammalian zygote, 5mC level of paternal pronucleus is substantially decreased before the first cleavage [34, 35]. The 5mC of paternal pronucleus is oxidized and converted to 5-hydroxymethylcytosine (5hmC) by *TET3* [36, 37] at the zygote stage. *Tet3* expression is high at zygote stage, but significantly reduced at 2-cell stage in the mouse [12]. On the other hand, high expression

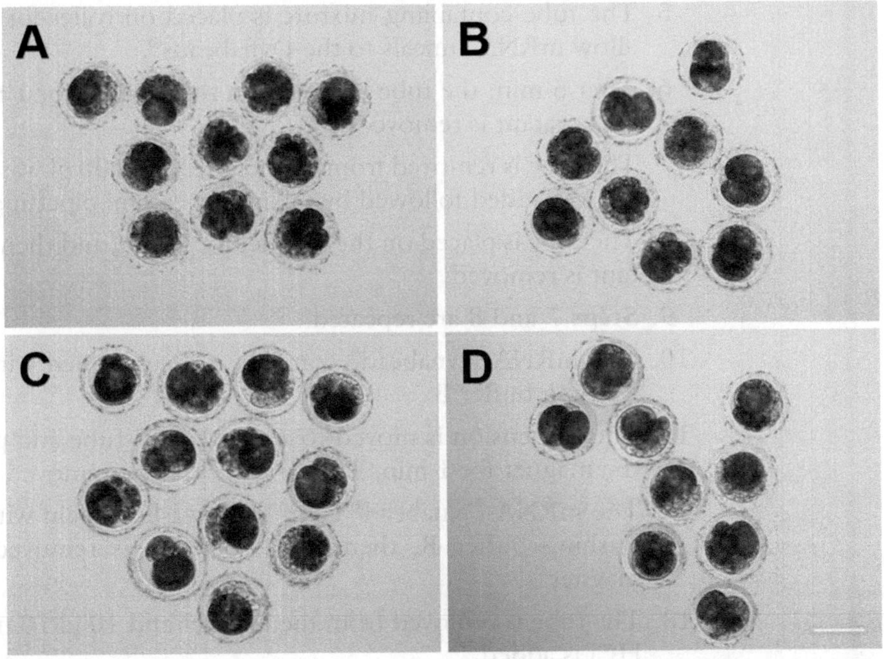

Fig. 3 Inhibition of RNA transcription, protein translation, and DNA replication in 2-cell stage embryos. Embryos at 2-cell stage are collected and cultured with inhibitors for 40 h. (**A**) Control embryos cultured without inhibitors developed normally beyond 4-cell stage. (**B**) Most of embryos could not develop to 4-cell stage when transcription is blocked by α-amanitin. (**C**) Embryos treated with cycloheximide, a translation inhibitor, could not reach 4-cell stage. (**D**) Embryo development is arrested at 2-cell stage when DNA replication is blocked by aphidicolin. Scale bar = 100 μm

of *TET3* is maintained until 2-cell stage, and subsequently decreased at 4-cell stage in pig embryos [13]. Interestingly, the embryonic stage when *TET3* expression is reduced coincides with the timing of ZGA in both species.

3.5.1 mRNA Isolation

1. mRNAs are isolated from the embryos (10–20 embryos per each treatment group) collected at the previous step (Subheading 3.4) using Dynabeads® mRNA DIRECT™ Micro Kit following manufacturer's instructions.
2. First, to wash and prepare the Dynabeads® Oligo $(dT)_{25}$, it is resuspended in the bottle by vortexing for 30 s and 20 μl (per one RNA isolation) is moved to nuclease-free 1.5 ml microcentrifuge tube, then 20 μl lysis buffer is added and mixed. The tubes are placed on a magnet for 1 min, then supernatant is removed.
3. Embryos are transferred to 100 μl lysis buffer in 1.5 ml tube and mixed by repetitive pipetting.
4. The prepared Dynabeads® Oligo $(dT)_{25}$ (20 μl) is added to the lysate and mixed.

5. The tube containing mixture is placed on roller for 5 min to allow mRNA anneals to the Dynabeads®.
6. After 5 min, the tube is placed on the magnet for 1 min, then supernatant is removed.
7. The tube is removed from the magnet and 100 μl washing buffer A is added followed by mixing by careful pipetting.
8. The tube is placed on the magnet for 1 min, and then supernatant is removed.
9. **Steps 7** and **8** are repeated.
10. The mRNA-Dynabead® complex is resuspended with 100 μl washing buffer B.
11. The suspension is moved to a new 1.5 ml tube and placed on the magnet for 1 min. Then, supernatant is removed.
12. The mRNA-Dynabead® complex is washed again with 100 μl washing buffer B, then the suspension is removed on the magnet.
13. The tube is removed from the magnet and 10 μl 10 mM Tris–HCl is added.
14. To isolate mRNA from the Dynabead®, the tube is incubated at 65–80 °C for 2 min and placed on the magnet, then supernatant is immediately transferred to a new nuclease-free tube.
15. The isolated mRNA is kept at −80 °C or immediately used in cDNA synthesis.

3.5.2 cDNA Synthesis

1. After mRNA isolation, the reverse transcriptase reaction is conducted using Superscript™ III First-Strand cDNA Synthesis System.
2. The isolated mRNA (8 μl) is mixed with 1 μl 50 μM oligo $(dT)_{20}$ and 1 μl 10 mM dNTPs in a PCR tube. Then, the mixture is incubated at 65 °C for 5 min and placed on ice for at least 1 min.
3. cDNA synthesis mix is prepared by mixing components in the following order: 2 μl 10× RT buffer, 4 μl 25 mM $MgCl_2$, 2 μl 0.1 M DTT, 1 μl RNaseOUT™ (40 U/μl), and 1 μl SuperScript® III RT (200 U/μl) in a PCR tube.
4. The prepared 10 μl of mRNA-primer mixture is added to the cDNA synthesis mix and incubated at 50 °C for 50 min followed by the termination of the reaction at 85 °C for 5 min. Then, the reaction is chilled on ice.
5. The reaction is collected with brief centrifugation and 1 μl of RNase H is added, and then incubated at 37 °C for 20 min.
6. cDNA is stored at −20 °C for later use or immediately used for real-time PCR.

3.5.3 Real-Time RT-qPCR

1. PCR is conducted using IQ SYBR Green Supermix and *YWHAG* or *GAPDH* is used as an internal control.
2. Primer validation is performed for each designed primer to verify that the amplification efficiencies are similar for each amplicon (*see* **Note 8**).
3. At least three biological and two experimental replications are conducted.
4. Considering the number of replication and reaction volume (e.g., 20 μl), sufficient amout of assay master mix for all reactions is prepared by adding IQ SYBR green supermix (2×), forward and reverse primers (final concentration is 100–500 nM each), and nuclease-free water, except cDNA.
5. cDNA is put into each well of a 96-well plate, then the master mix is added. For example, 1 μl cDNA and 19 μl master mix can be used per well when the reaction volume is 20 μl.
6. The 96-well plate is covered with film.
7. PCR is performed as the following condition: initial temperature of 94 °C for 2 min followed by 40 cycles of 5 s at 94 °C, 3 s at 60 °C, and 30 s at 72 °C.
8. Real-time fluorescence data is collected during the extension stage.
9. The abundance of *TET3* transcript is calculated relative to that of the internal controls, *YWHAG* or *GAPDH*. Relative abundance of *TET3* transcript between the embryo groups is calculated using the ΔΔCt method (*see* Fig. 4, *see* **Note 9**).

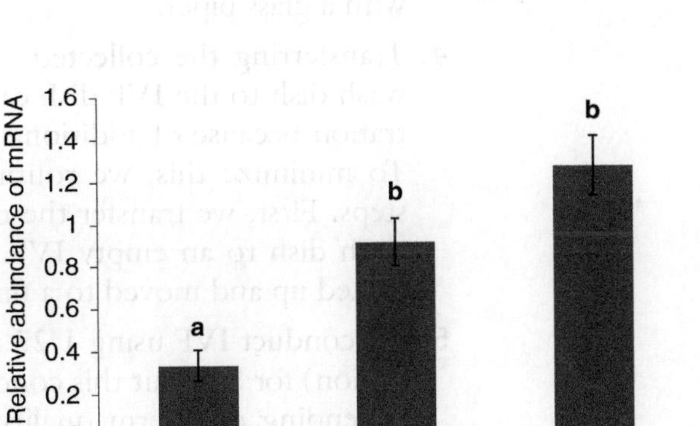

Fig. 4 Expression of *TET3* at 4-cell stage embryos after treatment of α-amanitin or cycloheximide (CHX). When transcription or translation was blocked using the inhibitors, the level of *TET3* remained high in 4-cell stage embryos. This suggests that an active mechanism is required for the degradation of *TET3* at the 4-cell stage. The difference was more dramatic when protein synthesis was inhibited using CHX. *GAPDH* was used to normalize the level of *TET3*. Different letters indicate significant difference ($p < 0.05$). Adapted from Lee et al. [13]

4 Notes

1. In pig IVM, serum or follicular fluid has generally been added to the system. Serum and follicular fluid contain hormones and amino acids which induce oocyte maturation. However, many of them are undefined factors and their composition is variable even among batches from a single source. To avoid irregular maturation rate of IVM and exclude the undefined factors, IVM is conducted with defined medium without serum and follicular fluid [38]. The average maturation rate is 70% using the defined medium.
2. The appropriate intensity of vortex mixer and time should be adjusted when denuding the COCs. Excessively high intensity or vortexing for too long can damage or even rupture the oocytes. Instead a repetitive pipetting in the presence of hyaluronidase can be alternatively used. An appropriate number of repetitive passages through a 1 ml pipette in the 1.5 ml tube can be effective in eliminating the cumulus cells. In this case, it requires attention during the pipetting to avoid making bubbles that obstruct vision when collecting the denuded oocytes afterwards.
3. The criterion for collecting matured oocytes is the extrusion of the first polar body, but this procedure can be laborious. To shorten the time for this procedure, first we remove oocytes without cellular membrane (Fig. 1c), and collect oocytes with obvious polar body (Fig. 1b). After then, we find the remaining matured oocytes with polar body by rolling the individuals with a glass pipet.
4. Transferring the collected oocytes directly from the 30 mm wash dish to the IVF dish can change the final sperm concentration because of additional medium inside the glass pipette. To minimize this, we routinely move the oocytes with two steps. First, we transfer the embryos from the second 30 mm wash dish to an empty IVF drop. Then, 25–30 embryos are picked up and moved to a final drop.
5. We conduct IVF using 1.25×10^5/ml of sperm (final concentration) for 5 h, but this concentration and time can be altered depending on sperm quality. Sperm quality is usually determined by morphology, viability, and motility. If cryopreserved sperm from same batch is used, the quality can be maintained constantly, but the quality of collected semen is variable even though the semen is collected from the same boar. To test the sperm quality, we add 50 μl of sperm (2.5×10^5/ml) to an empty IVF drop. If the viability and the motility are good, the prepared sperm is added to the IVF drops containing oocytes; however, higher concentration (2.5×10^5/ml of final concentration) is used when the quality is poor. Sometimes, IVF is

conducted for 6 h instead of increasing the sperm concentration or the combination.

6. We collect 2-cell stage embryos at 24–36 h after insemination. Every embryo shows different timing of the first cleavage usually between 24 h and 48 h after insemination. The previous study demonstrated that earlier-cleaving embryos are more developmentally competent than late-cleaving embryos [39], thus embryos that cleave after 36 h are excluded in this experiment.
7. The solubility of α-amanitin and cycloheximide is 1 mg/ml and 20 mg/ml in water, respectively. To minimize the alteration in composition of culture medium when the inhibitors are added, α-amanitin and cycloheximide stocks are prepared by dissolving them in PZM3 culture medium instead of using water. Aphidicolin is dissolved in DMSO because it is poorly soluble in water. Each aliquot of the inhibitor stocks is dissolved in pre-warmed PZM3 at final concentration right before the use.
8. We use the ΔΔCt method to compare the relative expression of *TET3* between the embryo groups. This method is based on the assumption that amplification efficiency is 100%. Primer validation is required before running of real-time PCR of target genes because error in fold change increases exponentially as primer efficiency is apart from the assumption. To obtain a primer efficiency value, a standard curve is needed. cDNA templates are serially diluted (i.e., 10 ng/µl, 100 ng/µl, µg/µl, and 10 µg/µl), then Ct values are measured. We graph Ct values obtained from the templates versus log of initial amount of templates (i.e., log 10 ng/µl = 1 and log 100 ng/µl = 2). The R^2 value should be greater than 0.98.
9. To compare the relative expression level of *TET3* between the embryo groups using the ΔΔCt method, ΔCt is calculated by subtracting Ct value of the internal control gene (*YWHAG* or *GAPDH*) from that of *TET3* in each group. ΔCt value of control group is used as a calibrator, thus ΔΔCt values are obtained by subtracting the ΔCt of the control group from that of other treatment groups. The fold differences in transcript abundance are obtained using the formula $2^{-\Delta\Delta Ct}$.

References

1. Li L, Lu X, Dean J (2013) The maternal to zygotic transition in mammals. Mol Asp Med 34(5):919–938. doi:10.1016/j.mam.2013.01.003
2. Bachvarova R, Cohen EM, De Leon V, Tokunaga K, Sakiyama S, Paynton BV (1989) Amounts and modulation of actin mRNAs in mouse oocytes and embryos. Development 106(3):561–565
3. Bouniol C, Nguyen E, Debey P (1995) Endogenous transcription occurs at the 1-cell stage in the mouse embryo. Exp Cell Res 218(1):57–62

4. Braude P, Bolton V, Moore S (1988) Human gene expression first occurs between the four- and eight-cell stages of preimplantation development. Nature 332(6163):459–461
5. Davis D (1984) Culture and storage of pig embryos. J Reprod Fertil Suppl 33:115–124
6. Frei R, Schultz G, Church R (1989) Qualitative and quantitative changes in protein synthesis occur at the 8–16-cell stage of embryogenesis in the cow. J Reprod Fertil 86(2):637–641
7. Crosby I, Gandolfi F, Moor R (1988) Control of protein synthesis during early cleavage of sheep embryos. J Reprod Fertil 82(2):769–775
8. Wang QT, Piotrowska K, Ciemerych MA, Milenkovic L, Scott MP, Davis RW, Zernicka-Goetz M (2004) A genome-wide study of gene activity reveals developmental signaling pathways in the preimplantation mouse embryo. Dev Cell 6(1):133–144
9. Golbus MS, Calarco PG, Epstein CJ (1973) The effects of inhibitors of RNA synthesis (α-amanitin and actinomycin D) on preimplantation mouse embryogenesis. J Exp Zool 186(2):207–216
10. Warner CM, Versteegh LR (1974) In vivo and in vitro effect of α-amanitin on preimplantation mouse embryo RNA polymerase. Nature 248(5450):678–680
11. Jarrell V, Day B, Prather R (1991) The transition from maternal to zygotic control of development occurs during the 4-cell stage in the domestic pig, *Sus scrofa*: quantitative and qualitative aspects of protein synthesis. Biol Reprod 44(1):62–68
12. Iqbal K, Jin S-G, Pfeifer GP, Szabó PE (2011) Reprogramming of the paternal genome upon fertilization involves genome-wide oxidation of 5-methylcytosine. Proc Natl Acad Sci 108(9):3642–3647
13. Lee K, Hamm J, Whitworth K, Spate L, K-w P, Murphy CN, Prather RS (2014) Dynamics of TET family expression in porcine preimplantation embryos is related to zygotic genome activation and required for the maintenance of NANOG. Dev Biol 386(1):86–95
14. Yoshioka K, Suzuki C, Tanaka A, Anas IM-K, Iwamura S (2002) Birth of piglets derived from porcine zygotes cultured in a chemically defined medium. Biol Reprod 66(1):112–119
15. Mukherjee AB (1972) Normal progeny from fertilization in vitro of mouse oocytes matured in culture and spermatozoa capacitated in vitro. Nature 237:397–398
16. Mattioli M, Bacci M, Galeati G, Seren E (1989) Developmental competence of pig oocytes matured and fertilized in vitro. Theriogenology 31(6):1201–1207
17. Kline D, Kline JT (1992) Repetitive calcium transients and the role of calcium in exocytosis and cell cycle activation in the mouse egg. Dev Biol 149(1):80–89
18. Wang W-H, Sun Q-Y, Hosoe M, Shioya Y, Day BN (1997) Quantified analysis of cortical granule distribution and exocytosis of porcine oocytes during meiotic maturation and activation. Biol Reprod 56(6):1376–1382
19. Abeydeera LR, Wang WH, Cantley TC, Rieke A, Prather RS, Day BN (1998) Presence of epidermal growth factor during in vitro maturation of pig oocytes and embryo culture can modulate blastocyst development after in vitro fertilization. Mol Reprod Dev 51(4):395–401
20. Kidson A, Colenbrander B, Verheijden J, Bevers M (2001) Polyspermia in the pig is dependent on both IVF medium and sperm dose during fertilization in vitro. In: Proceedings of the sixth international conference on pig reproduction, p 75
21. Martinez-Madrid B, Dominguez E, Alonso C, Diaz C, Garcia P, Sanchez R (2001) Effect of IVF medium and sperm concentration on fertilization parameters. In: Proceedings of the sixth international conference on pig reproduction, p 75
22. Petters R, Wells K (1992) Culture of pig embryos. J Reprod Fertil Suppl 48:61–73
23. Hamatani T, Carter MG, Sharov AA, Ko MS (2004) Dynamics of global gene expression changes during mouse preimplantation development. Dev Cell 6(1):117–131
24. Wang Q, Latham KE (1997) Requirement for protein synthesis during embryonic genome activation in mice. Mol Reprod Dev 47(3):265–270
25. Baranovskiy AG, Babayeva ND, Suwa Y, Gu J, Pavlov YI, Tahirov TH (2014) Structural basis for inhibition of DNA replication by aphidicolin. Nucleic Acids Res 42(22):14013–14021
26. Lo Y-S, Tseng W-H, Chuang C-Y, Hou M-H (2013) The structural basis of actinomycin D-binding induces nucleotide flipping out, a sharp bend and a left-handed twist in CGG triplet repeats. Nucleic Acids Res 41(7):4284–4294
27. Sobell HM (1985) Actinomycin and DNA transcription. Proc Natl Acad Sci 82(16):5328–5331
28. Brueckner F, Cramer P (2008) Structural basis of transcription inhibition by α-amanitin and implications for RNA polymerase II translocation. Nat Struct Mol Biol 15(8):811–818
29. Warfel AH, ELBERG S (1970) Specific inhibition of nuclear RNA polymerase II by a-amanitin. Science 170:447–449

30. Baumli S, Endicott JA, Johnson LN (2010) Halogen bonds form the basis for selective P-TEFb inhibition by DRB. Chem Biol 17(9):931–936
31. Chao S-H, Price DH (2001) Flavopiridol inactivates P-TEFb and blocks most RNA polymerase II transcription in vivo. J Biol Chem 276(34):31793–31799
32. Titov DV, Gilman B, He Q-L, Bhat S, Low W-K, Dang Y, Smeaton M, Demain AL, Miller PS, Kugel JF (2011) XPB, a subunit of TFIIH, is a target of the natural product triptolide. Nat Chem Biol 7(3):182–188
33. Schneider-Poetsch T, Ju J, Eyler DE, Dang Y, Bhat S, Merrick WC, Green R, Shen B, Liu JO (2010) Inhibition of eukaryotic translation elongation by cycloheximide and lactimidomycin. Nat Chem Biol 6(3):209–217
34. Mayer W, Niveleau A, Walter J, Fundele R, Haaf T (2000) Embryogenesis: demethylation of the zygotic paternal genome. Nature 403(6769):501–502
35. Santos F, Hendrich B, Reik W, Dean W (2002) Dynamic reprogramming of DNA methylation in the early mouse embryo. Dev Biol 241(1):172–182
36. Gu T-P, Guo F, Yang H, Wu H-P, Xu G-F, Liu W, Xie Z-G, Shi L, He X, S-g J (2011) The role of Tet3 DNA dioxygenase in epigenetic reprogramming by oocytes. Nature 477(7366):606–610
37. Wossidlo M, Nakamura T, Lepikhov K, Marques CJ, Zakhartchenko V, Boiani M, Arand J, Nakano T, Reik W, Walter J (2011) 5-Hydroxymethylcytosine in the mammalian zygote is linked with epigenetic reprogramming. Nat Commun 2:241
38. Kishida R, Lee E, Fukui Y (2004) In vitro maturation of porcine oocytes using a defined medium and developmental capacity after intracytoplasmic sperm injection. Theriogenology 62(9):1663–1676
39. Isom SC, Whitworth KM, Prather RS (2012) Timing of first embryonic cleavage is a positive indicator of the in vitro developmental potential of porcine embryos derived from in vitro fertilization, somatic cell nuclear transfer and parthenogenesis. Mol Reprod Dev 79(3):197–207

Chapter 14

Targeted Gene Knockdown in Early Embryos Using siRNA

Lu Zhang and Zoltan Machaty

Abstract

RNA interference is a convenient and highly effective technique to investigate the biological function of genes. Adequately designed RNA molecules introduced into an oocyte are able to bind specific endogenous mRNAs and trigger their degradation. Subsequent fertilization of these oocytes will result in the generation of embryos in which the expression of the gene of interest is downregulated, and following the degradation of maternal proteins the role of the gene product can be studied. Here, we describe the approach how post-transcriptional gene silencing can be achieved in oocytes and early embryos using siRNA.

Key words Oocyte, Embryo, siRNA, RNAi, Microinjection, Gene function, Downregulation, Knock down

1 Introduction

In recent years, RNA interference (RNAi) has become a popular method to analyze gene function. By harnessing an evolutionary conserved endogenous biological pathway, it allows for easy and effective downregulation of specific genes, and thus the identification of the corresponding change in the phenotype. RNA interference originally evolved as a defense mechanism against viruses and genetic parasites. It was first described in plants but the underlying mechanism of the process was clarified in the nematode *Caenorhabditis elegans* [1]. The mechanism jumps into action when a long double-stranded RNA (dsRNA) is detected inside the cell. Such molecules are produced by most replicating viruses and are not normally found in uninfected cells. An enzyme called Dicer cleaves the long dsRNA into 19–25 nucleotide-long fragments known as small interfering RNAs (siRNAs) [2]. Individual siRNA duplexes are then incorporated into RNA-induced silencing complexes known as RISCs. Part of the complex is a helicase that unwinds the duplex. The sense strand of the siRNA undergoes degradation, while the antisense strand serves as a guide to direct the RISC to additional copies of the long dsRNA [3]. Subsequently,

Kiho Lee (ed.), *Zygotic Genome Activation: Methods and Protocols*, Methods in Molecular Biology, vol. 1605,
DOI 10.1007/978-1-4939-6988-3_14,

the Argonaut (Ago) component of the RISC triggers site-specific cleavage of the offending RNA molecule [4]. The cleavage products are then released and the disengaged RISC targets additional complementary RNA molecules. Through the activation of this pathway, organisms such as plants, nematodes, and insects are able to convert viral dsRNAs into antiviral siRNAs.

RNAi can also be used under experimental conditions; in this case to downregulate the cells' own genes. In *C. elegans* and *D. melanogaster* it was observed that introducing long dsRNAs into cells led to the destruction of target RNAs [1, 5]. However, in somatic cells of mammals, dsRNAs longer than 30 bp activate the interferon response, a potent protein-based innate immune response [6]. This involves the stimulation of an RNase that is responsible for general RNA degradation and a global repression of protein synthesis. In such cells siRNAs can be used: these small double-stranded RNAs evade the interferon response by acting as Dicer products and entering the RNAi pathway further downstream [7]. Interestingly, the interferon response is absent in oocytes, and gene silencing is possible using both long dsRNAs and siRNAs [8–10]. The technique has been used successfully in a number of species including mouse [11], rat [12], human [13], cattle [14], and pig [15]. Below we are going to describe how microinjecting siRNA can be utilized to downregulate gene expression in early embryos. The description is about pig embryos but with appropriate modifications the same approach can be applied to other species as well.

2 Materials

2.1 Handling Oocytes, Sperm, and Embryos

1. Oocytes are obtained from pig ovaries collected at an abattoir.
2. Physiological saline: 0.9% NaCl.
3. Hepes-buffered Tyrode's Lactate (TL-Hepes) medium: 2.0 mM $CaCl_2$•$2H_2O$, 114.0 mM NaCl, 3.2 mM KCl, 2.0 mM $NaHCO_3$, 0.4 mM NaH_2PO_4, 10.0 mM Na Lactate (60% syrup), 0.5 mM $MgCl_2$•$6H_2O$, 10.0 mM Hepes, 12.0 mM sorbitol, 0.2 mM sodium pyruvate, 0.075 g/L penicillin, 0.05 g/L streptomycin, 1 ml phenol red (0.5%), 0.1 g/L polyvinyl alcohol (PVA); pH 7.4 [16].
4. In vitro maturation medium: Medium 199 supplemented with 1 mg/ml polyvinyl alcohol (PVA), 0.1 mg/ml cysteine, 10 ng/ml epidermal growth factor (EGF), 0.5 IU/ml porcine follicle stimulating hormone (FSH), and 0.5 IU/ml ovine luteinizing hormone (LH) [17].
5. Sperm wash medium: Dulbecco's phosphate-buffered saline (DPBS; without Ca^{2+} and Mg^{2+}) containing 0.1% bovine serum albumin (BSA), 75 μg/ml penicillin, and 50 μg/ml streptomycin; pH 7.2 [18].

6. Fertilization medium: 113.1 mM NaCl, 3 mM KCl, 7.5 mM $CaCl_2 \cdot 2H_2O$, 20 mM Tris (crystallized free base), 11 mM glucose, 5 mM sodium pyruvate, 0.1% bovine serum albumin (BSA), and 1 mM caffeine [18].
7. Porcine Zygote Medium-3 (PZM-3): 108.0 mM NaCl, 10.0 mM KCl, 0.35 mM KH_2PO_4, 0.4 mM $MgSO_4 \cdot 7H_2O$, 25.07 mM $NaHCO_3$, 0.2 mM Na-pyruvate, 2.0 mM Ca-(lactate)2•$5H_2O$, 1.0 mM l-Glutamine, 5.0 mM hypotaurine, 20.0 mM Basal Medium Eagle (BME), 10.0 mM Minimum Essential Medium (MEM), 0.065 mM penicillin G, 0.050 mM streptomycin, 3.0 mM bovine serum albumin (BSA); pH 7.2 [19].
8. 20 mg/ml hyaluronidase in TL-Hepes medium: Make 100 μl aliquots, freeze at −20 °C. Before use add 900 μl TL-Hepes to this stock after thawing.
9. Vortex mixer.
10. 1.5 ml microcentrifuge tubes.
11. 50 ml plastic conical tubes.
12. 100 × 15 mm plastic petri dishes.
13. 35 × 10 mm plastic petri dishes.
14. 60 × 15 mm plastic petri dishes.
15. Nunc™ 4-well culture dishes.
16. Mineral oil, embryo tested.
17. Slide warmer.
18. CO_2 incubator.
19. Stereo microscope.

2.2 Microinjection

1. Glass capillary with filament (e.g., BF100-78-15 thin-wall borosilicate tubing, outer diameter 100 μm, inner diameter 78 μm, from Sutter Instrument Company).
2. Glass capillary without filament (e.g., B100-75-15, outer diameter 100 μm, inner diameter 75 μm, also from Sutter).
3. Micropipette puller (e.g., P-87 from Sutter).
4. Microforge.
5. Inverted microscope equipped with 4×, 10×, and 20× objectives and Hoffman modulation contrast system.
6. Hydraulic micromanipulators.
7. Pipette holder for the injection pipette.
8. Syringe system for holding the oocytes.
9. Microinjector.
10. Vibration isolation table.

11. Micro-loader pipette tips.
12. Fluorescent dextran (e.g., fluorescein isothiocyanate-conjugated dextran).
13. siRNA (e.g., Custom Stealth siRNA from Life Technologies; *see* **Notes 1** and **2**).

3 Methods

3.1 Oocyte Collection

1. Obtain ovaries of sows or prepubertal gilts at an abattoir and place them in an insulated container containing 39 °C physiological saline. Transfer the container to the laboratory.
2. Harvest follicular fluid from medium-size (3–6 mm) follicles using a 20-gauge hypodermic needle attached to a 10 ml syringe. Make sure to apply only gentle suction to avoid damaging the cumulus investment of the oocytes. Collect follicular fluid in a 50 ml sterile conical tube.
3. After about 15 min, remove the supernatant containing mostly follicular fluid. Distribute the sediment into six to eight sterile petri dishes (100 mm in diameter) and add approximately 10 ml TL-Hepes medium. Place the dishes on a warming stage and allow the cells to settle for about 5 min.
4. Locate the cumulus-oocyte complexes using a stereo microscope. Collect those with evenly dark ooplasm and at least three intact layers of cumulus cells, and then transfer them into 3 ml of TL-Hepes medium in a sterile small petri dish (30 mm in diameter).

3.2 Oocyte Maturation

1. Rinse the cumulus-oocyte complexes in a small petri dish containing 3 ml of maturation medium. Repeat rinsing two more times.
2. Transfer the cumulus-oocyte complexes into 4-well Nunc dishes, 50 complexes into each well containing 500 µl of maturation medium covered with mineral oil. Place the dishes into an incubator, at 39 °C with maximum humidity, in an atmosphere of 5% CO_2 in air, for 44 h.
3. At the end of the maturation period remove the cumulus cells from the oocytes. For this purpose, transfer the cumulus-oocyte complexes into a 1.5 ml microcentrifuge tube containing 1 ml TL-Hepes medium with 2 mg/ml hyaluronidase. Shake the tube vigorously for ~5 min using a vortex mixer.
4. Transfer the medium from the tube into a small petri dish. Rinse the tube with additional TL-Hepes medium and add it to the petri dish as well.
5. Collect the oocytes with intact plasma membrane and evenly dark cytoplasm using a stereo microscope and transfer them into fresh TL-Hepes. Rinse the oocytes in 3 ml TL-Hepes medium two more times.

3.3 Microinjection

1. Prepare a holding pipette using a puller and borosilicate tubing without an inner filament. The setting for the Sutter Instrument puller we use is the following: P = 45, Heat = 925, Vel = 45, Time = 200. After pulling, break the front end of the capillary where it has an outer diameter of approximately 100 μm. Using a microforge, melt the cut end until the edges become smooth while making sure that the opening is retained (*see* **Note 3**).
2. Make an injection pipette with the puller, using borosilicate tubing with filament. The puller setting we use to make injection pipettes is: (1) P = 45, Heat = 805, Pull = 45, Vel = 30, Time = 200, (2) Heat = 800, Pull = 40, Vel = 35, Time = 200. These parameters result in a pipette with a short shank; at this point the tip of the pipette is still closed.
3. Fill the injection pipette with siRNA. Pick up ~1 μl RNA solution with the micro-loader, insert the micro-loader into the injection pipette through its back end, and expel the solution. It will run down along the filament and accumulate in the tip of the pipette.
4. Connect the siRNA-loaded injection pipette to the pipette holder that is connected to the microinjector. Attach the pipette holder to the micromanipulator.
5. Transfer oocytes in a drop of Ca^{2+}-free TL-Hepes medium covered with mineral oil in an upturned lid of a 60 mm petri dish. Arrange them so that they are positioned in the top third of the drop. Place the dish on the heated stage of the inverted microscope (*see* **Note 4**).
6. Connect the holding pipette to the Cell Tram Air pipette controller and attach it to the micromanipulator. Lower the pipette into the medium holding the oocytes. Fill the tip with medium by means of the pipette controller.
7. Lower the injection pipette into the medium. Open the tip by gently tapping it against the front of the holding pipette, creating an opening that is less than 1 μm in diameter. Apply a small amount of positive pressure to the back of the injection pipette by means of the microinjector to avoid the medium flowing into the pipette by capillary action.
8. Stabilize an oocyte with the holding pipette by gentle suction delivered via the controller.
9. Push the injection pipette through the zona pellucida, against the plasma membrane. The plasma membrane will bend and envelope the injection pipette but it will not rupture immediately. Advance the pipette further and when the membrane that envelops the pipette breaks, expel a small amount (approximately 40 pl solution) of siRNA by means of the microinjector. A slight but noticeable swelling of the oocyte is an indication that the pipette penetrated the oolemma and the injection occurred into the cytoplasm (*see* **Note 5**).

10. Remove the injection pipette from the oocyte, move the holding pipette to the lower section of the drop, and release the oocyte. Then pick up another noninjected oocyte, reposition it in the middle, and inject it as described above. Repeat with the rest of the oocytes (*see* **Notes 6–10**).

3.4 In Vitro Fertilization

1. Wash the oocytes in 500 μl of fertilization medium three times and place them (in groups of 30) into 50 μl droplets of fertilization medium covered with mineral oil in a small petri dish.
2. Wash boar semen (freshly collected or frozen-thawed) three times by centrifugation at 900 × *g* for 4 min in the sperm wash medium. Dilute the final sperm pellet with the fertilization medium.
3. Add the sperm suspension, at a final concentration of 5×10^5 cells/ml, to each 50 μl droplet containing the oocytes. Co-incubate the gametes for 5 h at 39 °C with maximum humidity in an atmosphere of 5% CO_2 in air.
4. At the end of the fertilization period transfer potential zygotes (in groups of 10) into 20 μl droplets of PZM-3 medium under mineral oil for culture.

3.5 Verification of Knockdown

1. To demonstrate that siRNA microinjection led to gene downregulation at the transcript level, RT-PCR should be performed. Isolate mRNA from 20 microinjected oocytes/embryos from each experimental group and generate cDNA by reverse transcription. Amplify a fragment of the target gene cDNA by means of PCR and separate PCR products by agarose gel electrophoresis. The amount of transcript in the siRNA-injected oocytes/embryos should be reduced compared to the control.
2. Downregulation at the protein level can be verified by Western blot. Collect approximately 500 microinjected oocytes per experimental group and lyse them in SDS sample buffer. Separate the proteins by SDS-PAGE and transfer them electrophoretically onto nitrocellulose membranes. Block nonspecific binding sites, incubate the membrane with the appropriate primary and secondary antibodies, and visualize the proteins by means of one of the commercially available methods.
3. The absence (or presence) of the knockdown protein can also be verified by immunocytochemistry. This approach requires fewer oocytes/embryos for verification compared to Western blot and it is also able to indicate cellular localization of the protein. In addition, if multiple antibodies are used in parallel, the effect of downregulation on other proteins can also be examined. For immunocytochemical verification of protein knockdown, the oocytes/embryos need to be fixed, permeabilized and the nonspecific binding sites blocked. The cells are

then treated with the appropriate primary and secondary antibodies, after which the proteins can be visualized by laser-scanning confocal microscopy (*see* **Note 11**).

4 Notes

1. The structure of siRNA to achieve successful target mRNA degradation has been clarified, which facilitates the design of efficient molecules for silencing [20, 21]. Maximum effectiveness calls for a 19-bp RNA duplex with a 2-nucleotide overhang on the 3′ ends (Fig. 1). The first nucleotide of the guide strand is considered the Anchor Site. Low stability at this terminus promotes separation from the passenger strand and loading of the guide strand into RISC, thus uridine or adenosine is preferred at this site [22]. The Seed Region spans positions 2–8 and is responsible for the initial base-pair formation during target recognition. The presence of four to seven adenosines/uridines in this region was shown to reduce unintended off-target effects [23]. The Central Region (nucleotides 9–12) overlaps with the catalytic part of the Ago protein. Low stability of the guide strand in this region promotes cleavage of the target mRNA, and uridine was suggested to be the preferential base at position 10 of the passenger strand. The 3′ supplementary region (positions 13–18) propagates duplex formation over the cleavage region of the Ago protein; adenosine at position 3 in the passenger strand provides the siRNA with superior efficacy. The 3′ end of the guide strand that spans nucleotides 19–21 contains a 2-nucleotide overhang that protrudes from the siRNA duplex. Although this unpaired structure has no role in target recognition, its presence is vital for efficacy of the siRNA [20]. A number of online programs are available to identify the most appropriate siRNA based on the mRNA sequence information of the protein to be knocked down. In addition, once the siRNA is selected, it is necessary to determine its specificity. This is done by blasting its target sequence against all known porcine (or the relevant species') genomic sequences to ensure that it is not homologous to other gene sequences, so that the siRNA will hybridize only to

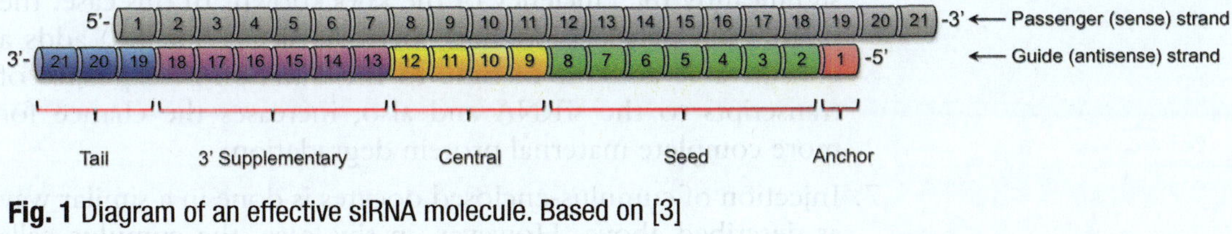

Fig. 1 Diagram of an effective siRNA molecule. Based on [3]

the mRNA of the protein of interest. Finally, it is important to experimentally evaluate and optimize siRNA molecules, as not all duplexes that meet the above-mentioned criteria are effective [24].

2. Remember to include appropriate controls in the experiments. At a minimum, such controls should include an endogenous positive control that is not injected with siRNA and a negative control in which oocytes are injected with scrambled siRNA.
3. Although microinjection is the most common way to introduce RNA into oocytes, electroporation, transfection, or expression from a transgene are also potential alternatives [10].
4. The holding medium should contain no Ca^{2+} to avoid inadvertent oocyte activation. Activation may occur as a result of Ca^{2+} contamination during the process of microinjection, when the injection pipette penetrates the plasma membrane. To make Ca^{2+}-free medium simply omit Ca^{2+} from the original recipe and use only sterile, plastic containers during preparation and storage.
5. In the case of species with very fragile oocytes (such as the mouse), the insertion of the injection pipette through the plasma membrane is typically assisted by a small electrical pulse delivered via an electrical amplifier. Such a method has been described in detail earlier [25].
6. Some experiments may require injection of oocytes prior to the metaphase II stage. To understand biological function of a gene product in the oocyte, it is necessary to deplete maternal proteins in the oocyte cytoplasm. The efficiency of RNAi-mediated gene knockdown is dependent on the time maternal transcripts are exposed to the siRNA and also, the half-life of the maternal protein already expressed in the oocyte. Proteins are continuously degraded and then replaced by newly synthesized ones and this constant protein turnover is one of the cell's strategies to maintain a functional proteome [26]. The rate of turnover varies markedly. Half-life of proteins in dividing mammalian cells ranges between 0.5 and 35 h, which in nondividing cells can increase up to 43 h, while protein half-life in brain cells can be as long as 9 days [27–29]. Mature oocytes can be injected with relative ease, but following siRNA injection they can be cultured for only a limited time (up to about 8 h) [30, 31], after which oocyte quality decreases markedly. Injecting germinal vesicle (GV)-stage oocytes increases significantly the efficiency of the knockdown. In this case, the maturation period (~44 h in the case of pig oocytes) adds a considerable amount of time to the duration of exposure of transcripts to the siRNA and also, increases the chance for more complete maternal protein degradation.
7. Injection of cumulus-enclosed oocytes is done in a similar way as described above. However, in this case, the cumulus cells

that cover the zona pellucida may make it difficult to confirm whether the siRNA solution was injected into the cytoplasm as intended or it was simply released into the perivitelline space because the injection pipette has not penetrated the plasma membrane. Verifying successful microinjection is possible by injecting a fluorescent dextran into the cytoplasm together with the siRNA solution. Dextrans are water-soluble polysaccharides that can be conjugated to different fluorophores. To visualize microinjection, load the injection pipette with a mix of siRNA solution and fluorescein isothiocyanate (FITC)-conjugated dextran (conc. 1%). Microinjection can be verified by a brief exposure to UV light: collect only the oocytes that display green fluorescence due to the presence of the fluorescent dextran in their cytoplasm.

8. When injecting cumulus-enclosed GV-stage oocytes, the cumulus cell may clog the injection pipette. In order to circumvent this problem, the injection pipette should be cleaned frequently by rubbing the tip against the holding pipette and by flushing it with maximum pressure from the microinjector. If none of this helps, replace the injection pipette with a new one.
9. The length of exposure to the siRNA and the time for protein breakdown can be further increased by the use of phosphodiesterase inhibitors such as 3-isobutyl-1-methylxanthine (IBMX). IBMX is known to raise intracellular cAMP levels and thus maintain the cell cycle at the GV stage. Incubation of the cumulus-oocyte complexes in 0.2 mM IBMX for 20–24 h after microinjection keeps the oocytes at the GV stage [32]; then after a thorough washing in IBMX-free maturation medium they can be transferred into the maturation dishes for the duration of the maturation period.
10. Depletion of certain maternal proteins may require even longer times due to their extended half-lives. To circumvent this problem, a method was developed in the mouse that involves the collection of ovarian follicles from 12-day-old mice and injection of the oocytes within the follicles with siRNA [33]. The injected follicles are then incubated until the oocytes reach their final size. At the end of the growth period the cumulus-oocyte complexes are harvested from the follicles and matured in vitro.
11. Gene silencing in mammalian embryos via the microinjection of long dsRNA or siRNA is transient [34]. As opposed to *C. elegans*, mammalian cells lack the RNA-dependent RNA polymerases that could amplify the RNA in the cytoplasm. Therefore, RNAi depends on the number of RNA molecules introduced into the cell. The duplexes are progressively diluted by consecutive cell divisions and the silencing effect lasts only for 2–7 days [4]. It is also important to remember that gene expression is knocked down rather than knocked out by RNAi.

References

1. Fire A, Xu S, Montgomery MK, Kostas SA, Driver SE, Mello CC (1998) Potent and specific genetic interference by double-stranded RNA in *Caenorhabditis elegans*. Nature 391:806–811
2. Bernstein E, Caudy AA, Hammond SM, Hannon GJ (2001) Role for a bidentate ribonuclease in the initiation step of RNA interference. Nature 409:363–366
3. Alagia A, Eritja R (2016) siRNA and RNAi optimization. Wiley Interdiscip Rev RNA 7(3):316–329. doi:10.1002/wrna.1337
4. Leung RK, Whittaker PA (2005) RNA interference: from gene silencing to gene-specific therapeutics. Pharmacol Ther 107:222–239
5. Mello CC, Conte D Jr (2004) Revealing the world of RNA interference. Nature 431:338–342
6. Cullen BR (2014) Viruses and RNA interference: issues and controversies. J Virol 88:12934–12936
7. Elbashir SM, Harborth J, Lendeckel W, Yalcin A, Weber K, Tuschl T (2001) Duplexes of 21-nucleotide RNAs mediate RNA interference in cultured mammalian cells. Nature 411:494–498
8. Svoboda P, Stein P, Hayashi H, Schultz RM (2000) Selective reduction of dormant maternal mRNAs in mouse oocytes by RNA interference. Development 127:4147–4156
9. Yang S, Tutton S, Pierce E, Yoon K (2001) Specific double-stranded RNA interference in undifferentiated mouse embryonic stem cells. Mol Cell Biol 21:7807–7816
10. Svoboda P (2008) RNA silencing in mammalian oocytes and early embryos. In: Paddison PJ, Vogt PK (eds) RNA interference. Curr Top Microbiol Immunol 320:225–256. Springer-Verlag, Berlin, Heidelberg
11. Wianny F, Zernicka-Goetz M (2000) Specific interference with gene function by double-stranded RNA in early mouse development. Nat Cell Biol 2:70–75
12. Lazar S, Gershon E, Dekel N (2004) Selective degradation of cyclin B1 mRNA in rat oocytes by RNA interference (RNAi). J Mol Endocrinol 33:73–85
13. Homer HA, McDougall A, Levasseur M, Murdoch AP, Herbert M (2005) RNA interference in meiosis I human oocytes: towards an understanding of human aneuploidy. Mol Hum Reprod 11:397–404
14. Nganvongpanit K, Müller H, Rings F, Hoelker M, Jennen D, Tholen E, Havlicek V, Besenfelder U, Schellander K, Tesfaye D (2006) Selective degradation of maternal and embryonic transcripts in in vitro produced bovine oocytes and embryos using sequence specific double-stranded RNA. Reproduction 131:861–874
15. Koh S, Lee K, Wang C, Cabot RA, Machaty Z (2009) STIM1 regulates store-operated Ca^{2+} entry in oocytes. Dev Biol 330:368–376
16. Hagen DR, Prather RS, Sims MM, First NL (1991) Development of one-cell porcine embryos to the blastocyst stage in simple media. J Anim Sci 69:1147–11450
17. Abeydeera LR, Wang WH, Prather RS, Day BN (1998) Maturation in vitro of pig oocytes in protein-free culture media: fertilization and subsequent embryo development in vitro. Biol Reprod 58:1316–1320
18. Abeydeera LR, Wang WH, Cantley TC, Rieke A, Prather RS, Day BN (1998) Presence of epidermal growth factor during in vitro maturation of pig oocytes and embryo culture can modulate blastocyst development after in vitro fertilization. Mol Reprod Dev 51:395–401
19. Yoshioka K, Suzuki C, Tanaka A, Anas IM, Iwamura S (2002) Birth of piglets derived from porcine zygotes cultured in a chemically defined medium. Biol Reprod 66:112–119
20. Mittal V (2004) Improving the efficiency of RNA interference in mammals. Nat Rev Genet 5:355–365
21. Reynolds A, Leake D, Boese Q, Scaringe S, Marshall WS, Khvorova A (2004) Rational siRNA design for RNA interference. Nat Biotechnol 22:326–330
22. Frank F, Sonenberg N, Nagar B (2010) Structural basis for 5′-nucleotide base-specific recognition of guide RNA by human AGO2. Nature 465:818–822
23. Ui-Tei K (2013) Optimal choice of functional and off-target effect-reduced siRNAs for RNAi therapeutics. Front Genet 4:107
24. Kumar R, Conklin DS, Mittal V (2003) High-throughput selection of effective RNAi probes for gene silencing. Genome Res 13:2333–2340
25. Swann K (2013) Measuring Ca^{2+} oscillations in mammalian eggs. Methods Mol Biol 957:231–248
26. Toyama BH, Hetzer MW (2013) Protein homeostasis: live long, won't prosper. Nat Rev Mol Cell Biol 14:55–61
27. Cambridge SB, Gnad F, Nguyen C, Bermejo JL, Krüger M, Mann M (2011) Systems-wide proteomic analysis in mammalian cells reveals conserved, functional protein turnover. J Proteome Res 10:5275–5284

28. Yen HC, Xu Q, Chou DM, Zhao Z, Elledge SJ (2008) Global protein stability profiling in mammalian cells. Science 322:918–923
29. Price JC, Guan S, Burlingame A, Prusiner SB, Ghaemmaghami S (2010) Analysis of proteome dynamics in the mouse brain. Proc Natl Acad Sci U S A 107:14508–14513
30. Okada Y, Yamagata K, Hong K, Wakayama T, Zhang Y (2010) A role for the elongator complex in zygotic paternal genome demethylation. Nature 463:554–558
31. Wossidlo M, Nakamura T, Lepikhov K, Marques CJ, Zakhartchenko V, Boiani M, Arand J, Nakano T, Reik W, Walter J (2011) 5-Hydroxymethylcytosine in the mammalian zygote is linked with epigenetic reprogramming. Nat Commun 2:241
32. Stein P, Svoboda P (2006) Microinjection of dsRNA into mouse oocytes and early embryos. CSH Protoc 2006(3)
33. Inoue A, Sunaga K, Aoki F, Zhang Y (2014) siRNA-mediated depletion of stable proteins in mouse oocytes. Protocol Exchange. doi:10.1038/protex.2014.024
34. Paddison PJ (2008) RNA interference in mammalian cell systems. In: Paddison PJ, Vogt PK (eds) RNA interference. Curr Top Microbiol Immunol 320:1–19. Springer-Verlag, Berlin, Heidelberg

Chapter 15

Generating Mouse Models Using Zygote Electroporation of Nucleases (ZEN) Technology with High Efficiency and Throughput

Wenbo Wang, Yingfan Zhang, and Haoyi Wang

Abstract

Mouse models with genetic modifications are widely used in biology and biomedical research. Although the application of CRISPR-Cas9 system greatly accelerated the process of generating genetically modified mice, the delivery method depending on manual injection of the components into the embryos remains a bottleneck, as it is laborious, low throughput, and technically demanding. To overcome this limitation, we invented and optimized the ZEN (Zygote electroporation of nucleases) technology to deliver CRISPR-Cas9 reagents via electroporation. Using ZEN, we were able to generate genetically modified mouse models with high efficiency and throughput. Here, we describe the protocol in great detail.

Key words Mouse model generation, CRISPR-Cas9, Electroporation, Mouse zygotes, ZEN

1 Introduction

The type II CRISPR-Cas9 system of *Streptococcus pyogenes* works efficiently in generating mouse models with various genetic modifications. Specifically, the system can introduce targeted mutations during embryogenesis [1–5]. This ability to introduce targeted modifications in embryos can have a dramatic impact on studies investigating events in early development such as zygotic genome activation because genetically modified embryos can be generated without breeding. Although this new technology is highly versatile and efficient, the delivery of CRISPR-Cas9 components is still mainly dependent on the microinjection, which is technically demanding and labor intensive.

To overcome this limitation, we developed and optimized the Zygote Electroporation of Nuclease (ZEN) method [6, 7]. We demonstrated that the CRISPR-Cas9 components, including Cas9 mRNA or Cas9 protein, single guide RNA (sgRNA), and a DNA oligo donor, could be efficiently delivered into mouse zygotes by electroporation to generate mice with targeted genetic modifications (Fig. 1).

Kiho Lee (ed.), *Zygotic Genome Activation: Methods and Protocols*, Methods in Molecular Biology, vol. 1605,
DOI 10.1007/978-1-4939-6988-3_15, © Springer Science+Business Media LLC 2017

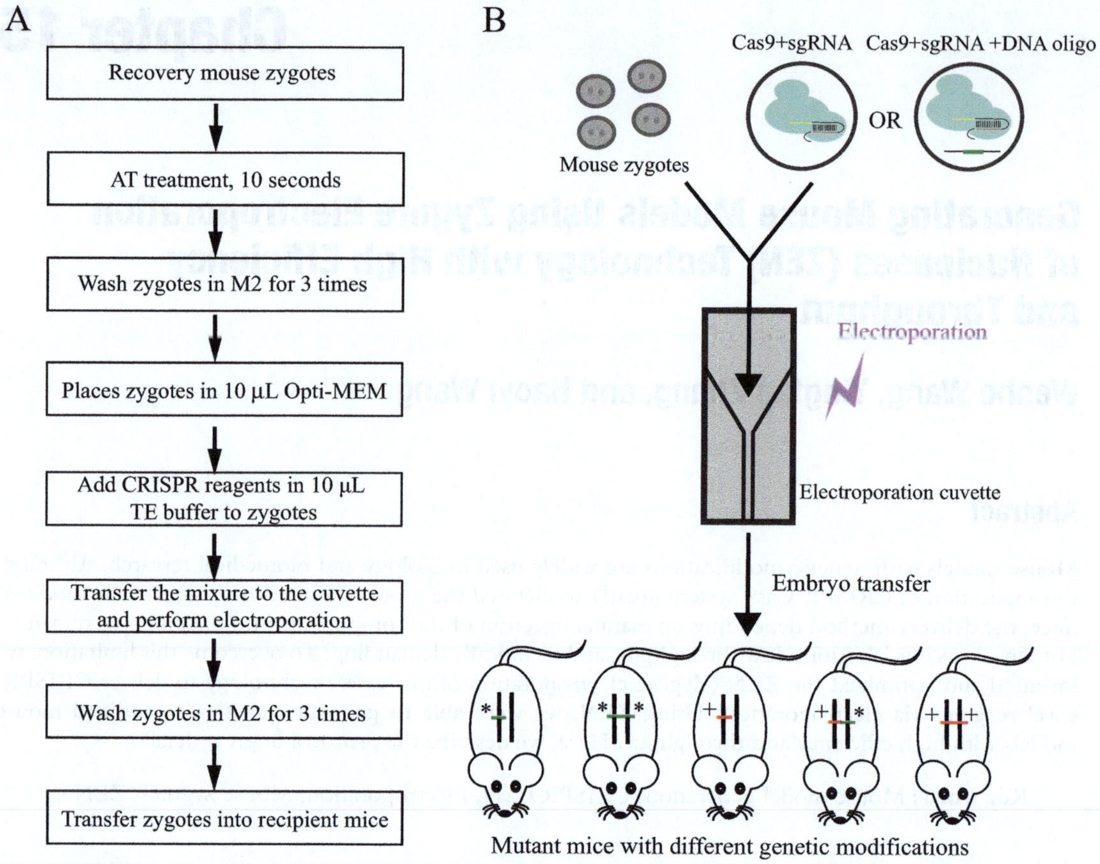

Fig. 1 ZEN Enables High-Throughput generation of genetically modified Mice. (**a**) Flow chart of ZEN method. (**b**) The process of generating genetically modified mice using ZEN. *Asterisk* represents precise modification through HDR, and *plus* indicates indel mutations resulted from NHEJ

Compared to microinjection, ZEN is easy to set up, less laborious, high-throughput, and has a gene-editing efficiency similar to microinjection [6, 7]. After we published our results, another group also reported that delivery of Cas9 RNP (Cas9:sgRNA ribonucleoprotein) through electroporation is efficient to generate genetic modification in the mouse genome, confirming the generality of this method [8].

In this protocol paper, we describe this method in detail.

2 Materials and Equipment

2.1 Mouse Embryo Preparation

1. Donor female mice (various strains, 12–15 g or 8–10 weeks, The Jackson Laboratory).
2. Stud male mice (various strains, 12–24 weeks, The Jackson Laboratory).
3. PMSG (pregnant mare's serum gonadotropin, ProSpec).

4. hCG (human Chorionic Gonadotropin, ProSpec).
5. Hyaluronidase.
6. M2 medium.
7. K-RVCL-50 Medium (COOK Medical).
8. Mineral oil (Sigma-Aldrich).
9. Needles: 25-gauge 5/8″; 30-gauge 1/2″ needle.
10. One microliter syringe.
11. Mouth micropipette.
12. Microdissecting instruments.
13. Tissue culture dishes: 60 mm; 100 mm.
14. MINC Benchtop Incubator (COOK Medical).
15. SteREO Discovery. V8 Microscope (Zeiss).

2.2 sgRNA Synthesis

1. Two universal primers and primers with various sgRNA sequences (Table 1) (Integrated DNA Technologies).
2. PrimeSTAR GXL DNA Polymerase (Clontech).
3. RNase-free PCR tubes.
4. RNase-free microfuge tubes (1.5 mL).
5. RNase AWAY™ Surface Decontaminant (Thermo Fisher Scientific).
6. MEGAshortscript T7 kit (Thermo Fisher Scientific).
7. QIAquick PCR Purification Kit (Qiagen).
8. SeaKem LE Agarose (Lonza).
9. 10× TAE buffer-4 L (Lonza).
10. GeneRuler 1 kb Plus DNA Ladder (Thermo Fisher Scientific).
11. 6× DNA loading dye.
12. Thermo cycler.
13. Electrophoresis equipment and supplies.
14. Manual Gel Documentation Systems (InGenius3, Syngene).
15. NanoDrop 2000 (Thermo Fisher Scientific).
16. Centrifuge.

Table 1
Primer information

Name	Sequence (5′ to 3′)
Universal Rev. Primer 1	Aaaaaagcaccgactcggtgccactttttcaagttgataacggactagccttatttaaacttgc tatgctgtttccagcatagctcttaaac
Universal Rev. Primer 2	aaaagcaccgactcggtgcc
Forward Primer	taatacgactcactatag-gRNA sequence-gtttAagagctatgctggaaac

2.3 Electroporation

1. Ultramer DNA oligos (Integrated DNA Technologies).
2. Cas9 protein (PNA Bio).
3. M2 medium.
4. OptiMEM reduced serum medium (Gibco).
5. Acidic Tyrode's solution (Sigma-Aldrich).
6. RNase-free microfuge tubes (1.5 mL).
7. Savant SpeedVac concentrator (Thermo Fisher Scientific).
8. ECM830 Square Wave Electroporation System (BTX Harvard Apparatus).
9. MINC Benchtop Incubator (COOK Medical).
10. SteREO Discovery.V8 Microscope (Zeiss).
11. Centrifuge.

2.4 Sample Collection and Genotyping

1. Mouse ear punch.
2. Scissors.
3. PCR tubes or 96 plates.
4. Genotyping primers (Integrated DNA Technologies).
5. PrimeSTAR GXL DNA Polymerase (Clontech).
6. Zero Blunt TOPO® PCR Cloning Kit (Invitrogen).
7. SeaKem LE Agarose (Lonza).
8. 10× TAE buffer-4 L (Lonza).
9. GeneRuler 1 kb Plus DNA Ladder (Thermo Fisher Scientific).
10. 6× DNA-loading dye.
11. Thermo cycler.
12. Electrophoresis equipment and supplies.
13. Manual Gel Documentation Systems (InGenius3, Syngene).
14. LB plates with specific antibiotics.

3 Methods

3.1 Zygote Isolation and Grading

All animal work must be approved by the local Institutional Animal Care and Use Committee and adhered to the standards of Guide for the Care and Use of Laboratory Animals set forth by the NIH. The embryos for electroporation can either be produced through natural breeding or from IVF (in vitro fertilization). Embryos from natural breeding usually have better quality, while IVF can produce large quantities of embryos. Here, we briefly introduce how embryos are produced through natural breeding [6, 7] (*see* **Note 1**).

3.1.1 Superovulation

1. Order female mice (6–8 weeks) of desired strain (*see* **Note 2**).
2. Inject mice intraperitoneally (i.p.) with 2.5 IU or 5 IU PMSG.
3. Inject mice i.p. with 2.5 IU or 5 IU hCG 46–50 h after administration of PMSG.

3.1.2 Mating

Immediately post administration of hCG, mate the female mice with stud males at a 1:1 ratio. The next day (22 h later), check for the presence of a copulation plug.

3.1.3 Harvest Embryos

1. Euthanize female mice with a copulation plug, and open the peritoneal cavity using dissecting instruments.
2. Remove the oviducts from the female mice.
3. Transfer the oviducts into M2 medium containing hyaluronidase (0.3 mg/mL).
4. Release the oocyte clutch by puncturing the ampulla with a pair of microdissecting forceps and allow them to incubate in the M2 medium containing hyaluronidase until the cumulus cells fall off and the embryos are disaggregated.
5. Wash the embryos with warm M2.

3.1.4 Embryos Grading

1. Transfer embryos to a new drop of fresh M2 medium and grade for fertilization and viability by examining the presence of the two pronuclei and integrity of the membranes enclosing the embryo.
2. Pass the graded zygotes through two washes of fresh M2 medium and then place in microdrops of K-RVCL-50 medium that have been equilibrated under mineral oil in an incubator (37 °C, 5% CO_2/5% O_2/Nitrogen).

3.2 sgRNA Preparation

Several sgRNA designing programs are freely available online. The consideration of sgRNA design and features of different programs are reviewed elsewhere [9, 10]. In this protocol, we use CRISPR Design (http://crispr.mit.edu/) as an example.

3.2.1 sgRNA Sequence Selection

1. Find the target region of the gene of interest. The location of the target region is dependent on the aim of the project. For example, if KO is required, the region should be located to the downstream of the start codon but relatively to the 5′ region of the gene. More detailed design principles are described in Qin et al. [10].
2. Copy about 100 nt within this region and paste into the sequence window on CRISPR Design webpage. Select the target genome, which is the model organism you are using and submit the mission. Select the available guides with the highest score indicating the lowest potential off-target efficiency.
3. Use the selected sgRNA sequences to design the PCR primers (Table 1).

3.2.2 PCR Amplification and PCR Products Purification

In this protocol, three primers are used in the PCR amplification: Two universal primers and one unique primer with specific sgRNA sequences are listed in Table 1. The T7 promoter sequence was added in front of the sgRNA sequence.

1. Set up the PCR reaction (50 μL):
   First, prepare the master mix:

Reagent	Volume (μL)
10× PCR reaction buffer	10
dNTP	4
Universal Rev. Primer 1 (10 μM)	3
Universal Rev. Primer 2 (10 μM)	3
PrimeStar GXL taq	1
H_2O (Molecular Grade)	26

Aliquot 47 μL of master mix per reaction and add 3 μL of sgRNA forward primer (10 μM) to each reaction. Mix well and load the PCR reactions to a thermo cycler.

2. PCR cycles:
   (1) 98 °C, 2 min; (2) 98 °C, 30 s; (3) 60 °C, 30 s; (4) 68 °C, 1 min; to **step 2**), 30 cycles; (5) 68 °C, 5 min; (6) 4 °C, hold.
3. Purify PCR products using QIAQuick PCR purification kit according to the manufacturer's instructions. Avoid all RNase contamination in subsequent steps. Elute PCR products with 30–50 μL RNase-free water. The final concentration of PCR products should be more than 50 ng/μL. Purified PCR products are the templates for the in vitro transcription (IVT) of sgRNAs.

3.2.3 sgRNA Synthesis

1. Set up sgRNA IVT reactions:

Reagent	Volume (μL)
T7 10x Reaction Buffer	2
ATP	2
CTP	2
GTP	2
UTP	2
T7 Enzyme Mix	2

Aliquot 12 μL of master mix per reaction and add 8 μL of purified PCR product. Mix well by pipetting.

2. Incubate for 4 h at 37 °C.

3. Add 1 μL of Turbo DNase (from MEGAshortscript kit) to each reaction and incubate for 15 min at 37 °C to degrade the template DNA.

3.2.4 sgRNA Purification

Purify sgRNAs by precipitation and rehydration:

1. Transfer 21 μL reaction to a new 1.5 mL RNase-free Eppendorf tube.
2. Add 115 μL nuclease-free water and 15 μL NH_4OAc stop solution from the MEGAshortscript kit to each reaction and mix well.
3. Add 300 μL of 100% EtOH and mix thoroughly.
4. Chill the mixture at −80 °C for at least 20 min.
5. Centrifuge at full speed (>10,000 × *g*) at 4 °C for 15 min.
6. Remove the supernatant carefully and add 1 mL of 70% EtOH to wash the RNA pellet.
7. Centrifuge at full speed (>10,000 × *g*) at 4 °C for 5 min.
8. Remove the supernatant and air-dry the pellet.
9. Resuspend the pelleted RNA in 50 μL of nuclease-free water or Ambion's Elution Solution.
10. Quantify the concentration of sgRNA using NanoDrop. The 260/280 ratio at 2.0 or above is indicative of a clean RNA preparation. The sgRNA can be diluted to 2 μg/μL and frozen in −80 °C for future use.

Alternatively, sRNA can be purified using MEGAclear kit following manufacturer's instruction.

3.2.5 sgRNA Quality Control

1. Transfer 2 μL of final sgRNA product (around 1–2 μg/μL) to a new PCR tube free of RNase.
2. Add 2 μL distilled RNase-free water.
3. Add 4 μL Gel Loading Buffer II from the Megashortscript kit.
4. Mix well and incubate at 65 °C for 10 min.
5. Directly put the samples on ice for at least 2 min.
6. Load the samples and 3 μL 1Kb plus DNA marker to 1% agarose gel.
7. Electrophorese the gel at 120 V until the bromophenol blue reaches the middle part of the gel.
8. Check the gel under UV and take images. There should be one single and sharp band for each sgRNA product at around 200 bp.

3.3 Electroporation Mixture Preparation

1. Prepare the reagent at 2× concentration in 10 μL volume in TE buffer:

   Cas9 protein: 500 ng/μL; sgRNA: 600 ng/μL; DNA: oligo donors 1 μg/μL.

2. Incubate at 37 °C for 15–30 min.
3. Put the samples on ice and perform electroporation immediately (*see* **Note 3** and **4**).

3.4 Zygote Electroporation and Embryo Transfer (see Note 5)

1. Place the mouse zygotes in the acidic Tyrode's solution for 10 s (IVF embryos do not need this treatment, since their zona pellucida has been weakened during IVF).
2. Remove the embryos and wash through three drops of 100 μL pre-warmed M2 media.
3. Place the embryos in 10 μL drop of Opti-MEM media that has been pre-warmed and equilibrated. For each sample, up to 150 zygotes can be electroporated together, and we routinely use 60 zygotes. Depending on the skill level of the personnel handling the embryos, different numbers of zygotes can be used.
4. Add 10 μL of CRISPR-Cas9 reagents, consisting of the Cas9 protein, sgRNA, and donor oligonucleotide reconstituted in TE buffer (pH 7.5), to the embryos in Opti-MEM medium.
5. Pipet up and down a few times gently, pick up the 20 μL of embryos suspended in CRISPR-Cas9 reagents with a 20 μL pipet tip, position the pipet tip between electrodes of the cuvette, and deposit the contents into the chamber of a 1 mm electroporation cuvette.
6. Load the cuvette to the electroporator (We use ECM830 Square Wave Electroporation System). The electroporation setting is the same for all the experiments using 1 ms pulse duration and two pulses with 100 ms pulse interval at 30 V.
7. Press the start button. One-time electroporation is quite efficient for generating mouse models we tested, but multiple electroporation can be used to further improve the efficiency by pressing start button four to six times, with 3 s interval.
8. Following the electroporation, deposit pre-warmed and pre-equilibrated 100 μL M2 medium into the cuvette using a sterile plastic pipette (comes with the cuvette).
9. Remove the medium containing zygotes from the cuvette and place into a Petri dish.
10. Rinse the cuvette with additional 100 μL of M2 and add to the embryos in Petri dish.
11. Transfer the zygotes to the culture dish and keep them in culture in a MINC benchtop incubator (COOK; 37 °C, 5% CO_2).
12. The embryos will be transferred into pseudopregnant female mice (CByB6F1/J).
13. Embryos at pronuclear stage will be transferred into each recipient mouse, following the standard embryo transfer protocol [11].
14. Alternatively, these embryos can be kept in culture and develop into blastocysts for genotyping.

3.5 Blastocysts Collection and Genomic DNA Extraction

It is useful to genotype the blastocysts in vitro developed from electroporated zygotes to evaluate the gene-editing efficiency before generating live born mice. Here, we describe the blastocysts genotyping method.

1. Use thin-wall strip PCR tubes with individual attached caps. Open the caps and put the PCR tubes under the inverted microscope.
2. Make several drops of M2 medium.
3. Transfer blastocysts in the culture dish to a drop of warm M2.
4. Wash the blastocysts in the drops of M2 medium to get rid of the mineral oil.
5. Load 8–16 blastocysts into the mouth micropipette.
6. Put one embryo in each PCR tube under the inverted microscope, keeping the volume of the liquid to the minimum.
7. Prepare blastocysts genomic DNA extraction buffer (10 mM Tris–HCl, pH 8.0; 2 mM EDTA; 2.5% Tween-20, 2.5% Triton-X 100; 100 μg/mL Proteinase K).
8. Add 6 μL of the extraction buffer to each PCR tube.
9. Perform the extraction reaction in a thermo cycler: 55 °C, 1 h; 95 °C, 10 min; RT, hold.
10. The extracted DNA can be stored in −20 °C or directly used for genotyping.

3.6 Mouse Tissue Collection and Genotyping

Pups at day 12 or after day 12 are ready for tissue collection. Either ear notch or tail tip can be used for genotyping. Ear notch and tail tip sampling can be done following the standard procedure.

3.6.1 Mouse Tissue Collection

3.6.2 Extraction of Genomic DNA from Mouse Tail Tip or Ear Notch

1. Put 2 mm tail tip or ear punch in a PCR tube or 96-well plate. Spin to collect the samples to the bottom of the wells.
2. Add 50 μL 50 mM NaOH and quick spin to make sure tissue samples are in the solution.
3. Perform the extraction reaction in a thermo cycler: 98 °C, 1 h; 25 °C, hold.
4. Add 60 μL neutralization buffer (167 mM Tris–HCl, pH 8.0, 42 mM HCl) to neutralize the reaction and mix well. This mixture will serve as the template for the following PCR amplification.

3.6.3 PCR Reaction and DNA Electrophoresis

1. Set up the PCR reaction:

Reagent	Volume (μL)
10× PCR reaction buffer	5
dNTP	2
Universal Rev. Primer 1	1.5
Universal Rev. Primer 2	1.5
PrimeStar GXL taq	0.5
H_2O (Molecular Grade)	13.5

Add 1 μL of genomic DNA sample in 25 μL reaction and mix well. The remaining DNA samples can be kept at 4 °C for a few weeks.

2. PCR cycles:
   (1) 98 °C, 2 min; (2) 98 °C, 30 s; (3) 60 °C, 30 s; (4) 68 °C, 1 min; to **step (2)**, 34 cycles; (5) 68 °C, 5 min; (6) 4 °C, hold.
3. Transfer 2 μL of the PCR reaction to a new tube.
4. Add 3 μL of water.
5. Add 1 μL of 6× loading dye and mix well.
6. Load the samples and 3 μL 1Kb plus DNA marker to 1% agarose gel.
7. Electrophorese the gel at 120 V until the bromophenol blue reaches the bottom of the gel.

3.6.4 Sanger Sequencing and Genotyping Strategies

1. Prepare required PCR products and desired primers following the required protocol for Sanger sequencing. Send the PCR products for sequencing. TIDE can be used to analyze different alleles existing in these PCR products [12].
2. T7E1 assay or Surveyor assay can be performed to confirm insertion or deletion (indel) formation following standard procedures.
3. If an enzymatic site (Knock-in experiment) is inserted, digest 4 μL PCR products with specific restriction enzymes. Run an agarose gel to confirm the insertion.

3.6.5 Sequencing Trace Analysis Using TIDE Online Software

1. Open the webpage of TIDE online analysis tool following the link: https://tide-calculator.nki.nl/.
2. Type in sample name and gRNA sequence.
3. Load control sequencing traces and sample sequencing traces (SCF or ABI chromatogram files).
4. Tick advanced settings and click update view. The alignment window, decomposition window, and indel size range can be adjusted.
5. Click update view.
6. The indels and the according frequency will be shown. If a small DNA insertion is expected, adjust the indel size range to

include the expected insert size. If the desired insert exists in the sample, there will be a peak at the expected insert size. The frequency of the alleles with the insert will be shown.

3.6.6 TOPO Blunt-End Cloning and Sanger Sequencing

To confirm that the genomic modifications are precise and correct, the PCR product needs to be subcloned into TOPO blunt-end cloning vectors.

1. Perform the TOPO cloning using Zero Blunt TOPO® PCR Cloning Kit following the manufacturer's protocol.
2. Pick and expand the single clones.
3. Purify the plasmids.
4. Sequence the insert by Sanger sequencing.

3.6.7 Southern Blot

The genomic DNA can also be used for southern analysis following the standard procedure.

4 Notes

1. Surgical training in the Jackson Laboratory. There is a workshop on Assisted Reproductive Technologies (ARTs) in the laboratory mouse every year in the Jackson Laboratory. We encourage people to get proper training from these useful workshops.
2. Mouse strain selection. We used C57B6/J and C57B6/NJ in our experiments. The targeting efficiency and the survival rate of the pups are both high. In addition, NSG mice can also be modified using this technology, as it is a more mild treatment for the embryos compared to the microinjection. This is very useful for generating genetic modifications in mouse strains that are sensitive to microinjection.
3. Cas9 RNP preparation. The Cas9 protein can be diluted with RNase-free water or TE, and kept at −80 °C. Multiple freeze-thaw cycles should be avoided. The electroporation solution can also be frozen in −80 °C before incubation at 37 °C. Thaw the frozen electroporation solution and incubate in 37 °C for 15–30 min before performing electroporation.
4. Cas9 mRNA can also be used in the ZEN technology. The final concentration of Cas9 mRNA will be 600 ng/μL in the electroporation solution. Please note that compared with Cas9 RNP, Cas9 mRNA has much lower targeting efficiency for several target genes we tested.
5. ZEN technology can be used to generate indel mutations, large DNA segment deletions, specific nucleotide changes, and small DNA fragment insertions. DNA segment deletion can be generated using two sgRNAs simultaneously. A donor

oligonucleotide can be used to generate precise nucleotides changes or small tag insertion (3× FLAG, His, V5, loxP, etc.). The oligonucleotide is designed such that the mutation is flanked by 40–60 nt long homology arms on each side. The intended mutation should disrupt (for tag or loxP insertion models) or replace (for point mutation incorporation models) the PAM or the guide sequence, particularly the PAM proximal sequences. Order the oligonucleotide from a vendor, considering using PAGE or HPLC to enrich for the full-length oligonucleotide.

References

1. Jinek M, Chylinski K, Fonfara I, Hauer M, Doudna JA, Charpentier E (2012) A programmable dual-RNA-guided DNA endonuclease in adaptive bacterial immunity. Science 337(6096):816–821. doi:10.1126/science.1225829. science.1225829 [pii]
2. Cong L, Ran FA, Cox D, Lin SL, Barretto R, Habib N, Hsu PD, Wu XB, Jiang WY, Marraffini LA, Zhang F (2013) Multiplex genome engineering using CRISPR/Cas systems. Science 339(6121):819–823. doi:10.1126/science.1231143
3. Mali P, Yang L, Esvelt KM, Aach J, Guell M, DiCarlo JE, Norville JE, Church GM (2013) RNA-guided human genome engineering via Cas9. Science 339(6121):823–826. doi:10.1126/science.1232033. science.1232033 [pii]
4. Wang H, Yang H, Shivalila CS, Dawlaty MM, Cheng AW, Zhang F, Jaenisch R (2013) One-step generation of mice carrying mutations in multiple genes by CRISPR/Cas-mediated genome engineering. Cell 153(4):910–918. doi:10.1016/j.cell.2013.04.025. S0092-8674(13)00467-4 [pii]
5. Yang H, Wang H, Shivalila CS, Cheng AW, Shi L, Jaenisch R (2013) One-step generation of mice carrying reporter and conditional alleles by CRISPR/Cas-mediated genome engineering. Cell 154(6):1370–1379. doi:10.1016/j.cell.2013.08.022. S0092-8674(13)01016-7 [pii]
6. Wang W, Kutny PM, Byers SL, Longstaff CJ, DaCosta MJ, Pang C, Zhang Y, Taft RA, Buaas FW, Wang H (2016) Delivery of Cas9 protein into mouse zygotes through a series of electroporation dramatically increases the efficiency of model creation. J Genet Genomics 43(5):319–327. doi:10.1016/j.jgg.2016.02.004. S1673-8527(16)30002-9 [pii]
7. Qin W, Dion SL, Kutny PM, Zhang Y, Cheng A, Jillette NL, Malhotra A, Geurts AM, Chen YG, Wang H (2015) Efficient CRISPR/Cas9-mediated genome editing in mice by zygote electroporation of nuclease. Genetics 200:423–430. doi:10.1534/genetics.115.176594. genetics.115.176594 [pii]
8. Chen S, Lee B, Lee AY, Modzelewski AJ, He L (2016) Highly efficient mouse genome editing by CRISPR ribonucleoprotein electroporation of zygotes. J Biol Chem 291(28):14457–14467. doi:10.1074/jbc.M116.733154. M116.733154 [pii]
9. Wiles MV, Qin W, Cheng AW, Wang H (2015) CRISPR-Cas9-mediated genome editing and guide RNA design. Mamm Genome 26(9–10):501–510. doi:10.1007/s00335-015-9565-z
10. Qin W, Kutny PM, Maser RS, Dion SL, Lamont JD, Zhang Y, Perry GA, Wang H (2016) Generating mouse models using CRISPR-Cas9-mediated genome editing. Curr Protoc Mouse Biol 6(1):39–66. doi:10.1002/9780470942390.mo150178
11. Byers SL, Payson SJ, Taft RA (2006) Performance of ten inbred mouse strains following assisted reproductive technologies (ARTs). Theriogenology 65(9):1716–1726. doi:10.1016/j.theriogenology.2005.09.016. S0093-691X(05)00405-X [pii]
12. Brinkman EK, Chen T, Amendola M, van Steensel B (2014) Easy quantitative assessment of genome editing by sequence trace decomposition. Nucleic Acids Res 42(22):e168. doi:10.1093/nar/gku936. gku936 [pii]

Chapter 16

CRISPR/Cas9-Mediated Gene Targeting during Embryogenesis in Swine

Junghyun Ryu and Kiho Lee

Abstract

Ability to disrupt genes is essential in elucidating gene function. Unlike rodents or amphibians, it has been difficult to generate gene-targeted embryos in large animals. Therefore, studies of early embryo development have been hampered in large animals. A recent technology suggests that targeted mutations can be successfully introduced during embryogenesis, thus by-passing the need of breeding to produce gene-targeted embryos. This is particularly important in large animal models because of longer gestation period and higher animal cost. Here, we describe a specific approach to disrupt up to two genes simultaneously during embryogenesis using the CRISPR/Cas9 technology in swine. The approach can help understand the mechanism of zygotic genome activation in large animals.

Key words Pig embryos, CRISPR/Cas9, Zygotic genome activation, Gene targeting, Early development

1 Introduction

Ability to generate gene-targeted embryos is essential in elucidating gene function. In mammals, the technology has been available mostly in rodents, especially in the mouse. Development of embryonic stem (ES) cells and gene targeting using homologous recombination mechanism have made the mouse a leading species to study early development [1, 2]. The mouse models have contributed greatly in biomedicine, and various signaling pathways have been elucidated using the models [1]. However, because of the differences in physiology, rodent models are not always the best model to recapitulate human biology [3].

Swine models, on the other hand, can closely represent human physiology. If the emphasis is on early development, swine embryos have similar development trajectory as human embryos. Both human and swine embryos have major zygotic genome activation at the 4-cell stage [4, 5]. In addition, changes in the level of DNA methylation after fertilization are also observed in both species

Kiho Lee (ed.), *Zygotic Genome Activation: Methods and Protocols*, Methods in Molecular Biology, vol. 1605, DOI 10.1007/978-1-4939-6988-3_16, © Springer Science+Business Media LLC 2017

[6, 7]. However, swine has not been considered to be a useful model to study early development because of difficulty in modifying its genome.

Conventionally, in swine, targeted modification was introduced into somatic cells, then embryos/animals were generated through somatic cell nuclear transfer (SCNT) [8, 9]. This approach allowed us to by-pass the need of ES cells in generating gene-targeted embryos/animals. However, because the process involves SCNT, i.e., cloning, animals generated through the method often presented developmental abnormalities associated with the cloning. Additional breeding should be the solution to this problem, but considering gestation period being close to 4 months (114 days), and housing and animal cost being high, the use of swine in biomedicine was limited.

A recent technology demonstrates that gene-targeted embryos can be generated without cloning or breeding by introducing clustered regularly interspaced short palindromic repeats (CRISPR)/Cas9 system into early embryos [10, 11]; reported efficiency of the targeting was as high as 100%. A recent report from our group demonstrates that the technology can also efficiently disrupt two genes simultaneously. Ability to introduce targeted modifications during embryogenesis will allow swine to be a model to study early embryo development. Here, we describe a specific approach to introduce targeted modifications during embryogenesis in swine.

2 Materials

2.1 Materials Needed for in Vitro Maturation

2.1.1 Lab Equipment

1. 5% CO_2 incubator.
2. Stereo microscope with warm plate.
3. Microneedles (18-gauge).
4. Syringes (10 ml).
5. Centrifuge tubes (50 ml).
6. Petri dishes (30 × 10 mm and 100 × 25 mm).
7. 4-Well culture dishes.
8. Glass capillary tubes.
9. Captrol III® micropipet (Drummond Scientific).
10. Portable pipet-aid.
11. 10 ml serological pipets.

2.1.2 Reagents

1. Saline: 0.9% NaCl, supplemented with 100 U/ml penicillin-streptomycin.
2. Hepes-buffered Tyrode's Lactate (TL-Hepes) medium: 2.0 mM $CaCl_2 \cdot 2H_2O$, 114.0 mM NaCl, 3.2 mM KCl, 2.0 mM $NaHCO_3$, 0.4 mM NaH_2PO_4, 10.0 mM Na Lactate (60% syrup), 0.5 mM $MgCl_2 \cdot 6H_2O$, 10.0 mM Hepes, 12.0 mM

sorbitol, 0.2 mM sodium pyruvate, 0.075 g/L penicillin, 0.05 g/L streptomycin, 1 ml phenol red (0.5%), 0.1 g/L polyvinyl alcohol (PVA); pH 7.4.

3. IVM medium: Medium 199 supplemented with 3.05 mM glucose, 0.91 mM sodium pyruvate, 0.57 mM cysteine, 10 ng/ml epidermal growth factor (EGF), 0.5 μg/ml luteinizing hormone (LH), 0.5 μg/ml follicle stimulating hormone (FSH), 10 ng/ml gentamicin, and 0.1% polyvinyl alcohol (PVA); pH 7.4.
4. IVM wash medium: IVM medium without LH and FSH; pH 7.4.
5. Embryo culture graded Mineral oil.

2.2 Materials Needed for in Vitro Fertilization

2.2.1 Lab Equipment

1. CO_2 incubator.
2. O_2/CO_2 incubator.
3. Stereo microscope with warm plate.
4. Benchtop centrifuge.
5. Microcentrifuge tubes (1.5 ml).
6. Vortex mixer.
7. Petri dishes (30 × 10 mm).
8. Glass capillary tubes.
9. Captrol III® micropipet (Drummond Scientific).
10. Portable pipet-aid.
11. 10 ml serological pipets.

2.2.2 Reagents

1. IVF medium (mTBM): modified Tris-buffered medium with 113.1 mM NaCl, 3 mM KCl, 7.5 mM $CaCl_2$, 11 mM glucose, 20 mM Tris, 2 mM caffeine, 5 mM sodium pyruvate, and 2 mg/ml BSA; pH 7.4.
2. Manipulation medium: medium 199 supplemented with 0.6 mM $NaHCO_3$, 2.9 mM Hepes, 30 mM NaCl, 10 ng/ml gentamicin, and 3 mg/ml BSA; pH 7.4.
3. Denuding medium: 0.3 M mannitol, 0.001% BSA, 0.03% hyaluronidase, 5% TL-Hepes medium in distilled water; pH 7.4.
4. AndroPRO® Plus semen extension medium (MOFA®).
5. Sperm wash medium: DPBS with 0.1% BSA and 10 ng/ml gentamicin.
6. Embryo culture graded Mineral oil.

2.3 Materials Needed for in Vitro Culture

2.3.1 Lap Equipment

1. O_2/CO_2 incubator.
2. Stereo microscope with warm plate.
3. Petri dishes (30 × 10 mm).
4. Glass capillary tubes.
5. Captrol III® micropipet (Drummond Scientific).

2.3.2 Reagents

1. PZM3 culture medium [12].
2. Embryo culture graded Mineral oil.

2.4 Materials Needed for Microinjection of CRISPR/Cas9 System into Presumable Zygotes

2.4.1 Lab Equipment

1. Heat block (adjustable to 65–80 °C).
2. Thermocycler.
3. 37 °C bacteria culture incubator.
4. 37 °C bacteria culture shaker.
5. Water bath.
6. UV trans illuminator.
7. FemtoJet (Eppendorf).
8. Micro manipulator.
9. O_2/CO_2 incubator.
10. Stereo microscope with warm plate.
11. Petri dishes (100 × 10 mm).
12. Glass capillary tubes.
13. Captrol III® micropipet (Drummond Scientific).
14. Nikon microscope with warm plate.
15. Microcapillary Puller (Shutter).
16. Benchtop centrifuge.
17. Microcentrifuge tubes (1.5 ml).
18. Electrophoresis.

2.4.2 Reagents

1. pX330 vector (Addgene).
2. BbsI enzyme (NEB).
3. T4 ligase and buffer.
4. mMESSAGE mMACHINE® T7 Ultra Kit (Ambion).
5. MEGAshortscript™ Kit (Ambion).
6. Poly(A) Tailing Kit (Ambion).
7. mMESSAGE mMACHINE® T7 Ultra Kit (Ambion).
8. Phusion High-Fidelity DNA Polymerase (ThermoFisher).
9. GeneJET PCR Purification Kit (ThermoFisher).
10. GeneJET Gel Extraction Kit (ThermoFisher).
11. GeneJET Plasmid Miniprep Kit (ThermoFisher).
12. PZM3 culture media.
13. Manipulation media.
14. Embryo graded mineral oil.
15. LB.
16. LB agar plate.

17. Ampicillin.
18. DNA ladder.
19. RNA ladder.
20. Agarose.
21. Chemical competent cells.

2.5 Genotyping of CRISPR/Cas9 Injected Embryos

2.5.1 Lab Equipment

1. Thermocycler.
2. Stereo microscope with warm plate.
3. Petri dishes (100 × 10 mm).
4. Glass capillary tubes.
5. Captrol III® micropipet (Drummond Scientific).
6. PCR tube.
7. Microcentrifuge tubes (1.5 ml).
8. Bioedit program.
9. DNA eletrophoesis units.

2.5.2 Reagents

1. DPBS with 1% BSA media (pH 1.98).
2. DPBS with 1% BSA media.
3. Embryo lysis buffer: 50 mM KCl, 1.5 mM $MgCl_2$, 10 mM Tris–HCl pH 8.5, 0.5% Nonidet P40, 0.5% Tween-20 and 200 μg/ml proteinase K.
4. Platinum taq (ThermoFisher).
5. GeneJET PCR Purification Kit (ThermoFisher).
6. Agarose.
7. 100 bp DNA ladder.

3 Methods

3.1 In Vitro Maturation (IVM)

1. In the morning, IVM and IVM washing dishes are prepared and incubated for 4–6 h at 38.5 °C/5% CO_2/100% humidity. The 4-well dishes containing 500 μl IVM medium are covered with 450 μl mineral oil per well and two 30 mm IVM wash dishes with 3 ml IVM wash medium per dish.
2. Porcine ovaries of pre-pubertal gilts or sows are obtained from a local slaughterhouse and transported to the laboratory.
3. The ovaries are washed with pre-warmed saline at 38.5 °C. Immature oocytes are aspirated from follicles (>2 mm in diameter) using an 18-guage microneedle attached to a 10 ml syringe, and harvested into 50 ml centrifuge tube (*see* **Note 1**).
4. After 20 min incubation, the supernatant is removed and then pre-warmed TL-Hepes medium is added for washing.

5. Washed aspirates are placed in the 100 mm petri dish.
6. Cumulus-oocyte complexes (COCs) with evenly granulated cytoplasm and intact surrounding cumulus cells are collected using a finely drawn glass pipet under a microscope (*see* **Note 2**).
7. The collected COCs are washed twice with pre-incubated IVM wash medium in the 30 mm dish.
8. Approximately 50 COCs are maturated in a well (4-well dish) containing 500 μl IVM medium and incubated for 42–44 h at 38.5 °C, 5% CO_2, and 100% humidity.

3.2 In Vitro Fertilization

In pigs, modified Tris-buffered medium (mTBM) is known to be proper to minimize polyspermy not affecting the penetration rate [13, 14]. For successful IVF, viability of sperm needs to be monitored prior to the IVF.

3.2.1 Preparation of Oocytes

1. IVF medium (20 ml) is incubated at 38.5 °C/5% CO_2/100% humidity 2 days prior to the IVF.
2. IVF drops (each 50 μl) covered with mineral oil and two 3 ml mTBM washing medium are prepared in 30 mm petri dishes using the incubated IVF medium on the day before the IVF. These plates and the rest of media are incubated at the same condition until the IVF day.
3. After maturing COCs for 42–44 h, the COCs are transferred to 1.5 ml tube containing 1 ml denuding medium pre-warmed at 38.5 °C, and the expanded cumulus cells surrounding oocytes are removed by vortexing for 3 min.
4. Three manipulation dishes are prepared during the vortexing: A 30 mm dish with 2.5 ml and another two dishes with 3.5 ml manipulation medium pre-warmed at 38.5 °C.
5. Denuded oocytes are moved to the first manipulation dish (2.5 ml) and collection of matured oocytes with a visible polar body and washing are sequentially conducted in other two dishes (*see* **Note 3**).
6. After washing the collected oocytes with the incubated IVF medium (two 30 mm dishes), 25–30 collected oocytes are placed in 50 μl IVF droplets.

3.2.2 Sperm Preparation

1. Fresh semen collected from a boar is diluted into an extender and stored at 17 °C. The semen can be sorted for up to 10 days.
2. For IVF, 1 ml semen is washed in 9 ml sperm wash medium by centrifugation at 750 × *g* for 3 min.
3. The sperm pellet is resuspended and washed two more times with 10 ml wash medium at the same centrifugation condition. Washed sperm pellet is resuspended and diluted to 2.5×10^5 using the incubated IVF medium (*see* **Note 4**).

4. Then, 50 μl sperm suspension is added to the prepared IVF droplets containing oocytes (final sperm concentration is 1.25×10^5) and incubated for 5 h at 38.5 °C/5% CO_2/100% humidity.

3.3 In Vitro Culture

1. PZM3 droplets, 20 μl each placed on 30 mm dish, are covered with mineral oil and another two 30 mm wash dishes are filled with 3 ml PZM3, followed by incubation at 38.5 °C, 5% O_2, and 5% CO_2 in humidified air during the 5 h of IVF.
2. Oocytes fertilized for 5 h are transferred to the incubated wash dish containing PZM3 using a glass pipet.
3. Excess sperms attached to the oocyte surface (zona pellucida) are removed by repetitive passage of medium through glass pipet in the wash dish.
4. The oocytes are moved to the second wash dish.
5. 20–25 oocytes are placed in the pre-incubated PZM3 droplets and incubated at 38.5 °C/5% O_2/5% CO_2 in humidified air.

3.4 Microinjection of sgRNA and Cas9 mRNA into Presumable Zygotes

Introducing sgRNA and Cas9 mRNA into developing embryos can induce random insertion or deletion mutations (indel mutations) on a target sequence [15]. The random indel mutations can introduce a pre-mature stop codon by disrupting the order of amino acids, thus disrupting the gene function.

3.4.1 In Vitro Transcription of Cas9 mRNA

1. Using the pair of primers (Table 1) Cas9 mRNA sequence is amplified from pX330 vector using Phusion taq following the manufacturer's protocol. The PCR condition is as follows, initial denature at 98 °C for 2 min, denature at 98 °C for 30 s, annealing at 62 °C for 30 s, and extension at 72 °C for 3 min for 29 cycles, 72 °C for 5 min, and hold at 4 °C.
2. The PCR product is loaded on a 0.8% agarose gel for electrophoresis (*see* **Note 5**).
3. Once the size of the PCR product is confirmed, the rest of PCR products are purified using a PCR purification kit following the manufacturer's protocol.
4. In vitro transcription kit is used to generate Cas9 mRNA from the PCR product. Assembly of the transcription reaction is: 1 μg of PCR product, 10 μl of 2× NTP/CAP, 2 μl of 10× buffer, 2 μl of enzyme mix, and use nuclease-free water to bring the reaction volume to 20 μl.
5. The reaction mixture is incubated at 37 °C for 1 h.
6. After incubation, add 1 μl of TURBO DNase into the reaction and incubate at 37 °C for 15 min.
7. The in vitro transcribed product is placed into a fresh 1.5 ml tube with 20 μl of 5× E-PAP buffer, 10 μl of ATP (10 mM),

Table 1
Primers used to generate template DNAs for in vitro transcription and genotype CRISPR/Cas9 injected embryos for their modifications on *RAG2* and *IL2RG*

To generate template DNA for in vitro transcription	
Cas9 mRNAF	TAA TAC GAC TCA CTA TAG GGA GAA TGG ACT ATA AGG ACC ACG AC
Cas9 mRNAR	GCG AGC TCT AGG AAT TCT TAC
T7 RAG2 F	TTA ATA CGA CTC ACT ATA GGT ATA GTC GAG GGA AAA GTA
T7 IL2RG F	TTA ATA CGA CTC ACT ATA GGG AAA CGG TTG AGA GTC CCA
T7 sgRNA R	AAA AGC ACC GAC TCG GTG CC
To genotype for RAG2 and IL2RG mutations	
RAG2 F	AAG GAT TCC TGC TAC CTT CCT CCT
RAG2 R	AGA TAG CCC ATC TTG AAG TTC TGG
IL2RG F	CTG GAC TAT TAG AAG GAT GTG GGC
IL2RG R	ATA TAG TGG GAA GCC TGG GAT GCT

10 μl of MnCl2 (25 mM), 36 μl of nuclease-free water, and 4 μl of E-PAP enzyme to add additional poly A tail to the mRNA. The final volume should be 100 μl. Incubate the reaction at 37 °C for 1 h.

8. To purify the mRNA, the entire reaction product (100 μl) is mixed with 350 μl of binding buffer from the RNA purification kit and add 250 μl of 100% ethanol. The total of 700 μl is then loaded into a filter cartridge. Centrifuge for 1 min at 15,000 × *g*.
9. Wash the filter cartridge two times with 500 μl of washing buffer.
10. To completely remove the traces of wash buffer, centrifuge for 1 min.
11. To elute the Cas9 mRNA, place the filter into a new collection tube and add 50 μl of elution buffer. Then, place the filter into a heating block at 65 °C for 5 min.
12. Centrifuge for 1 min at 15,000 × *g*.
13. Load the purified Cas9 mRNA to a RNase-free gel for electrophoresis (*see* Fig. 1).

3.4.2 sgRNA Preparation

1. Digest 1 μg of pX330 vector with BbsI enzyme. Reaction mixture contains 2 μl of 10× buffer, 1 μl of BbsI enzyme, 1 μg of pX330 vector, and distilled water up to 20 μl. Mixture is incubated at 37 °C for 1 h.
2. sgRNA sequences are designed using a web-based program (http://crispr.mit.edu/). Then, sequences with the highest score are selected. The target sequences are blasted against the entire pig genome to verify their specificity (*see* **Note 6**).

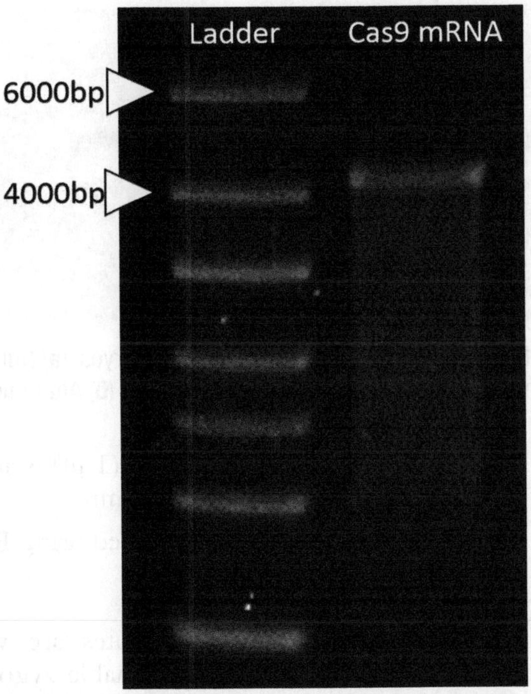

Fig. 1 Image of Cas9 mRNA with additional poly A tail. In vitro transcribed Cas9 mRNA was loaded on a 2% agarose gel. The mRNA was denatured prior to the loading. Only a single product was detected on the gel

3. The selected sgRNA sequence is introduced into the linearized pX330 by following a standard protocol (http://www.addgene.org/crispr/zhang/) [16].
4. Using the plasmids carrying correct sgRNA as template, a pair of primers is used to amplify the sgRNA (Table 1). Phusion taq is used for the PCR reaction following the manufacturer's protocol. The PCR condition is as follows, initial denature at 98 °C for 2 min, denature at 98 °C for 30 s, annealing at 64 °C for 30 s and extension at 72 °C for 30 s for 34 cycles, 72 °C for 5 min, and hold at 4 °C.
5. The PCR product is loaded on a 2% agarose gel for electrophoresis.
6. After verifying size of the amplicons, the rest of PCR products are purified using PCR purification kit following manufacturer's protocol.
7. For in vitro transcription, the following reaction mixture is assembled: 2 μl of 10× buffer, 2 μl of ATP solution, 2 μl of CTP solution, 2 μl of GTP solution, 2 μl of UTP solution, 2 μl of enzyme, and 8 μl of PCR product. Then, the mixture is incubated at 37 °C for 3 h.

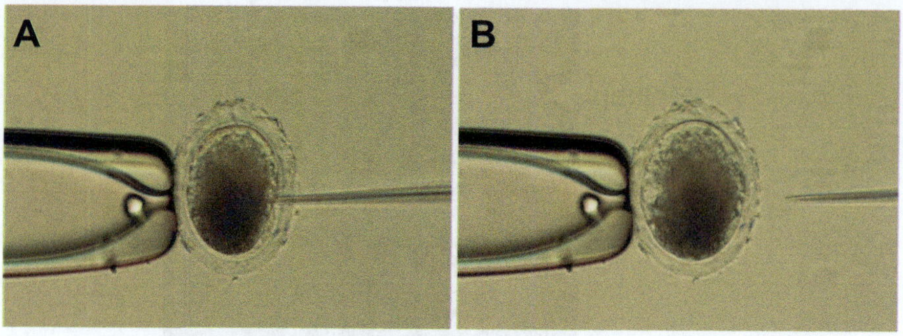

Fig. 2 Image of microinjection of CRISPR/Cas9 system into pig embryos. (**a**) During microinjection, sgRNA and Cas9 mRNA are injected into the cytoplasm of presumable zygotes. (**b**) After microinjection

8. After 3 h, TURBO DNase (1 μl) is added to the reaction and incubation at 37 °C for 15 min.
9. The sgRNAs are then purified using RNA purification kit then used for microinjection.

3.4.3 Microinjection

1. After 2 h post-IVF, zygotes are washed in manipulation medium. Then, the presumable zygotes are transferred to an injection dish (manipulation medium covered with mineral oil). Recommended concentrations of RNAs are 10 ng of sgRNA and 20 ng of Cas9 mRNA. The RNAs are injected into cytoplasm of the presumable zygotes using FemtoJet microinjector (*see* **Notes 7** and **8**). The microinjection is conducted on a heated plate at 37 °C (Fig. 2).
2. After microinjection, the zygotes are washed two times in PZM3 medium.
3. The zygotes are cultured in PZM3 at 38.5 °C, 5% CO_2, and 5% O_2 incubator for additional 7 days.

3.5 Genotyping of CRISPR/Cas9 Injected Embryos

DNAs extracted from individual blastocyst are used to identify mutations on target sites, introduced by the CRISPR/Cas9 system.

3.5.1 DNA Isolation and PCR

1. Seven days after IVF, blastocysts are collected for genotyping.
2. The blastocysts are placed in DPBS with 1% BSA adjusted pH 1.98 media drops and gently pipetted a few times to completely remove zona pellucida and sperms attached to the zona pellucida.
3. Genomic DNA from the blastocysts is extracted using embryo lysis buffer. An individual blastocyst is placed into a PCR tube with 12 μl of embryo lysis buffer and then incubated at 65 °C for 30 min followed by 95 °C for 10 min (*see* **Note 9** and **10**).
4. The genomic regions flanking CRISPR/Cas9 target region are amplified using Platinum Taq DNA polymerase. PCR conditions

are as follows, initial denature at 95 °C for 2 min, denature at 95 °C for 30 s, annealing at 55 °C for 30 s and extension at 72 °C for 30 s for 39 cycles, 72 °C for 5 min, and holding at 4 °C (*see* **Note 11**).

5. The PCR products are loaded on a 2% gel for electrophoresis.
6. Then the rest of the PCR products are purified using PCR purification kit and used for Sanger sequencing using the forward primer (Table 1).

3.5.2 Analysis of the Genotyping

1. Results from the Sanger sequencing are opened with the Bioedit program (Chromas lite or other software can also be used). The sequencing results are blasted against wild-type genomic sequence to verify modifications induced by the CRISPR/Cas9 system.
2. The blast results can indicate mutations introduced by the CRISPR/Cas9 system. To clearly distinguish the type of modifications (heterozygous, homozygous, biallelic, or mosaic), chromatogram from mutated embryos needs to be thoroughly compared with chromatogram peaks from wild-type control (Fig. 3, *see* **Note 12**).

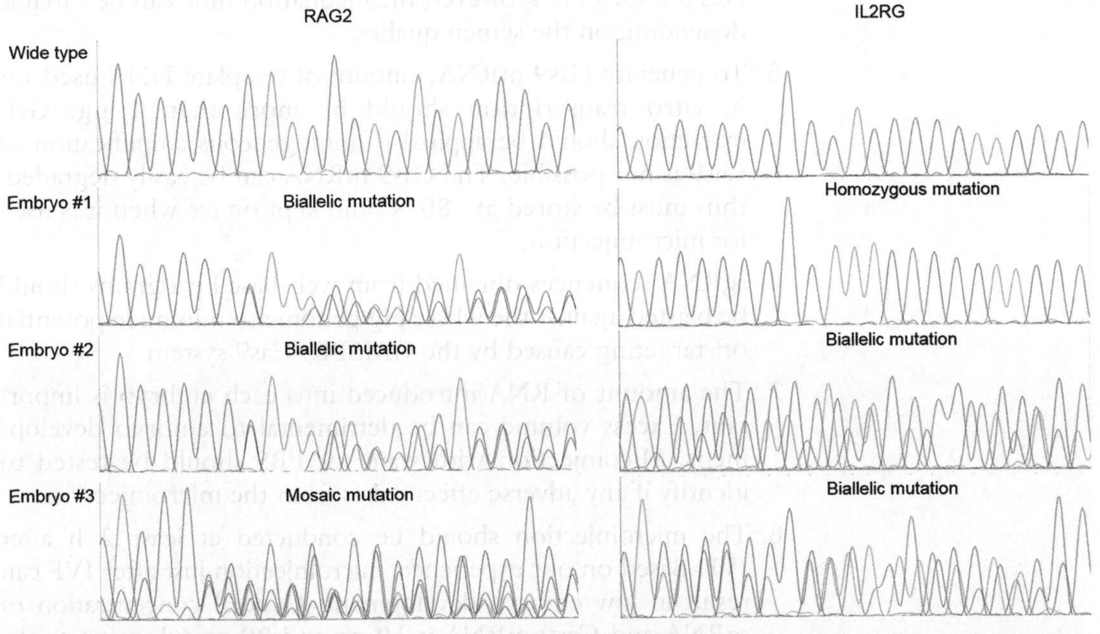

Fig. 3 Sequencing results from *RAG2* and *IL2RG* double knock-out embryos. Sequencing readings from embryos carrying homozygous mutation have mutated single peaks. Bialleic mutations can be identified by having two polymorphic sequencing peaks but no wild type matching sequences. More than two sequencing peaks indicate the embryos have mosaic mutation; all embryos carrying mosaic genotypes do not have matching wild type sequence. The colors indicate each nucleotide; red – thymidine, black – guanine, green – adenine, and blue – cytosine

4 Notes

1. During aspiration, avoid blood contamination from the follicle. The blood may cause low maturation rate.
2. Good quality oocytes should consistently have two or more layers of cumulus cells. To minimally disturb the cumulus cells while transferring COCs, the size of glass capillaries should be larger than 200 μm.
3. After freeing the oocytes from cumulus cells, searching for maturated oocytes is an important step for successful IVF. The extrusion of the first polar body is commonly used as an indicator of successful maturation. However, the presence of the first polar body is not always apparent as the polar body can be located behind the cytoplasm. Use of manipulation medium can help with the process because the medium has higher osmolality compared to other media. The diameter of the glass capillary used to transfer oocytes should be narrower (120–150 μm) as cumulus cells are no longer attached to the oocytes.
4. The age and condition of boars can dramatically affect semen quality. We recommend to use fresh semen from a boar aging between 1 to 4 years post-puberty. We generally incubate gametes 5 h for IVF. However, the incubation time can be variable depending on the semen quality.
5. To generate Cas9 mRNA, amount of template DNA used for in vitro transcription should be more than 1 μg. Gel-extraction should be applied if homogeneous amplification of Cas9 is not possible. The Cas9 mRNA can be easily degraded, thus must be stored at −80 °C and kept on ice when it is used for microinjection.
6. sgRNA sequences obtained from web-based programs should be blasted against the whole pig genome to minimize potential off-targeting caused by the CRISPR/Cas9 system.
7. The amount of RNA introduced into each embryo is important. Excess volume can be detrimental to embryo development. Micoinjection with water or PBS should be tested to identify if any adverse effect exists from the microinjection.
8. The microinjection should be conducted at least 2 h after IVF. Based on our experience, microinjection just after IVF can result in low embryo development. General concentration of sgRNA and Cas9 mRNA is 10 ng and 20 ng/μl, respectively. However, the concentration depends on target locations. According to our experience, it is possible to completely disrupt a gene by using 2.5 ng/μl of sgRNA and 5 ng/μl of Cas9 mRNA. On the other hand, some genes require higher concentration of CRISPR/Cas9 system. Therefore, it is important to conduct an optimization experiment to identify a working concentration.

9. The lysis buffer should be freshly made each time. Based on our experience, old buffer leads to lower genomic DNA yield.
10. Excess sperms, attached to zona pellucida, should be completely removed from the blastocyst to ensure accurate genotyping. Contamination of the sperm DNA can result in the appearance of wild-type sequences in the genotyping results.
11. PCR amplification from an individual blastocyst is challenging because of extremely low amount of DNA. Number of PCR cycles used here should be at least 39 cycles. However, high cycle number may cause nonspecific amplifications, especially in the negative control. To prevent nonspecific amplifications, specificity and efficiency of primers should be tested using low amount of genomic DNA as PCR template. We use 40 pg of standard wild-type genomic DNA to test the primers prior to using genomic DNA isolated from individual blastocyst.
12. It is difficult to get clear genotyping results by sequencing only the PCR products. Biallelic mutation should have double peaks in the sequencing results without carrying the wild-type sequence. Following each chromatogram peak will allow you to examine the presence of wild-type sequence. Mosaic mutation typically have more than two peaks in each nucleotide position. It is difficult to determine the presence of wild-type sequence in mosaic embryos. Alternatively, the PCR products can be cloned into a cloning vector and sequenced to identify modifications on each allele. Mosaic mutations will have more than two genotypes from the sequencing analysis.

Acknowledgment

This work was supported by NIH grant R21OD019934.

References

1. Capecchi MR (2005) Gene targeting in mice: functional analysis of the mammalian genome for the twenty-first century. Nat Rev Genet 6(6):507–512. doi:10.1038/nrg1619
2. Thomas KR, Capecchi MR (1987) Site-directed mutagenesis by gene targeting in mouse embryo-derived stem cells. Cell 51(3):503–512
3. Wilke M, Buijs-Offerman RM, Aarbiou J, Colledge WH, Sheppard DN, Touqui L, Bot A, Jorna H, de Jonge HR, Scholte BJ (2011) Mouse models of cystic fibrosis: phenotypic analysis and research applications. J Cyst Fibros 10(Suppl 2):S152–S171. doi:10.1016/s1569-1993(11)60020-9
4. Braude P, Bolton V, Moore S (1988) Human gene expression first occurs between the four- and eight-cell stages of preimplantation development. Nature 332(6163):459–461. doi:10.1038/332459a0
5. Prather RS (1993) Nuclear control of early embryonic development in domestic pigs. J Reprod Fertil Suppl 48:17–29
6. Santos F, Hyslop L, Stojkovic P, Leary C, Murdoch A, Reik W, Stojkovic M, Herbert M, Dean W (2010) Evaluation of epigenetic marks in human embryos derived from IVF and ICSI. Hum Reprod 25(9):2387–2395. doi:10.1093/humrep/deq151
7. Deshmukh RS, Ostrup O, Ostrup E, Vejlsted M, Niemann H, Lucas-Hahn A, Petersen B, Li J, Callesen H, Hyttel P (2011) DNA methylation in porcine preimplantation embryos developed

in vivo and produced by in vitro fertilization, parthenogenetic activation and somatic cell nuclear transfer. Epigenetics 6(2):177–187

8. Lai L, Kolber-Simonds D, Park KW, Cheong HT, Greenstein JL, Im GS, Samuel M, Bonk A, Rieke A, Day BN, Murphy CN, Carter DB, Hawley RJ, Prather RS (2002) Production of alpha-1,3-galactosyltransferase knockout pigs by nuclear transfer cloning. Science (New York, NY) 295(5557):1089–1092. doi:10.1126/science.1068228
9. Dai Y, Vaught TD, Boone J, Chen SH, Phelps CJ, Ball S, Monahan JA, Jobst PM, McCreath KJ, Lamborn AE, Cowell-Lucero JL, Wells KD, Colman A, Polejaeva IA, Ayares DL (2002) Targeted disruption of the alpha1,3-galactosyltransferase gene in cloned pigs. Nat Biotechnol 20(3):251–255. doi:10.1038/nbt0302-251
10. Whitworth KM, Lee K, Benne JA, Beaton BP, Spate LD, Murphy SL, Samuel MS, Mao J, O'Gorman C, Walters EM, Murphy CN, Driver JP, Mileham A, McLaren D, Wells KD, Prather RS (2014) Use of the CRISPR/Cas9 system to produce genetically engineered pigs from in vitro-derived oocytes and embryos. Biol Reprod 91:78. doi:10.1095/biolreprod.114.121723
11. Lei S, Ryu J, Wen K, Twitchell E, Bui T, Ramesh A, Weiss M, Li G, Samuel H, Clark-Deener S, Jiang X, Lee K, Yuan L (2016) Increased and prolonged human norovirus infection in RAG2/IL2RG deficient gnotobiotic pigs with severe combined immunodeficiency. Sci Rep 6:25222. doi:10.1038/srep25222
12. Yoshioka K, Suzuki C, Tanaka A, Anas IM, Iwamura S (2002) Birth of piglets derived from porcine zygotes cultured in a chemically defined medium. Biol Reprod 66(1):112–119
13. Abeydeera LR, Wang WH, Cantley TC, Rieke A, Prather RS, Day BN (1998) Presence of epidermal growth factor during in vitro maturation of pig oocytes and embryo culture can modulate blastocyst development after in vitro fertilization. Mol Reprod Dev 51(4):395–401. doi:10.1002/(sici)1098-2795(199812)51:4<395::aid-mrd6>3.0.co;2-y
14. Lee K, Redel BK, Spate L, Teson J, Brown AN, Park KW, Walters E, Samuel M, Murphy CN, Prather RS (2013) Piglets produced from cloned blastocysts cultured in vitro with GM-CSF. Mol Reprod Dev 80(2):145–154. doi:10.1002/mrd.22143
15. Wang H, Yang H, Shivalila CS, Dawlaty MM, Cheng AW, Zhang F, Jaenisch R (2013) One-step generation of mice carrying mutations in multiple genes by CRISPR/Cas-mediated genome engineering. Cell 153(4):910–918. doi:10.1016/j.cell.2013.04.025
16. Ran FA, Hsu PD, Wright J, Agarwala V, Scott DA, Zhang F (2013) Genome engineering using the CRISPR-Cas9 system. Nat Protoc 8(11):2281–2308. doi:10.1038/nprot.2013.143

Chapter 17

Potential Involvement of SCF-Complex in Zygotic Genome Activation During Early Bovine Embryo Development

Veronika Benesova, Veronika Kinterova, Jiri Kanka, and Tereza Toralova

Abstract

Proper timing of degradation of maternal protein reserves is important for early embryonic development. The major modification that triggers proteins to degradation is ubiquitination, mediated by ubiquitin-proteolytic system. We focus here on Skp 1-Cul 1-F-box complex (SCF-complex), E3 ubiquitin-ligase, a part of ubiquitin-proteolytic system, which transfer ubiquitin to the substrate protein. We describe in this chapter the methods for the characterization of the expression profile of mRNA and protein of invariant members of SCF-complex and for the definition of SCF-complex activity.

Key words SCF-complex, Preimplantation development, qRT-PCR, Western blot, Immunofluorescence, In situ proximity ligation assay

1 Introduction

Early embryo development is driven by maternal mRNAs and proteins. However, during the development these reserves are gradually degraded and replaced with mRNAs produced by newly activated embryonic genome. The degradation of mRNAs peaks around embryonic genome activation and results in the degradation of the vast majority of maternal mRNAs [1]. However, little information about the protein degradation is available. It is thought that a large proportion of the degradation of maternal proteins is mediated by the ubiquitin-proteolytic system [2, 3]. Ubiquitination of proteins is managed by the cooperation of three enzyme complexes: E1—ubiquitin activating enzyme, E2—ubiquitin conjugating enzyme, and E3—ubiquitin ligase [4].

Previously, we characterized the SCF-complex (Skp 1-Cullin 1-F-box), a modular RING type E3 ubiquitin-ligase, in bovine preimplantation embryos [5]. This complex consists of three invariable components—Skp1, Cullin 1 (Cul1) and Rbx1, and one of many F-box proteins, which determines the substrate specificity

Kiho Lee (ed.), *Zygotic Genome Activation: Methods and Protocols*, Methods in Molecular Biology, vol. 1605,
DOI 10.1007/978-1-4939-6988-3_17, © Springer Science+Business Media LLC 2017

[6–8]. It is thought that up to 20% of ubiquitinated proteins are triggered for degradation by the SCF-complex [9].

Technologies and methods used to characterize the expression profile of SCF-complex in bovine embryos are presented below. The profile of the mRNA was measured by the quantitative RT-PCR. To determine at which stage the transcription of each mRNA from the embryonic genome begins, the embryos were cultured with α-amanitin, specific inhibitor of RNA polymerase II. For the expression profile and localization of proteins, the western blot and immunofluorescence were used. To define the SCF-complex activity during bovine preimplantation development in situ proximity ligation assay was used. This method enables detection of the protein-protein interaction and moreover the localization of this interaction.

2 Materials

2.1 Oocytes and Embryos Preparation

Abattoir-derived ovaries from cows and heifers were collected and transported in thermocontainers in sterile saline at about 33 °C.

2.2 α-Amanitin Treatment

This treatment is used to block the RNA polymerase II-dependent transcription.

1. Add α-amanitin (Sigma-Aldrich, St. Louis, MO) to the culture medium at a final concentration of 100 μg/ml for either from the 1-cell stage to 2-cell stage (20–34 hpf), from the 1-cell stage to 4-cell stage (20–44 hpf), from the 4-cell stage to early 8-cell stage (44–72 hpf) or from the 4-cell stage to late 8-cell stage (44–96 hpf).
2. Collect the embryos at required stage. Collect the control embryos (without added α-amanitin) at the same time interval.
3. Wash embryos in PBS.
4. Place it on ice or store at −80 °C.
5. Use the embryos for Subheading 3.3.

2.3 Buffers

1. PBS/BSA/sap: phosphate-buffered saline (PBS) supplemented with 0.3% (w/v) BSA and 0.05% (w/v) saponin.
2. TBS-Tween: 20 mM Tris, pH 7.4, 137 mM NaCl, and 0.5% Tween 20.
3. 10% Towbin: 5.82 g 25 mM Tris, 2.93 g 192 mM glycin, pH 9.2, 100 ml Methanol, adjust volume to 1 L dH_2O.
4. Modified Parker medium (MPM): 16.8 ml H 199 10× concentrated, 7.8 ml 7.5% $NaHCO_3$, 100 mg Ca l-Lactate, 40 mg

Na pyruvate, 300 mg HEPES, 20 mg l-Glutamine, 50 U/ml Penicillin K-salt, 50 U/ml Streptomycin, 125 ng/ml Amphotericin B, 15 U/ml P.G. 600 (Intervet, Boxmeer, Holland), 10% oestrus bovine serum, adjust to 200 ml dH_2O.

5. Fertilization medium [10].
6. Menezo B2 medium [11, 12].

2.4 Antibodies

1. Rabbit anti-cullin 1 (Abgent, San Diego, CA).
2. Mouse anti-Skp 1 (Abcam, Cambridge, UK).
3. Rabbit anti-ROC 1 (Abcam).
4. HRP-conjugated donkey anti-rabbit antibody (Jackson Immuno Research, Suffolk, UK).
5. HRP-conjugated donkey anti-mouse antibody (Jackson Immuno Research).
6. Goat anti-rabbit antibody conjugated with FITC (Santa Cruz Biotechnology, Santa Cruz, TX).
7. Goat anti-mouse conjugated with Alexa Fluor 594 (Invitrogen Dynal AS, Eugene, OR).

3 Methods

3.1 Oocyte Collection and in Vitro Maturation

1. Wash ovaries briefly in ethanol.
2. Wash ovaries in PBS at 25–30 °C two times.
3. Dissect the follicles (size 5–9 mm) with fine scissors.
4. Puncture the follicles and transfer the cumulus-oocytes complexes (COCs) to 4-well dishes (Nunc, Roskilde, Denmark) with 0.5 ml MPM media without a paraffin overlay and cultivate under a humidified atmosphere at 39 °C with 5% CO_2 for 24 h [10].
5. After 24 h denude part of the MII oocytes by gentle pipetting.
6. Wash MII oocytes in PBS.
7. Use the oocytes for Subheading 3.6 or freeze at −80 °C and use them for Subheading 3.3 or 3.5.
8. Use the rest of COCs for Subheading 3.2.

3.2 In Vitro Fertilization and Embryo Culture

1. After 24 h of maturation wash the rest of COCs in PBS four times.
2. Wash COCs in fertilization medium [10].
3. Transfer up to 40 of COCs to 4-well dishes (Nunc) with 250 μl of fertilization medium.
4. Wash the viable spermatozoa in fertilization medium.

5. Centrifuge spermatozoa at 100 × *g* for 5 min.
6. Count the spermatozoa in a hemocytometer and dilute in the appropriate volume of fertilization medium to concentration of 2 × 10^6 spermatozoa/ml.
7. Add 250 μl of spermatozoa suspension to fertilization well to obtain a final concentration of 1 × 10^6 spermatozoa/ml.
8. Incubate plates for 20 h at 39 °C in an atmosphere of 5% CO_2, 5% O_2, 90% N_2.
9. At 20 h post fertilization denude presumptive zygotes by gentle pipetting.
10. Transfer groups of 25 zygotes to 25 μl of Menezo B2 medium [11, 12] and cultivate them under the same conditions as in **step 8**.
11. Collect the embryos at 34, 44, 72, 96, 156, and 180 h post fertilization at required stages 2-cell, 4-cell, 8-cell, morula stage, blastocyst stage, and hatched blastocyst stage.
12. Wash embryos in PBS.
13. Use them for Subheading 3.6 or freeze it at −80 °C and use it for Subheading 3.3 or 3.5.

3.3 Poly(A)+ mRNA Extraction

1. Prepare pools of 20 oocytes and embryos at each stage of development.
2. Use Dynabeads mRNA DIRECT Micro Kit (Invitrogen). Unless indicated, the chemicals are included in the kit. Prepare the magnetic rack (PerkinElmer Chemagen Technologie GmbH, Baesweiler, Germany).
3. Bring all buffers except the 10 mM Tris–HCl to room temperature (RT) before use. Store Tris–HCl at 4 °C before use. Warm the incubator to 95 °C.
4. Add 1 pg of luciferase mRNA (Promega, Madison, WI) per oocyte/embryo on ice as an external standard (*see* **Note 1**).
5. Adjust volume to 100 μl/tube by Lysis/Binding buffer and gently mix by vortex.
6. Keep oocyte and embryos in Lysis/Binding buffer for 20 min on ice. Control embryos when they are lysed.
7. *Preparation of Dynabeads:* Shake the bottle with Dynabeads and transfer 20 μl/reaction into new tubes.
8. Place the tube on a magnet for 30 s and discard the supernatant.
9. Remove the tube from magnet, add an equivalent volume of Lysis/Binding Buffer, and resuspend Dynabeads.
10. Place the tube on a magnet for 30 s and discard the supernatant.
11. Remove the tube from magnet, add an equivalent volume of Lysis/Binding Buffer, and resuspend Dynabeads.

12. Transfer 20 μl of Dynabeads/reaction into new tubes.
13. Vortex oocytes/embryos and shortly centrifuge.
14. Transfer the lysed oocytes/embryos to Dynabeads and place tubes on a rotator for 5 min at RT.
15. Place the tubes on the magnet for 30 s and discard the supernatant.
16. Remove the tubes from the magnet and resuspend the samples in 100 μl Washing Buffer A by gentle pipetting.
17. Place the tubes on the magnet for 30 s and discard the supernatant.
18. Remove the tubes from the magnet and resuspend the samples in 100 μl Washing Buffer A by gentle pipetting.
19. Place the tubes on the magnet for 30 s and discard the supernatant.
20. Remove the tubes from the magnet and resuspend the samples in 100 μl Washing Buffer B by gentle pipetting.
21. Transfer the suspension to a new tube.
22. Place the tubes on the magnet for 30 s and discard the supernatant (*see* **Note 2**).
23. Remove the tubes from the magnet and resuspend the samples in 100 μl Washing Buffer B by gentle pipetting.
24. Place the tubes on the magnet for 30 s and discard the supernatant.
25. Remove the tubes from the magnet and resuspend the samples in 3 μl of ice-cold Tris–HCl/embryo.
26. Place the tubes into incubator for 2 min at 95 °C.
27. Place the tubes on the magnet and immediately transfer the supernatant to new tube and place it on ice or store at −80 °C.

3.4 Quantitative RT-PCR

1. Design primers using Beacon Designer 7 (Premier Biosoft, Palo Alto, CA). *See* Table 1.
2. Unless indicated, the chemicals are included in OneStep RT-PCR Kit (Qiagen, Hilden, Germany).
3. Thaw template mRNA, primer solutions dNTP Mix, OneStep RT-PCR buffer and RNase-free water, and place them on ice.
4. Prepare the master mix into one tube composed of: Nuclease-free water (4.55 μl/reaction; Life Technologies, Carlsbad, CA), OneStep Buffer 1× (0.5 μl/reaction), dNTP Mix 400 μM of each (0.5 μl/reaction), forward and reverse primers (both 400 μM; 0.25 μl/reaction), Sybr Green I (1:50,000 of 1000× stock solution, 0.25 μl/reaction; Invitrogen), RNasin Ribonuclease Inhibitor (Promega; 0.2 μl/reaction), OneStep Enzyme Mix (0.5 μl/reaction) (*see* **Note 3**).

Table 1
Information of primers used in this study

Primer	Sequences	Annealing temperature (°C)	Amplicon size (bp)
Cul 1-like (XM_589507.3)	5′-CGG ACT GGA GCC AGA ATC CCA-3′ 5′-GTC TGG GCT TGA GGG GAC ACA-3′	60	178
Cul 1 (NM_001193233.1)	5′-AAC CCC CAC GGA CTC AAG CAG A-3′ 5′-GCC CCT CGA GCT TGG TTT GAC T-3′	60	173
Skp 1 (NM_001034781)	5′-GCC ATC TCC TTG AGC CCT AC-3′ 5′-CAT TTG GCA AGG GGA CTG GA-3′	55	172
Rbx 1 (NM_001034781)	5′-CAG GCG TCC GCT ACT TCT G-3′ 5′-TGT TTT GAG CCA GCG AGA GA-3′	63	93
Luciferase	5′-ACT TCG AAA TGT CCG TTC GG-3′ 5′-ACT TCG AAA TGT CCG TTC GG-3′	55	633

Table 2
RT and PCR conditions used in this study

	Temperature (°C)	Time
Gene	Cul1-like, Cul1/Skp1/Rbx1	Cul1-like, Cul1/Skp1/Rbx1
Reverse transcription	50	30 min
Initial activation	95	15 min
Cycling:		
Denaturation	94	15 s
Annealing	60/55/63	20 s
Extension	72	30 s
Final extension	72	10 min

5. Mix the master mix thoroughly and distribute appropriate volume (9.0 μl) into the PCR microtubes.
6. Add template mRNA to final volume 12 μl (1 pg–2 μg/reaction). All reactions are prepared in duplicates.
7. Place the tubes into RotorGene 3000 (Corbett Research, Morthlake, Australia). Reaction conditions are as in Table 2.

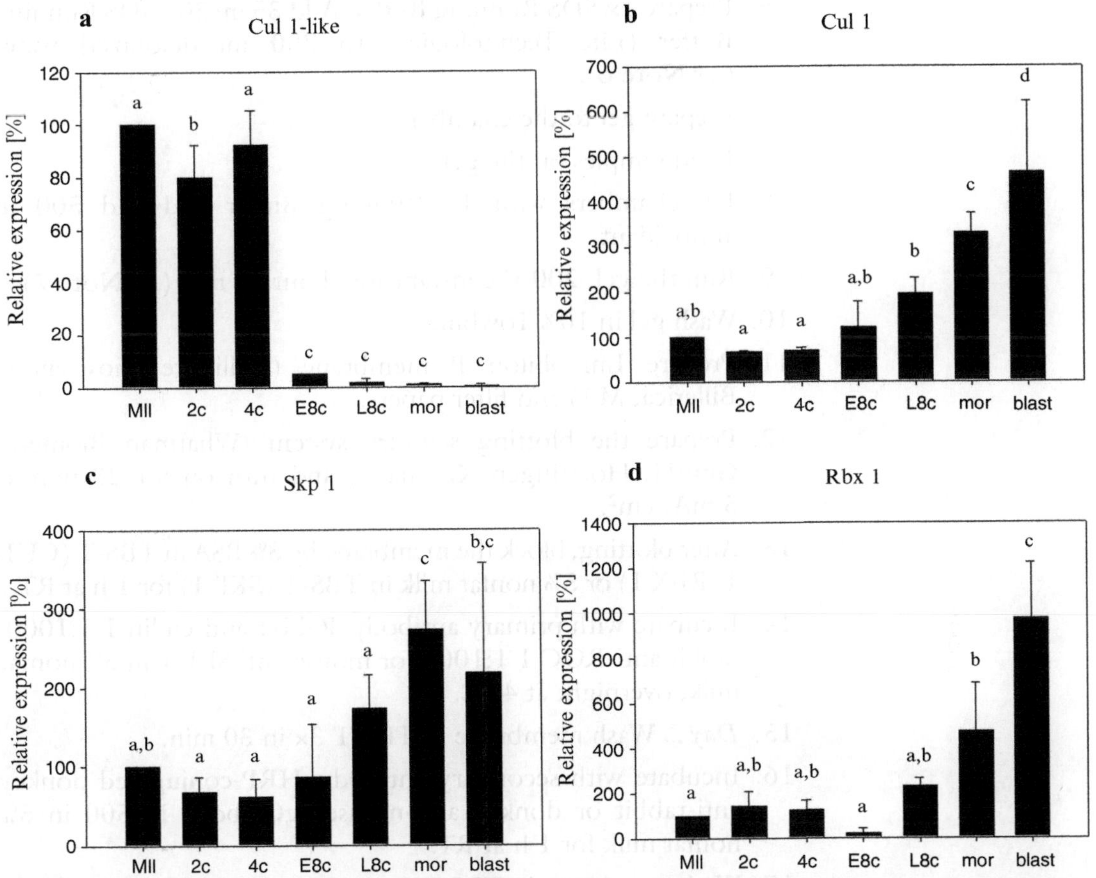

Fig. 1 Relative mRNA expression of invariant members of SCF complex. Untreated embryos. The data were normalized according to the relative concentration of the external standard (luciferase mRNA, 1 pg per oocyte/embryo). (**A**) Cul 1-like, (**B**) Cul1, (**C**) Skp1, (**D**) Rbx1. Bars show ±S.D. ^{a,b,c,d}Values with different superscripts indicate statistical significance ($P < 0.05$). MII MII stage oocyte, 2c 2-cell stage embryo, 4c 4-cell stage embryo, E8c early 8-cell embryo, L8c late 8-cell stage, mor morula, bl blastocyst

8. Determine the relative concentration of the template in the different samples by comparative quantification in analysis software (Corbett Research), as described in [13]. Normalize the results according to the relative concentration of the external standard (Luciferase) (Fig. 1).

3.5 Western Blot

1. *Day 1*. Prepare gel 4–12% Bis–Tris Gel (*see* **Note 4**).
2. Preparation of samples. Add 2.5 μl LDS Sample Buffer, 1 μl Reducing agent (both Life Technologies), 6 μl deionized water to sample (*see* **Note 5**). Use 25 oocytes/embryos per extract.
3. Boil samples for 3–5 min in distilled water.
4. Transfer to ice.

5. Prepare 1× SDS Running Buffer. Add 35 ml 20× SDS Running Buffer (Life Technologies) to 700 ml deionized water (*see* **Note 6**).
6. Prepare gel to the chamber.
7. Load samples on the gel.
8. Fill chambers with 1× Running Buffer and add 500 μl antioxidant.
9. Run the gel. 200 V constant for about 35 min (*see* **Note 7**).
10. Wash gel in 10% Towbin.
11. Prepare Immobilon P membrane (Millipore Biosciences, Billerica, MA) and filter papers.
12. Prepare the blotting semidry system (Whatman Biometra GmbH, Hoettingen, Germany) and turn on for 28 min at 5 mA/cm^2.
13. After blotting, block the membrane by 3% BSA in TBS-T (CUL 1, RBX 1) or 5% nonfat milk in TBS-T (SKP 1) for 1 h at RT.
14. Incubate with primary antibody. Rabbit anti-cullin 1 1:1000, rabbit anti-ROC 1 1:1000, or mouse anti SKP 1 in 5% nonfat milk, overnight at 4 °C.
15. *Day 2*. Wash membrane in TBS-T 3× in 30 min.
16. Incubate with secondary antibody. HRP-conjugated donkey anti-rabbit or donkey anti-mouse IgG, both 1:7500 in 5% nonfat milk for 1 h at RT.
17. Wash membrane in TBS-T 3× in 30 min.
18. Visualize the proteins with Luminata crescendo Western HRP (Merck Millipore) in dark room.

3.6 Immunofluorescence Analysis

1. *Day 1*. Fix oocyte/embryos in 4% paraformaldehyde for 50 min at 4 °C.
2. Wash in PBS (*see* **Note 8**).
3. Transfer embryos to 0.5% TritonX-100 for 15 min (*see* **Note 9**).
4. Wash embryos in PBS/BSA/sap three times.
5. Block with 2% normal goat serum for 1 h at RT.
6. Wash in PBS/BSA/sap three times.
7. Incubate with primary antibody. Anti-cullin 1 1:100 or anti-SKP1 1:100 or anti-ROC1 1:100, overnight at 4 °C (*see* **Note 10**).
8. *Day 2*. Wash in PBS/BSA/sap 12 times in 1 h.
9. Incubate with secondary antibody goat anti-rabbit with FITC 1:350 or goat anti-mouse with Alexa Fluor 594 1:800 for 1 h at RT in the dark.
10. Wash in PBS/BSA/sap 12 times in 1 h.

11. Wash in PBS.
12. Mount with Vectashield Hardset Mounting Medium with DAPI (Vector Laboratories, Peterborough, UK) (*see* **Note 11**).
13. Check and make a photo of the samples on a fluorescent/confocal microscope (Fig. 2).

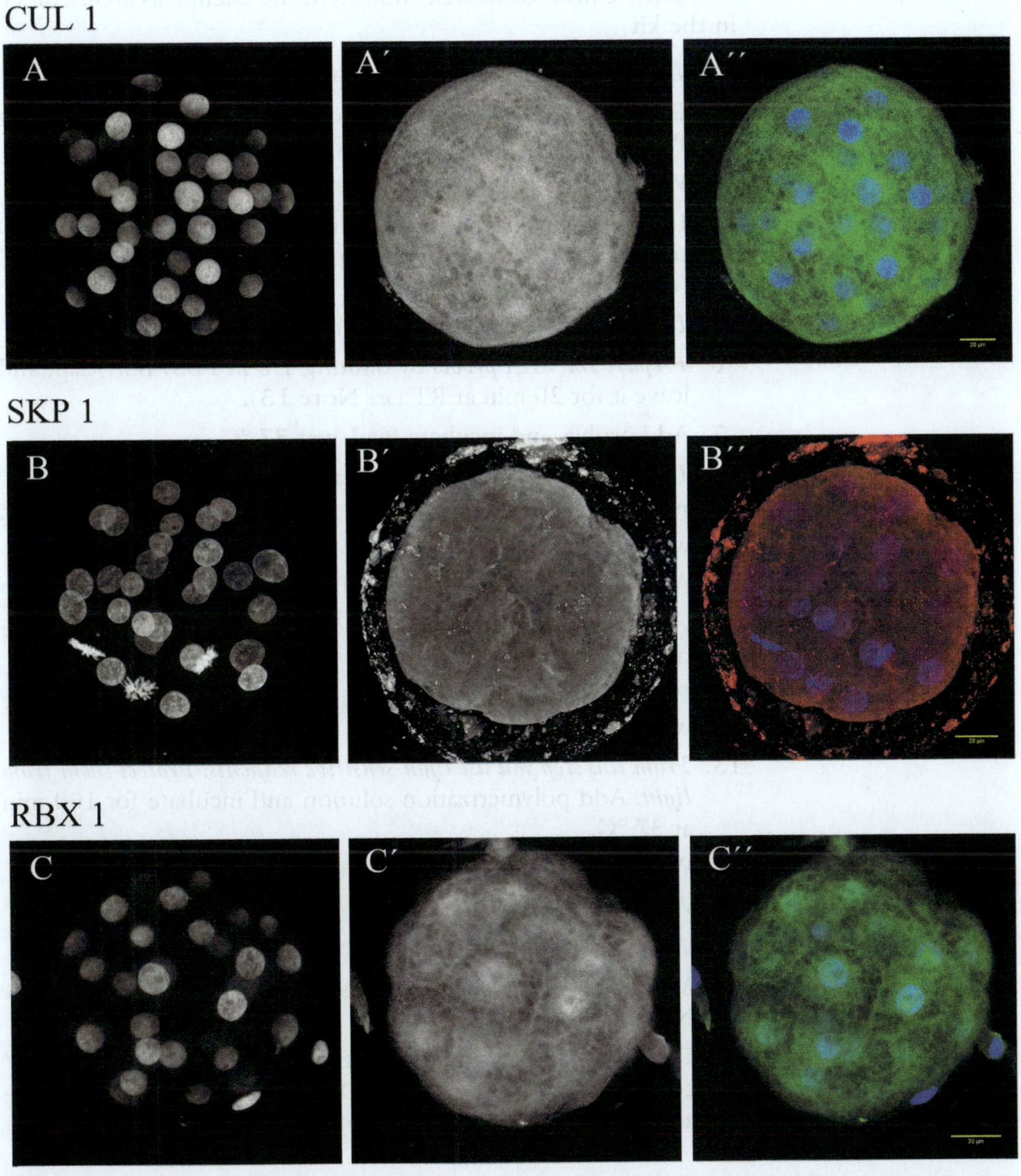

Fig. 2 Confocal laser scanning microscopy of CUL 1, SKP 1, RBX 1 of morula stage embryo. The embryo was labeled with protein-specific antibodies (**A′**) CUL 1, (**B′**) SKP 1, (**C′**) RBX 1, and the nuclei were stained with DAPI (**A–C**). In overlaid images CUL 1 (**A″**) and RBX 1 (**C″**) are green, SKP 1 (**B″**) is red, and DAPI (**A″–C″**) blue

3.7 In Situ Proximity Ligation Assay (PLA)

The PLA method enables detection and visualization of the protein-protein interaction. If the proteins are in close proximity (<40 nm), it causes cascade of reactions that leads to fluorescent signal. In our case, the interaction of SKP 1 and CUL 1 was observed, which indicates SCF-complex activity [14].

For PLA use Duolink kit (Olink Bioscience, Uppsala, Sweden). Unless otherwise indicated, the chemicals are included in the kit.

1. *Day 1.* Prepare embryos as in Subheading 3.4, **steps 1–4**.
2. Block with 2% normal donkey serum (NDS; Santa Cruz Biotechnology, Santa Cruz, TX), 1 h at RT.
3. Wash in PBS/BSA/sap solution three times.
4. Incubate with primary antibodies—rabbit anti-cullin 1 1:100 together with mouse anti-Skp 1 1:100 overnight at 4 °C (*see* **Note 12**).
5. *Day 2.* Wash embryos in PBS/BSA/sap 12 times in 1 h.
6. *Prepare the PLA probes* by diluting 1:5 in PBS/BSA/sap, and leave it for 20 min at RT (*see* **Note 13**).
7. Add probes and incubate for 1 h at 37 °C.
8. *Ligation.* Wash in PBS/BSA/sap 12 times in 1 h.
9. Prepare the Ligation solution by diluting 1:5 in RNase-free water (*see* **Note 14**).
10. Add ligation solution and incubate for 30 min at 37 °C.
11. *Amplification.* Wash in 1× (Washing Buffer A) WBA three times in 15 min.
12. Prepare polymerization solution by diluting 1:5 in RNase-free water (*see* **Note 15**).
13. *From this step you use light-sensitive reagents. Protect them from light.* Add polymerization solution and incubate for 100 min at 37 °C.
14. Wash six times in 1× WBB in 30 min.
15. Wash in 0.01× WBB (*see* **Note 16**).
16. Wash in PBS.
17. Mount by Vectashield Hardset medium with DAPI (*see* **Note 17**).
18. Check and make a photo of the samples on a fluorescent/confocal microscope (Fig. 3). Dots should represent the interaction of CUL 1 and SKP 1 proteins (*see* **Note 18**).

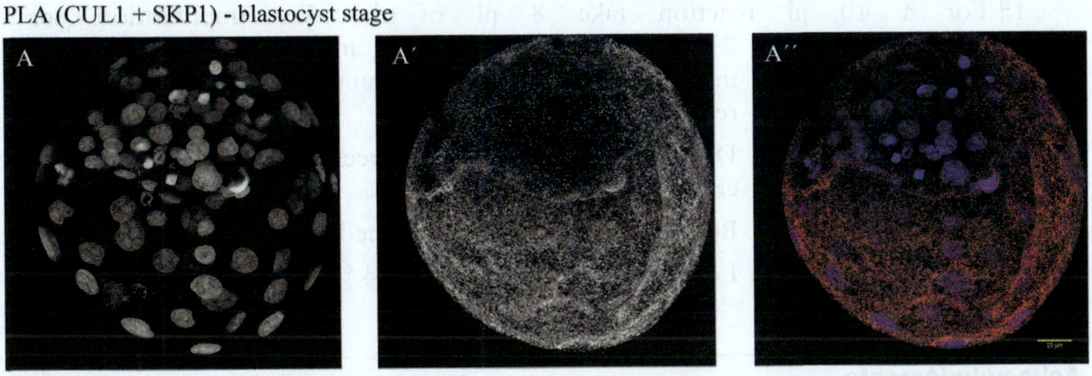

Fig. 3 Confocal laser scanning microscopy of bovine blastocyst after Duolink in situ PLA analysis. PLA signal indicates Cul1-Skp1 interaction (**A′**) and the nuclei were stained with DAPI (**A**). In overlaid (**A″**), PLA signal is red and DNA blue

4 Notes

1. Dilute luciferase in Lysis/Binding buffer.
2. Carefully, in contrast to previous steps, the Dynabeads are not tightly touched.
3. Prepare a volume of master mix 10% greater than required for total number of reactions.
4. We use NuPAGE Bis–Tris Mini Gels (Life Technologies).
5. It is better to prepare the solution together for every sample and then add the appropriate volume to each sample.
6. Use cold water, at 4 °C.
7. 100 V for 10 min at the beginning, then 175–200 V.
8. Fixed embryos can be stored in PBS for up to 3 weeks at 4 °C.
9. Dilute TritonX-100 in PBS.
10. Dilute in PBS/BSA/sap.
11. Before the use, centrifuge the medium for 10 min at 10,000 × *g*.
12. Prepare controls by omitting one or both primary antibodies or using another species-specific secondary conjugate; dilute primary antibodies in PBS/BSA/sap.
13. For a 40 μl reaction take 8 μl of PLA probe MINUS, 8 μl of PLA probe PLUS, and 24 μl of PBS/BSA/sap.
14. For a 40 μl reaction take 8 μl of the 5× Ligation stock, 31 μl of RNAse-free water, and add 1 μl of Ligase immediately before addition to the samples.

15. For a 40 μl reaction take 8 μl of the 5× Amplification stock, 31.5 μl of RNAse-free water, and add 0.5 μl of polymerase immediately before the addition to the samples; light-sensitive reagents.
16. Dilute 1× WBB in RNase-free water; be careful that the embryos tend to stick in WBB.
17. Before the use, centrifuge the medium for 10 min at 10,000 × *g*.
18. Leave it for at least 20 min at 4 °C before checking.

Acknowledgments

We thank R. Sucha and A. Susor for their helpful comments and expert assistance during the experiments and M. Kopcikova, J. Kankova, J. Supolikova, and M. Rakocyova for their technical assistance. This work was supported by the Czech Science Foundation (GACR 13-24730P), Grant Agency of Charles University in Prague (GAUK 1280214) and RVO 67985904.

References

1. Yokoi H, Natsuyama S, Iwai M, Noda Y, Mori T, Mori KJ, Fujita K, Nakayama H, Fujita J (1993) Non-radioisotopic quantitative RT-PCR to detect changes in mRNA levels during early mouse embryo development. Biochem Biophys Res Commun 195:769–775. doi:10.1006/bbrc.1993.2112
2. DeRenzo C, Seydoux G (2004) A clean start: degradation of maternal proteins at the oocyte-to-embryo transition. Trends Cell Biol 14:420–426. doi:10.1016/j.tcb.2004.07.005
3. Mtango NR, Latham KE (2007) Ubiquitin proteasome pathway gene expression varies in rhesus monkey oocytes and embryos of different developmental potential. Physiol Genomics 31:1–14. doi:10.1152/physiolgenomics.00040.2007
4. Glickman MH, Ciechanover A (2002) The ubiquitin-proteasome proteolytic pathway: destruction for the sake of construction. Physiol Rev 82:373–428. doi:10.1152/physrev.00027.2001
5. Benesova V, Kinterova V, Kanka J, Toralova T (2016) Characterization of SCF-complex during bovine preimplantation development. PLoS One 11:e0147096. doi:10.1371/journal.pone.0147096
6. Bai C, Sen P, Hofmann K, Ma L, Goebl M, Harper JW, Elledge SJ (1996) SKP1 connects cell cycle regulators to the ubiquitin proteolysis machinery through a novel motif, the F-box. Cell 86:263–274
7. Skowyra D, Craig KL, Tyers M, Elledge SJ, Harper JW (1997) F-box proteins are receptors that recruit phosphorylated substrates to the SCF ubiquitin-ligase complex. Cell 91:209–219
8. Zheng J, Yang X, Harrell JM, Ryzhikov S, Shim EH, Lykke-Andersen K, Wei N, Sun H, Kobayashi R, Zhang H (2002) CAND1 binds to unneddylated CUL1 and regulates the formation of SCF ubiquitin E3 ligase complex. Mol Cell 10:1519–1526
9. Soucy TA, Smith PG, Milhollen MA, Berger AJ, Gavin JM, Adhikari S, Brownell JE, Burke KE, Cardin DP, Critchley S, Cullis CA, Doucette A, Garnsey JJ, Gaulin JL, Gershman RE, Lublinsky AR, McDonald A, Mizutani H, Narayanan U, Olhava EJ, Peluso S, Rezaei M, Sintchak MD, Talreja T, Thomas MP, Traore T, Vyskocil S, Weatherhead GS, Yu J, Zhang J, Dick LR, Claiborne CF, Rolfe M, Bolen JB, Langston SP (2009) An inhibitor of NEDD8-activating enzyme as a new approach to treat cancer. Nature 458:732–736. doi:10.1038/nature07884
10. Pavlok A, Lucas-Hahn A, Niemann H (1992) Fertilization and developmental competence of bovine oocytes derived from different categories of antral follicles. Mol Reprod Dev 31:63–67. doi:10.1002/mrd.1080310111
11. Menezo Y (1976) Synthetic medium for gamete survival and maturation and for culture of

fertilized eggs. C R Acad Sci Hebd Seances Acad Sci D 282:1967–1970
12. Staessen C, Janssenswillen C, Clerck ED, Steirteghem AV (1998) Controlled comparison of commercial media for human in-vitro fertilization: Ménézo B2 medium versus Medi-cult universal and BM1 medium. Hum Reprod 13:2548–2554. doi:10.1093/humrep/13.9.2548
13. Kanka J, Kepková K, Nemcová L (2009) Gene expression during minor genome activation in preimplantation bovine development. Theriogenology 72:572–583. doi:10.1016/j.theriogenology.2009.04.014
14. Liu J, Furukawa M, Matsumoto T, Xiong Y (2002) NEDD8 modification of CUL1 dissociates p120(CAND1), an inhibitor of CUL1-SKP1 binding and SCF ligases. Mol Cell 10:1511–1518

Chapter 18

Use of Histone K-M Mutants for the Analysis of Transcriptional Regulation in Mouse Zygotes

Keisuke Aoshima, Takashi Kimura, and Yuki Okada

Abstract

Histone modifications are dramatically altered during the pronuclear (PN) stage of zygotes, and more markedly in paternal than maternal pronuclei. Among various types of histone modifications, lysine methylation exhibits the most dynamic changes in the PN stage. To analyze the physiological functions of histone methylations, it is therefore important to elucidate the mechanism of epigenetic reprogramming. However, loss-of-function approaches using mutant histones whose lysine residues have been substituted with arginine residues are unable to erase histone modifications at all levels, since they are incapable of entirely replacing endogenous histones. To solve this problem, we used an alternative histone mutant whose lysine residues were substituted with methionine (K-M mutants). This mutant cannot be methylated itself but also prevents methylation of endogenous histones. We also developed a simple method for analyzing global transcription levels in early preimplantation embryos, involving using a commercial kit to examine the involvement of histone methylation in zygotic gene activation.

Key words K-M mutant, Zygotic gene activation, Histone methylation, Microinjection

1 Introduction

Epigenetic reprogramming is a process through which epigenetic properties such as DNA methylation and histone modifications are dynamically altered to regulate gene expression patterns from differentiated to undifferentiated. In zygotes, epigenetic reprogramming occurs in the pronuclear (PN) stage. At that point, there are two distinct nuclei, derived from the sperm and the oocyte and termed the paternal and maternal pronuclei, respectively [1–3]. Epigenetic properties are altered preferentially in the paternal pronucleus, which implies that reprogramming in the paternal pronucleus is more dynamic than the maternal one [4–8]. Among the epigenetic properties, histone modifications in paternal pronuclei are gradually established as the PN stage advances, but those in maternal pronuclei maintain their originally modified state throughout the PN stage [9, 10]. Since histone modifications such

Kiho Lee (ed.), *Zygotic Genome Activation: Methods and Protocols*, Methods in Molecular Biology, vol. 1605,
DOI 10.1007/978-1-4939-6988-3_18, © Springer Science+Business Media LLC 2017

as H3K4 methylation and H3K9 methylation are known to be strongly associated with transcriptional regulation, the dynamic alteration of histone modifications in the PN stage seems to regulate transcription in zygotes.

In mice, there are two distinct post-fertilization events involving a change in gene expression pattern from maternal to zygotic: minor zygotic gene activation (minor ZGA) at the late PN stage, and major zygotic gene activation (major ZGA) at the 2-cell stage [11, 12]. Interestingly, minor ZGA occurs preferentially in paternal pronuclei, which coincides with the asymmetrical epigenetic alteration at the PN stage. The level of involvement in minor ZGA of the dynamic alteration of histone modifications in paternal pronuclei has not yet been clarified because of the lack of an effective method for artificially controlling histone modifications in the PN stage. Although histone mutants whose lysine residues had been replaced by arginine residues have been used to analyze the effect of histone modifications in zygotes, these mutants are not effective at a global level because of their inability to replace all endogenous histones, even when overexpressed. It has therefore been difficult to verify the effect of complete loss of specific histone modifications in zygotes.

To overcome this problem, we used an alternative histone mutant whose lysine (K) residue had been replaced by methionine (M). This mutant, termed the K-M mutant, was first reported by Lewis et al. from pediatric glioma patients carrying the K to M mutation at the 27th lysine of their histone H3.3 gene (H3.3-K27M) [13]. H3.3-K27M cannot itself be methylated, but abrogates endogenous H3K27 methylation via inhibition of PRC2, an H3K27 methyltransferase. This effect has also been observed in H3.3-K4M, H3.3-K9M and H3.3-K36M, which can respectively prevent the methylation of the endogenous histones H3K4, H3K9, and H3K36. We used these mutants to examine the effect of histone methylation at the PN stage, and found that H3K4 methylation was required for early embryonic development to control transcriptional activation during minor ZGA [14]. We also demonstrated that knockdown of *Mll3* and *Mll4* (histone methyltransferases), achieved by microinjecting the siRNAs into zygotes, caused effects similar to H3.3-K4M overexpression. We preferentially used GV oocytes for the siRNA microinjection and allowed them to grow until the MII stage, followed by intracytoplasmic sperm injection (ICSI). This gave the siRNA sufficient time (>12 h) to become functional in the zygotes. To monitor the alteration of zygotic transcription, we used the recently developed "click chemistry" methodology to visualize and quantify the transcriptional level using a commercial kit (although a similar method without a commercial kit has also been reported [15]). These techniques can provide a simple way to analyze the effects of histone modifications in early preimplantation embryos.

2 Materials

2.1 Animals

1. 4–8-week-old BDF1 mice (C57BL/6 female × DBA/2 strain male; CLEA Japan), for collecting MII oocytes.
2. 4–8-week-old C57BL6 × 129 mice, for collecting GV oocytes (CLEA Japan).

2.2 General Supplies

1. 35 mm Petri dish.
2. 60 mm Petri dish.
3. 50 μl calibrated micropipettes (Drummond Scientific).
4. 1.5 ml tube.
5. 15 ml tube.
6. 96-well Terasaki plate (Watson Bio Lab).
7. Glass-based dishes (Iwaki).

2.3 Reagents

1. Pregnant mare's serum gonadotropin (PMSG; Asuka Pharmaceutical).
2. Human chorionic gonadotropin (hCG; Asuka Pharmaceutical).
3. Bovine serum albumin (BSA; Sigma Aldrich).
4. Mineral oil for embryo culture (Irvine Scientific).
5. HTF medium (Irvine Scientific) [16, 17].
   - Add 3 mg/ml BSA to prevent oocyte from sticking to the bottom of Petri dishes.
6. KSOM (ARK Resource) [18].
7. M2 medium (ARK Resource) [19].
8. Polyvinylpyrrolidone (PVP) solution (Irvine Scientific).
9. TYH medium (LSI Medience) [20].
10. Hyaluronidase solution (Irvine Scientific).
11. mTaM medium (a 1:1 mixture of TYH medium and MEMα supplemented with 5% FBS and 2 mM L-carnitine) [21].
12. RiboMAX Large Scale RNA Production System T7 (Promega).
13. Ribo m7G Cap Analog (Promega).
14. PCI (Phenol (pH 4–5):Chloroform:Isoamyl alcohol = 125:24:1).
15. CI (Chloroform:Isoamyl alcohol = 24:1).
16. MicroSpin G-25 (GE Healthcare).
17. VECTASHIELD mounting medium with DAPI (Vector).
18. Phosphate-buffered saline containing 0.1% BSA (PBS/BSA).
19. Tyrode's solution (Sigma Aldrich).

20. 4% paraformaldehyde.
21. Methanol.
22. TBS containing 0.5% Triton X-100.
23. TBS containing 0.05% Triton X-100 (TBS-T).
24. Click-iT RNA Alexa Fluor 488 Imaging Kit (Thermo Fisher Scientific).
25. Immersion oil (Olympus).

2.4 Instruments

1. Stereoscopic microscope (Olympus SZ60).
2. Inverted microscope (Olympus IX71).
3. Pneumatic microinjector (Narishige).
4. Piezo Micro Manipulator (Prime Tech).
5. CO_2 incubator.
6. Clean bench.
7. Micropipette puller (P-1000; Sutter Instrument).
8. Microforge (MF-900; Narishige).
9. Confocal microscope (CV1000; Yokogawa Electric Corporation).
10. ImageJ software.

2.5 Plasmids

1. pcDNA3.1-H3.3-poly(A)83.

   The cDNA of Flag-HA-tagged H3.3 obtained from pOZ-e-H3.3 [22] is subcloned into a pcDNA3.1-poly(A)83 vector [23].
2. pcDNA3.1-H3.3K4M-poly(A)83.

3 Methods

3.1 mRNA Preparation

1. Perform in vitro transcription to pcDNA3.1-H3.3WT-poly(A)83 and pcDNA3.1-H3.3K4M- poly(A)83 using the RiboMAX Large Scale RNA Production System T7, following the instructions for using Ribo m7G Cap Analog (*see* **Note 1**).
2. Extract RNAs by means of PCI and CI treatment.
3. Purify RNAs with MicroSpin G-25 (*see* **Note 2**).

3.2 mRNA Microinjection

1. Treat female BDF1 mice intraperitoneally with 7.5 international unit (IU) PMSG.
2. 48 h later, treat the mice intraperitoneally with 7.5 IU hCG.
3. Prepare drops as indicated in Fig. 1.
4. Equilibrate the drops at 37 °C, 5% CO_2 until use.

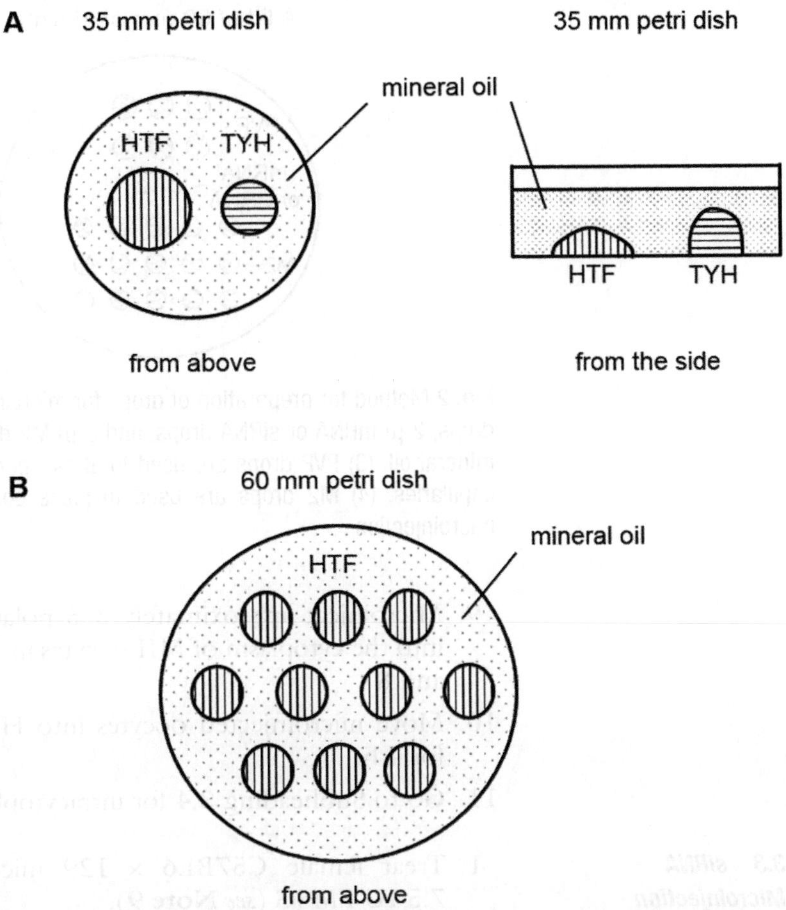

Fig. 1 Method for preparation of drops for oocyte and sperm cultures. (**a**) For oocyte and sperm cultures: (1) Place a 100 μl HTF drop and a 50 μl TYH drop on a 35 mm Petri dish. (2) Cover the drops with mineral oil. (3) Add an additional 50 μl TYH to the TYH drop. (**b**) To wash the oocytes: (1) Place as many 30 μl HTF drops as possible on a 60 mm Petri dish. (2) Cover the drops with mineral oil

5. After approximately 15 h, humanely euthanize hormone-treated mice and collect MII oocytes by puncturing their fallopian tubes with fine forceps.
6. Put the oocytes into a 100 μl drop of HTF medium and add 10 μl hyaluronidase solution, followed by culturing at 37 °C, 5% CO_2 for 30 min (*see* **Note 3**).
7. After oocytes are denuded (cumulus cells are detached by hyaluronidase treatment), wash them in HTF medium using a mouth pipette until all cumulus cells are removed (*see* **Note 4**).
8. Prepare an "injection dish" for microinjection, using the lid of a Petri dish as indicated in Fig. 2 (*see* **Notes 5** and **6**).

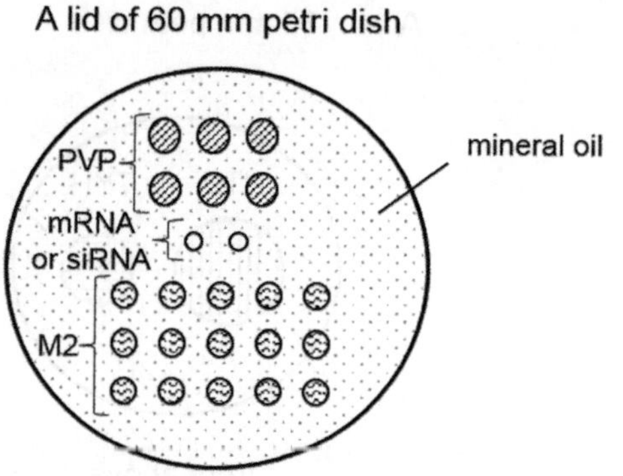

Fig. 2 Method for preparation of drops for microinjection. (1) Prepare 10 μl PVP drops, 2 μl mRNA or siRNA drops, and 5 μl M2 drops. (2) Cover the drops with mineral oil. (3) PVP drops are used to allow sperms to swim or to wash glass capillaries. (4) M2 drops are used to place oocytes or zygotes followed by microinjection

9. Microinject approximately 3–5 polar body size (pl) mRNAs into the cytoplasm of MII oocytes in M2 medium (*see* **Notes 7** and **8**).
10. Move microinjected oocytes into HTF medium and culture for 4 h.
11. Go to Subheading 3.4 for intracytoplasmic sperm injection.

3.3 siRNA Microinjection

1. Treat female C57BL6 × 129 mice intraperitoneally with 7.5 IU PMSG (*see* **Note 9**).
2. Prepare drops as indicated in Fig. 3 one day before collecting GV oocytes.
3. After 48 h, humanely euthanize hormone-treated mice and collect GV oocytes from their ovaries by puncturing antral follicles using a 23-gauge needle. Culture oocytes in a 100 μl drop of mTaM medium with 10 μl hyaluronidase solution for about 2 h, until GV oocytes develop into MI oocytes.
4. Prepare an "injection dish" as indicated in Fig. 2.
5. Microinject approximately 3–5 pl siRNAs into the cytoplasm of MI oocytes in M2 medium (*see* **Note 10**).
6. Move microinjected oocytes into mTaM medium and culture for about 15 h.
7. After microinjected oocytes mature to MII oocytes, go to Subheading 3.4 for intracytoplasmic sperm injection.

3.4 Intracytoplasmic Sperm Injection (ICSI)

1. Humanely euthanize male mice and collect sperm from their cauda epididymis and culture in TYH medium for 1 h at 37 °C, 5% CO_2.

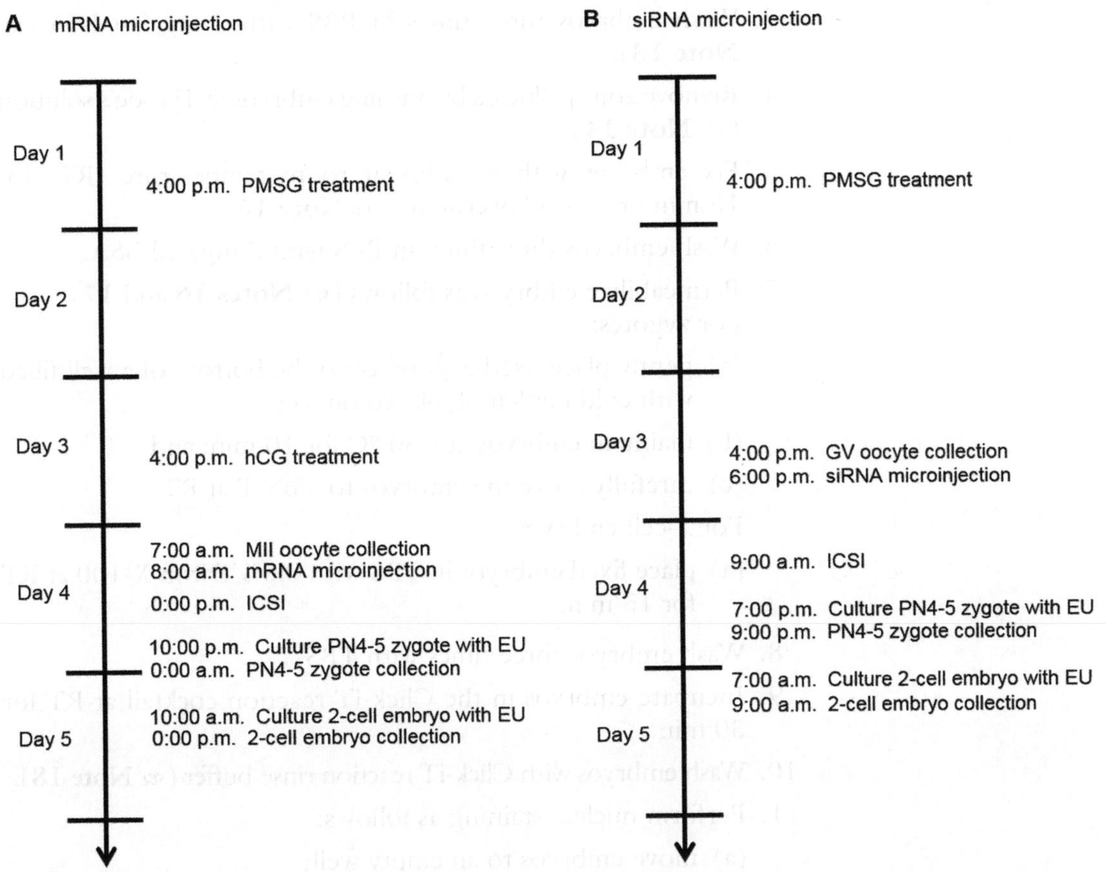

Fig. 3 Example experimental schedule. (**a**) For mRNA microinjection. (**b**) For siRNA microinjection

2. Add 1 μl TYH medium, containing the sperm, to a PVP solution in an injection dish.
3. Move mRNA- or siRNA-microinjected oocytes into M2 medium in the injection dish.
4. Inject a sperm from the tail and hitch the neck to the tip of a capillary.
5. Give Piezo pulse to the sperm.
6. Pick up sperm heads detached from the tail and inject them into the cytoplasm of the oocytes.
7. Culture the oocytes in HTF medium at 37 °C, 5% CO_2 (*see* **Note 11**).

3.5 Analyze Minor or Major ZGA

1. Use a Click-iT RNA Alexa Fluor 488 Imaging Kit.
2. Incubate PN4–5 zygotes or 2-cell embryos with 2 mM 5-ethynyl uridine (EU, supplied with the kit) at 37 °C for 2 h (*see* **Note 12**).

3. Wash embryos three times in PBS with 3 mg/ml BSA (*see* **Note 13**).
4. Remove zona pellucida by washing embryos in Tyrode's solution (*see* **Note 14**).
5. Fix embryos with 4% PFA at room temperature (RT) for 15 min or at 4 °C overnight (*see* **Note 15**).
6. Wash embryos three times in PBS with 3 mg/ml BSA.
7. Permeabilize embryos as follows (*see* **Notes 16** and **17**).
   For zygotes:
   (a) gently place fixed zygotes onto the bottom of a well filled with cold methanol, placed on ice;
   (b) maintain embryos at −30 °C for 10 min; and
   (c) carefully move the embryos to TBS-T at RT.
   For 2-cell embryos:
   (a) place fixed embryos in TBS with 0.5% Triton X-100 at RT for 15 min.
8. Wash embryos three times with TBS-T.
9. Incubate embryos in the Click-iT reaction cocktail at RT for 30 min.
10. Wash embryos with Click-iT reaction rinse buffer (*see* **Note 18**).
11. Perform nuclear staining as follows:
    (a) move embryos to an empty well;
    (b) remove extra buffer with a mouth pipette; and
    (c) slowly add 10 μl VECTASHIELD (*see* **Note 19**).
12. Wash embryos three times with TBS-T.
13. Wash embryos three times with TBS.
14. Prepare 5 μl drops of TBS on a glass-bottomed dish.
15. Move embryos into the drops.
16. Cover the drops with immersion oil.
17. Capture fluorescence images of each 2 μm section using an inverted confocal microscope.

3.6 Data Quantification

1. Import captured images to ImageJ software (*see* **Note 20**).
2. Subtract background.
3. Manually outline the pronucleus, nucleus, or a cytosol region of the same size as the nucleus (*see* **Note 21**).
4. Measure fluorescence intensities in the outlined areas.
5. Calculate total intensities for each whole pronucleus, nucleus, or cytosol region.
6. Calculate relative intensities as follows (Fig. 4):

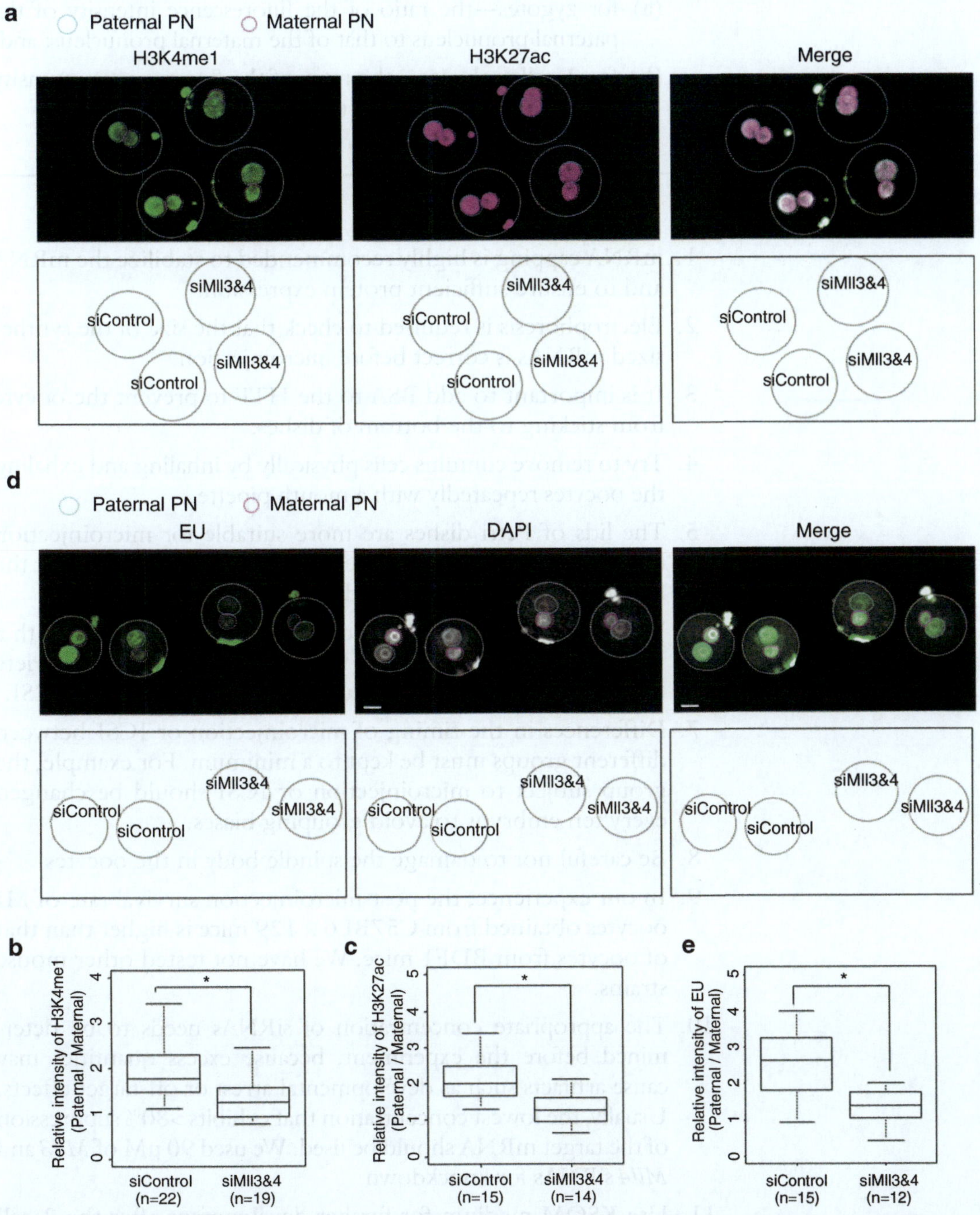

Fig. 4 Double knockdown of Mll3&4 causes paternal PN-specific reduction of global H3K4me1, H3K27ac, and global transcription. (**a**, **b**) Immunostaining against H3K4me1 and H3K27ac (**a**) and EU treatment (**b**) were performed for PN zygotes. In each figure, top pictures are representative images, middle schemes indicate which type of siRNA was injected to embryos corresponding to the representative images, and bottom graphs are boxplots about relative intensities of H3K4me1 and H3K27ac or EU for each sample from three independent experiments. For statistical analyses, Steel Dwass test (**a**) and Tukey-Kramer test (**b**, **c**) were performed. Bars = 20 μm. *P < 0.01. Adapted from Aoshima et al. [14]

(a) for zygotes—the ratio of the fluorescence intensity of the paternal pronucleus to that of the maternal pronucleus; and
(b) for 2-cell embryos—the ratio of the fluorescence intensity of the nucleus to that of the cytosol.

4 Notes

1. mRNA capping is highly recommended to stabilize the mRNA and to ensure sufficient protein expression.
2. Electrophoresis is required to check that the size of the synthesized mRNAs is correct before microinjection.
3. It is important to add BSA to the HTF to prevent the oocyte from sticking to the bottom of dishes.
4. Try to remove cumulus cells physically by inhaling and exhaling the oocytes repeatedly with a mouth pipette.
5. The lids of Petri dishes are more suitable for microinjection than the actual dishes because their low rims do not inhibit the action of microinjection needles and glass capillaries.
6. We used injection needles and glass capillaries prepared with a P-1000 Micropipette Puller (Sutter Instrument). Use a variety that is convenient for you to use for microinjection and ICSI.
7. Differences in the timing of microinjection or ICSI between different groups must be kept to a minimum. For example, the group subject to microinjection or ICSI should be changed every ten embryos to avoid grouping biases.
8. Be careful not to damage the spindle body in the oocytes.
9. In our experience, the post-microinjection survival rate of MI oocytes obtained from C57BL6 × 129 mice is higher than that of oocytes from BDF1 mice. We have not tested other mouse strains.
10. The appropriate concentration of siRNAs needs to be determined before the experiment, because excess quantities may cause artifacts such as developmental arrest or off-target effects. Usually, the lowest concentration that exhibits >80% suppression of the target mRNA should be used. We used 90 μM of *Mll3* and *Mll4* siRNAs for knockdown.
11. Use KSOM medium for further development after the 2-cell stage.
12. PN4–5 zygotes can be obtained about 12 h after ICSI. Check the stage morphologically (i.e., based on the size and distance of the two pronuclei) using a stereoscopic microscope.
13. Use a low-well or Terasaki plate after this step.

14. Denuded embryos easily stick to the bottom of the plate. Be careful, therefore, not to let them sink to the bottom.
15. Fixation with 4% PFA is necessary, even for methanol permeabilization, to prevent deformation of the embryos.
16. Methanol should be chilled to −30 °C before use.
17. Place zygotes gently onto the bottom and keep the methanol cold. Otherwise, the zygotes will move around vigorously in the methanol.
18. Be careful not to allow the embryos to stick to the bottom of the plate.
19. Add VECTASHIELD before embryos are dried.
20. We originally tried to use Volocity (PerkinElmer) to calculate the fluorescence intensities of the whole pronucleus, since that software can work with 3D images. For the analysis of PN4–5 zygotes, however, the distance between the paternal and maternal pronuclei was too small for the software to recognize them as separate pronuclei and calculate their fluorescence intensities individually.
21. Use DAPI signals to crop nuclei if the EU signal cannot be detected because of low fluorescence intensities.

References

1. Adenot PG, Mercier Y, Renard JP et al (1997) Differential H4 acetylation of paternal and maternal chromatin precedes DNA replication and differential transcriptional activity in pronuclei of 1-cell mouse embryos. Development 124:4615–4625
2. Santos F, Hendrich B, Reik W et al (2002) Dynamic reprogramming of DNA methylation in the early mouse embryo. Dev Biol 241:172–182
3. Reik W, Dean W, Walter J (2001) Epigenetic reprogramming in mammalian development. Science 293:1089–1093
4. Arney KL, Bao S, Bannister AJ et al (2002) Histone methylation defines epigenetic asymmetry in the mouse zygote. Int J Dev Biol 46:317–320
5. Lepikhov K, Walter J (2004) Differential dynamics of histone H3 methylation at positions K4 and K9 in the mouse zygote. BMC Dev Biol 4:12
6. Torres-Padilla ME, Bannister AJ, Hurd PJ et al (2006) Dynamic distribution of the replacement histone variant H3.3 in the mouse oocyte and preimplantation embryos. Int J Dev Biol 50:455–461
7. Santenard A, Ziegler-Birling C, Koch M et al (2010) Heterochromatin formation in the mouse embryo requires critical residues of the histone variant H3.3. Nat Cell Biol 12:853–862
8. Boskovic A, Bender A, Gall L et al (2012) Analysis of active chromatin modifications in early mammalian embryos reveals uncoupling of H2A.Z acetylation and H3K36 trimethylation from embryonic genome activation. Epigenetics 7:747–757
9. Morgan HD, Santos F, Green K et al (2005) Epigenetic reprogramming in mammals. Hum Mol Genet 14(Spec No 1):R47–R58
10. Cantone I, Fisher AG (2013) Epigenetic programming and reprogramming during development. Nat Struct Mol Biol 20:282–289
11. Latham KE, Solter D, Schultz RM (1991) Activation of a two-cell stage-specific gene following transfer of heterologous nuclei into enucleated mouse embryos. Mol Reprod Dev 30:182–186
12. Bouniol C, Nguyen E, Debey P (1995) Endogenous transcription occurs at the 1-cell stage in the mouse embryo. Exp Cell Res 218:57–62

13. Lewis PW, Muller MM, Koletsky MS et al (2013) Inhibition of PRC2 activity by a gain-of-function H3 mutation found in pediatric glioblastoma. Science 340:857–861
14. Aoshima K, Inoue E, Sawa H et al (2015) Paternal H3K4 methylation is required for minor zygotic gene activation and early mouse embryonic development. EMBO Rep 16:803–812
15. Aoki F, Worrad DM, Schultz RM (1997) Regulation of transcriptional activity during the first and second cell cycles in the preimplantation mouse embryo. Dev Biol 181:296–307
16. Quinn P, Kerin JF, Warnes GM (1985) Improved pregnancy rate in human in vitro fertilization with the use of a medium based on the composition of human tubal fluid. Fertil Steril 44:493–498
17. Nakagata N (1996) Use of cryopreservation techniques of embryos and spermatozoa for production of transgenic (Tg) mice and for maintenance of Tg mouse lines. Lab Anim Sci 46:236–238
18. Ho Y, Wigglesworth K, Eppig JJ et al (1995) Preimplantation development of mouse embryos in KSOM: augmentation by amino acids and analysis of gene expression. Mol Reprod Dev 41:232–238
19. Whitten W (1971) Nutrient requirements for the culture of preimplantation mouse embryos in vitro. Adv Biosci 6:129–141
20. Toyoda Y, Yokoyama M, Hosi T (1971) Studies on fertilization of mouse eggs in vitro. I. In vitro fertilization of eggs by fresh epididymal sperm. Jpn J Anim Reprod 16:147–151
21. Miki H, Ogonuki N, Inoue K et al (2006) Improvement of cumulus-free oocyte maturation in vitro and its application to microinsemination with primary spermatocytes in mice. J Reprod Dev 52:239–248
22. Tagami H, Ray-Gallet D, Almouzni G et al (2004) Histone H3.1 and H3.3 complexes mediate nucleosome assembly pathways dependent or independent of DNA synthesis. Cell 116:51–61
23. Yamagata K, Yamazaki T, Yamashita M et al (2005) Noninvasive visualization of molecular events in the mammalian zygote. Genesis 43:71–79

Index

Kiho Lee (ed.), *Zygotic Genome Activation: Methods and Protocols*, Methods in Molecular Biology, vol. 1605, DOI 10.1007/978-1-4939-6988-3, © Springer Science+Business Media LLC 2017

9781493969869